An Advanced Course in Nuclear Engineering

Series Editors: Yoshiaki Oka, Tokyo, Japan
Haruki Madarame, Tokyo, Japan
Mitsuru Uesaka, Tokyo, Japan

For further volumes:
http://www.springer.com/series/10746

An Advanced Course in Nuclear Engineering

Since the beginning of the twenty-first century, expectations have been rising for nuclear power generation in the world because the need for a safe and stable energy supply is increasing against the background of global environmental issues and the depletion of oil-based energy sources. The situation is calling for the development of human resources with advanced knowledge and techniques of nuclear energy. The role of nuclear energy remains unchanged in the world even after Fukushima Daiichi Nuclear Power Plant accident. Needs for education and human resource development of nuclear professionals who understand and manage the nuclear power have increased worldwide.

This series of books is really the renaissance of publication covering the full fields of, and reflecting recent advances in, nuclear engineering and technology in these three past decades in the world.

Anyone who tries to solve and counter practical problems and unknown application problems encountered in using nuclear energy will face the need to return to the basics. This requires the capability to envision various images of the subject. The systematic acquisition of knowledge of nuclear energy is fundamental to nuclear R&D and the development of the nuclear industry. This series of books is designed to serve that purpose.

We have been publishing this series as a set of standard books for systematically studying nuclear energy from the basics to actual practice, with the aim of nurturing experts and engineers who can act from a global perspective. While being aware of the situation surrounding nuclear energy, students need not only to familiarize themselves with basic knowledge but also to acquire practical expertise, including flexible knowledge and first-hand techniques that allow them to understand field practices without experiencing any gap.

This series consists of three courses on nuclear energy: the basic course, the frontier course, and the practical course. The content is based on the education at the Nuclear Professional School and the Department of Nuclear Engineering and Management, both established by the University of Tokyo with the aim of developing high-level human resources specializing in nuclear energy to cope with the new century of nuclear energy. The books were written mainly by faculty members of the University of Tokyo and researchers at the Japan Atomic Energy Agency and in related industries.

We would like to extend our gratitude to all those who have kindly taken the time to contribute to or cooperate in the creation and publication of this book series.

Yoshiaki Oka, Haruki Madarame, and Mitsuru Uesaka
Department of Nuclear Engineering and Management, and Nuclear Professional School,
School of Engineering, The University of Tokyo

Yoshiaki Oka • Katsuo Suzuki
Editors

Nuclear Reactor Kinetics and Plant Control

Editors
Yoshiaki Oka
Cooperative Major in Nuclear Energy
Graduate School of Advanced Science and Engineering
Waseda University
Tokyo, Japan

Katsuo Suzuki
Japan Nuclear Energy Safety Organization
Tokyo, Japan

Content editor
Takashi Kiguchi
Japan Nuclear Energy Safety Organization
Tokyo, Japan

Original Japanese edition
Genshiryoku Kyokasho: Genshiro Doutokusei to Plant Seigyo
By Yoshiaki Oka, Katsuo Suzuki Copyright (c) 2008 Published by Ohmsha, Ltd.
3-1 Kanda Nishikicho, Chiyoda-ku, Tokyo, Japan

ISSN 2195-3708 ISSN 2195-3716 (electronic)
ISBN 978-4-431-54693-1 ISBN 978-4-431-54195-0 (eBook)
DOI 10.1007/978-4-431-54195-0
Springer Tokyo Heidelberg New York Dordrecht London

Printed on acid-free paper

Springer is part of Springer Science+Business Media (www.springer.com)

Yoshiaki Oka • Katsuo Suzuki
Editors

Nuclear Reactor Kinetics and Plant Control

Editors
Yoshiaki Oka
Cooperative Major in Nuclear Energy
Graduate School of Advanced
Science and Engineering
Waseda University
Tokyo, Japan

Katsuo Suzuki
Japan Nuclear Energy Safety Organization
Tokyo, Japan

Content editor
Takashi Kiguchi
Japan Nuclear Energy Safety Organization
Tokyo, Japan

Original Japanese edition
Genshiryoku Kyokasho: Genshiro Doutokusei to Plant Seigyo
By Yoshiaki Oka, Katsuo Suzuki Copyright (c) 2008 Published by Ohmsha, Ltd.
3-1 Kanda Nishikicho, Chiyoda-ku, Tokyo, Japan

ISSN 2195-3708 ISSN 2195-3716 (electronic)
ISBN 978-4-431-54693-1 ISBN 978-4-431-54195-0 (eBook)
DOI 10.1007/978-4-431-54195-0
Springer Tokyo Heidelberg New York Dordrecht London

Printed on acid-free paper

Springer is part of Springer Science+Business Media (www.springer.com)

Preface

A nuclear reactor is a device controlling fission chain reactions. The world's first fission chain reaction was achieved in December 1942, in a reactor created at the University of Chicago by Enrico Fermi, a genius Italian physicist. We human beings can control the reactor because not all neutrons generated by fission are generated simultaneously with the fission, but some neutrons (delayed neutrons), although extremely small in number, are generated after the fission with a delay. The delay gives us time to control the reactor. The existence of the delayed neutrons may be a gift granted to human beings by the grace of God who intends us to use fission energy. Having an understanding of reactor kinetics and the control of nuclear power plants is necessary in order to understand reactor operation and safety issues—for example, how the reactor behaves in an accident or when in an abnormal state. Despite this requirement, we can hardly find books, in Japan or in other countries, that systematically provide the basics of reactor kinetics and plant control. In addition, the last three to four decades have seen great progress in the use of nuclear power, and this progress has led us to feel the necessity of describing new important matters concerning the practical use of nuclear power. Digital computers and information-processing technologies using digital devices have also developed tremendously in these decades. In nuclear power plants, for example, digital controls have largely replaced analog controls. Today we can easily carry out plant stability and other analyses by using digital computers. We believe that publication of a book describing the actual control of a nuclear power plant, as well as the basic control theory of reactor kinetics and the PID control technology still widely used for plant controls, is of great significance. The following outlines the structure of this book.

Part I describes reactor kinetics, a temporally changing characteristic of reactor reactivity or power.

Chapter 1 explains the concept of the fission chain reaction, a basis of reactor kinetics, and the neutron multiplication factor and reactivity.

Chapter 2 describes the point reactor kinetics equation and its solution in order to understand variation of power affected by the input size of reactivity and the time behavior after the input. Approximate solutions of the point reactor kinetics equation are frequently used for solving practice questions and are also useful for better understanding of physical behavior of reactor kinetics. The last part of this chapter describes the kinetics of subcriticality. All cases of reactivities handled in this chapter are those without feedback.

Chapter 3 explains feedback of the reactivity changed by variation of reactor temperature and other factors. This chapter first explains the Doppler reactivity coefficient, a main factor of the reactivity temperature coefficient and also important for immediate feedback effect. The chapter then describes reactivity coefficients of moderator and coolant. The reactivity coefficient can be turned to negative by designing a value of the moderator-to-fuel volume ratio slightly favorable on the under-moderating side. The last part of this chapter describes the analytical model of temperature feedback.

Chapter 4 describes measuring and experimenting reactivity and other kinetics parameters. The critical approach experiment, control rod calibration, and different measurement techniques of reactivity are described in this chapter. The chapter is important for deepening the understanding of their relation to practical use and application and for the understanding of reactor kinetics.

Part II describes the actual operation control of nuclear plants currently operating in Japan.

Chapter 5 first organizes basic PID control technologies concerning the stability of control systems, evaluation of controllability, and calibration method of control constants. Today, computers have been developed and many programs are available for control system design, enabling us to easily obtain the root of the denominator polynomial of a transfer function and to determine stability directly. In view of the current environment, we allow ourselves to omit description of the Routh–Hurwitz criterion and other methods to determine stability that have been explained in other standard books for control engineering. We do not omit, however, the Nyquist method to determine stability, because it provides visual understanding of frequency-dependent stability of a control system. In addition, a practical design of a PID control device is illustrated in this chapter with a simple structure, and the design process is explained in detail to enable understanding of how the control technologies are of actual use.

Chapter 6 is a description of reactor stability. This chapter first describes a transfer function of the reactor in each state of subcriticality, criticality, and high power with feedback reactivity and discusses features of the reactor in each state from the point of view of frequency characteristics. The chapter then provides the PID control design of a stable power control system for high-power reactors, so that the outline of the design process can be understood. Also described in this chapter are the problems for actual BWRs: first the issue of reactor power stability caused by the nuclear thermal-hydraulic feedback and next the kinetics of reactor power and space oscillation caused by poisonous xenon-135.

Chapters 7–10 describe actual operation controls of the BWR, PWR, FBR, and HTTR. Each chapter, using to a maximum extent the design data of actual reactors and the data of actual operation controls, relies largely on the following format: First, it describes the outline of each reactor plant with attention to the operation control characteristic of the plant. Second, it organizes the plant parameters to be controlled, selecting from temperature, pressure, flow rate, and so on in order to safely operate each phase of startup, steady operation, and shutdown. Our basic approach for controlling the parameters and required performance is also described. These chapters then describe instrumentation and control systems enabling the controls. Lastly, the chapters provide information about the instrumentation and control systems used to shut down the reactor plant safely in the case of an accident or occurrence of an abnormal state.

Chapter 11 introduces the optimal regulator that represents modern controls, the $H\infty$ control of robust control design theory, and controls using the expert system, neural network, fuzzy logic, or other artificial intelligence that we expect to be applied to future reactors.

February 2008 Yoshiaki Oka and Katsuo Suzuki

Contents

Contributors

Satoshi Hanada Mitsubishi Heavy Industries, Ltd., Tokyo, Japan

Toshihide Inoue Mitsubishi Heavy Industries, Ltd., Tokyo, Japan

Takashi Kiguchi Japan Nuclear Energy Safety Organization, Tokyo, Japan

Koichi Kondo Toshiba Corporation, Tokyo, Japan

Yuji Koshi Toshiba Corporation, Tokyo, Japan

Masahiko Kuroki Toshiba Corporation, Tokyo, Japan

Shuhei Miyake Mitsubishi Heavy Industries, Ltd., Tokyo, Japan

Kunihiko Nabeshima Japan Atomic Energy Agency, Tokai-mura, Japan

Shigeaki Nakagawa Japan Atomic Energy Agency, Tokai-mura, Japan

Yoshiaki Oka Tokyo University, Tokyo, Japan

Hiroshi Ono Toshiba Corporation, Tokyo, Japan

Yasuo Ota Toshiba Corporation, Tokyo, Japan

Katsuo Suzuki Japan Nuclear Energy Safety Organization, Tokyo, Japan

Masayoshi Tahira Toshiba Corporation, Tokyo, Japan

Hidetaka Takahashi Japan Atomic Energy Agency, Tokai-mura, Japan

Kiyoshi Tamayama Japan Atomic Energy Agency, Tokai-mura, Japan

Part I
Nuclear Reactor Kinetics

Chapter 1
Delayed Neutron and Nuclear Reactor Kinetics

Yoshiaki Oka

Introduction Reactor power changes when the temperature and position of the control rods of a nuclear reactor are changed. This change is unique to each reactor, and its characteristics are called "nuclear reactor kinetics."

The control rods are made of strong neutron-absorbing materials, and when they are inserted into the reactor, the reaction rate of neutron absorption increases. The reactor becomes subcritical and its power decreases. Conversely, the reaction rate of neutron absorption decreases when the control rods are withdrawn; the reactor becomes supercritical and its power increases. The reaction rate of neutron absorption changes when the reactor temperature is changed and, therefore the reactor power changes.

The reactor power is proportional to the number of fission reactions per second in the nuclear reactor. As fission reactions are caused by neutrons, the number of their reactions is proportional to the total number of neutrons in the reactor. However, the number of neutrons varies depending on the neutron production rate due to the fission reactions, the rate of neutron absorption by the nuclear fuel and reactor structure materials, and the rate of neutron leakage from the reactor.

1.1 Fission Chain Reactions

When a neutron collides with a heavy nucleus such as ^{235}U or ^{239}Pu, it splits the nucleus into two nuclei which are called fission fragments. In this process, multiple neutrons, several gamma rays, and neutrinos are also released. Multiple neutrons are produced by a single fission reaction. If one of these neutrons causes another fission event, it results in another emission of neutrons, followed by more fission events, and so on. This is called a fission chain reaction. Figure 1.1 illustrates the concept of the fission chain reaction.

Y. Oka and K. Suzuki (eds.), *Nuclear Reactor Kinetics and Plant Control*,
An Advanced Course in Nuclear Engineering,
DOI 10.1007/978-4-431-54195-0_1, © Springer Japan 2013

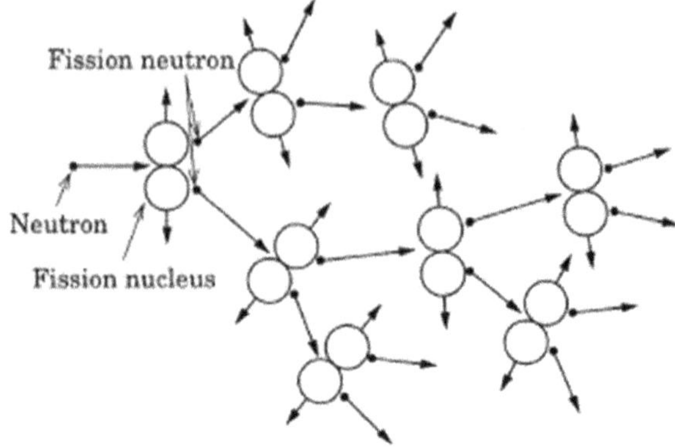

Fig. 1.1 Conceptual diagram of fission chain reaction

Because neutrons play a key role in maintaining the fission chain reaction, it is important to understand their behavior with regard to designing a reactor and analyzing its characteristics. A multiplication factor is used to analyze the fission chain reaction quantitatively. The effective neutron multiplication factor, k, is defined as the ratio between neutron production and loss (absorption plus neutron leakage from the reactor). In an infinitely sized reactor, the neutron loss consists of only neutron absorption, and k is expressed as the infinite neutron multiplication factor, $k\infty$. However, the actual reactor size is limited and, in addition to the neutron absorption, neutron leakage from the reactor needs to be considered. In this case, the multiplication factor is expressed as the effective neutron multiplication factor, k_{eff}.

$$k_{\mathrm{eff}} = \frac{\text{Neutron production rate}}{\text{Neutron loss rate (the sum of absorption rate and leakage rate)}} \quad (1.1)$$

If $k_{\mathrm{eff}} = 1$, the number of neutrons is constant in the reactor. In other words, the fission rate is constant and the constant energy release continues. At this state, the reactor is critical. If $k_{\mathrm{eff}} < 1$, the number of neutrons decreases gradually in the reactor with progression of the fission chain reaction. In this state, the reactor is subcritical. However, if "$k_{\mathrm{eff}} > 1$," the chain reaction rate increases and the reactor is supercritical.

As the concept of the fission chain reaction of Fig. 1.1 shows, neutrons are generated by nuclear fission and are lost when captured by nuclei or when leaked to the outside of reactor. When the generated neutrons are absorbed by a fissile nuclide (^{235}U, for example) or by a fissionable nuclide (^{238}U, for example) a certain rate of neutron generation triggers a fission. This produces new neutrons. If generation of a neutron is thought of as its "birth" and its loss is a "death," the fission chain reaction is a process whereby "child" neutrons are born of "parent" neutrons. The effective multiplication factor can also be defined as a ratio of the number of neutrons between two consecutive generations.

$$k_{\mathrm{eff}} = \frac{\text{Number of neutrons of a generation}}{\text{Number of neutrons of the preceeding generation}} \quad (1.2)$$

Actually, however, it is difficult to determine the length of a generation of neutrons. This is because some neutrons trigger fission reactions immediately after they are reproduced but other neutrons trigger after their moderation to thermal neutrons. Neutron capture occurs randomly, and so does leakage of neutrons to the outside of the reactor. The following discussion uses Eq. (1.1) to define the effective multiplication factor.

When a reactor operates with a constant power, its effective multiplication factor is equal to 1. The reactor is "critical". When the reactor is shut down, it is "subcritical". During startup, the reactor is controlled so that it becomes "supercritical"; the neutron production rate is increased above the neutron loss rate, and the number of neutrons in the reactor is gradually increased. When the reactor reaches the required power rating, it is returned to the critical state and is operated with a constant output. If the reactor is required to change from low power to high power, it is controlled to reach the supercritical state. If the reactor is required to return to low power, it is controlled to reach the subcritical state; when it reaches the required power, the reactor is returned to critical.

The nuclear reactor kinetics usually explains an increase or decrease in the number of neutrons in the entire core. In other words, the spatial distribution of neutrons is not considered in the core. Here, a point-wise reactor approximation is used, where the core is represented by one point and no space variable is considered. Reactor kinetics that considers the spatial distribution is called space–time kinetics.

1.2 Change in Multiplication Factor and Nuclear Reactor Kinetics

In order to operate a nuclear reactor at constant power, the effective multiplication factor (k_{eff}) must be equal to 1; that is, the neutron production by fission must be accurately balanced with the neutron loss due to the neutron absorption and leakage. Although the k_{eff} value is changed by various factors, they can be grouped by the times needed for them to occur which are as follows:

- Short period (from seconds to minutes): startup, shutdown, and disturbances during operation (including changes of temperature, pressure, and moderator density)
- Intermediate period (from hours to days): generation and decay of fission products (xenon and samarium) having strong neutron absorption
- Long period (from months to years): burnup (consumption) of nuclear fuel and accumulation of fission products

It is important to estimate of a change in the number of neutrons and a change in the power with time that occur when k_{eff} changes. This is covered by the reactor kinetics and nuclear plant dynamics. The nuclear reactor kinetics covers the change in the number of neutrons and the change in power due to a short-period change in multiplication factor. Over a long period, the change in multiplication factor is compensated for by control rods, chemical shim, and burnable poison. Details for this are presented in Part II Chaps. 3–6 of this book. In an intermediate period, the change in multiplication factor is estimated using the generation and decay model of xenon and samarium. It is different from the nuclear reactor kinetics model.

1.3 Prompt Neutron and Delayed Neutron

Most neutrons (99.35 % for ^{235}U fission by thermal neutrons) are emitted immediately by a nuclear fission event. These are called "prompt neutrons." A few neutrons are emitted a little after nuclear fission occurs and they are called "delayed neutrons."

The delayed neutrons are primarily produced from the decay of the fission products emitting neutrons. The fission products that emit delayed neutrons are called delayed neutron precursors. There are many delayed neutron precursors such as ^{87}Br, ^{88}Br, ^{137}I, ^{138}I, and ^{139}I and they have different half-lives. The delayed neutron precursors are treated in six groups with different half-lives for analysis of nuclear reactor kinetics. Table 1.1 shows data of delayed neutrons that are generated by thermal fission of ^{235}U.

The delayed neutron precursor of the longest half-life is ^{87}Br. Delayed neutron data differ from those of the fission nuclides and are between the data of thermal fission and fast fission neutrons. The data need to be used correctly depending on the type of the fuel and neutron spectrum of the reactor. Table 1.2 lists the delayed neutron fraction for each fission nuclide.

The delayed neutrons have approximately 0.4-MeV average energy, which is lower than the approximate 2-MeV average energy of the prompt neutrons. Therefore, the fraction of delayed neutrons that is leaked outside the reactor and lost disappear is slightly smaller than that of the prompt neutrons. The fraction of delayed neutrons that contributes to the fission chain reactions is slightly larger than that of the prompt neutrons. This effect is considered in the analysis of nuclear reactor kinetics. A slightly larger delayed neutron fraction is used than the absolute value "β" depending on the reactor and the effect is shown as "β_{eff}." If the reactor has a large core volume, neutron leakage is very small during moderation and there is almost no difference between them. The "β_{eff}" value depends on the reactor size and neutron spectrum. Although the delayed neutron fraction is low, it slows down the transient change of the reactor and, therefore, it plays a very important role in the reactor control.

The distribution of prompt neutron energies can be expressed by the following function:

$$\chi(E) = 0.453\,\mathrm{e}^{-1.012E} \sin h\sqrt{2.19E} \tag{1.3}$$

where, E is the neutron energy in MeV.

Table 1.1 Data of a delayed neutron generated by thermal fission of uranium-235

Group	Half-life (s)	Decay constant, λ_i (s^{-1})	Delayed neutron fraction, β_i
1	55.72	0.012 4	0.000 215
2	22.72	0.030 5	0.001 424
3	6.22	0.111	0.001 274
4	2.30	0.301	0.002 568
5	0.610	1.14	0.000 748
6	0.230	3.01	0.000 273

Total delayed neutron fraction β: 0.006 5

Table 1.2 Delayed neutron fraction of nuclides

Nuclide	Total delayed neutron fraction, β
^{232}Th	0.020 3[a]
^{233}U	0.002 6
^{235}U	0.006 5
^{238}U	0.014 8[a]
^{239}Pu	0.002 0[a]

[a]Generated by fast fission

1.4 Kinetic Parameters

This section explains the parameters commonly used for description of reactor kinetics. The definition of effective multiplication factor, k_{eff}, has been given by Eq. (1.1) or (1.2).

The "reactivity" indicates a degree of deviation from the critical state, and it can be defined as follows:

$$\rho = \frac{k_{eff} - 1}{k_{eff}} \tag{1.4}$$

If the reactor is supercritical, $k_{eff} > 1$ and the value of ρ is positive. If the reactor is subcritical, $k_{eff} < 1$ and the value of ρ is negative. ρ takes a value within the range of $-\infty < \rho < 1$.

The reactivity is expressed as a numerical value or a percentage. It is shown in the French unit of "pcm" (10^{-5}), in the English unit of "milli-k" (10^{-3}), or in the American unit of "dollars" ($) and "cents" (¢). A dollar is equal to the value ρ divided by generation rate β of delayed neutrons, and 1 $ is equal to 100 ¢.

If $k_{eff} = 1$, that is, if $\rho = 0$, it is strictly said to be the "delayed critical" state. The generation of neutrons in the reactor (including the generation of delayed neutrons caused by the decay of delayed neutron precursors) is equal to the loss of neutrons. When the reactivity is 1 $, the generation of prompt neutrons is equal to its loss. This is called the "prompt critical" state. The state above the prompt critical is called the "prompt supercritical."

The prompt neutron lifetime can be defined by the following equation:

$$l \equiv \frac{\text{Total number of neutrons in reactor}}{\text{Extinction rate of neutrons}} \tag{1.5}$$

Table 1.3 Diffusion time and slowing-down time of various moderators

Moderator	Diffusion time (ms)	Slowing-down time (μs)
Light water (H_2O)	0.205	1.0
Heavy water (D_2O)	100[a]	8.1
Beryllium (Be)	3.46	9.3
Graphite	13.0	23

[a]Depends on the purity of heavy water

The prompt neutron lifetime can be expressed as an average time from generation of a prompt neutron to its absorption. Because the neutron slowing-down time is much shorter than the time when the neutron is diffused and absorbed, the prompt neutron lifetime of the thermal reactor is almost the same as the diffusion time of a thermal neutron. The diffusion time and slowing-down time for various moderators are shown in Table 1.3.

Because the neutrons are not moderated to become thermal neutrons in the fast reactor, the prompt neutron lifetime is an order of 10^{-5}–10^{-7} s.

The prompt neutron generation time can be defined by the following equation:

$$\Lambda \equiv \frac{\text{Total number of neutrons in reactor}}{\text{Neutron generation rate}} \tag{1.6}$$

The prompt neutron generation time is equal to the prompt neutron lifetime divided by the effective multiplication factor.

Among the kinetics parameters described here, the denominator is the neutron loss rate obtained in Eqs. (1.1) and (1.5) for effective multiplication factor k_{eff} and prompt neutron lifetime ℓ and the denominator is the neutron generation rate for equation of reactivity ρ and prompt neutron generation time Λ. If $k_{\text{eff}} = 1$, values ℓ and Λ become equal to each other. In the kinetics equations described below, a pair of k_{eff} and ℓ values or a pair of ρ and Λ values should be used.

Chapter 2
Point Reactor Kinetics

Yoshiaki Oka

2.1 Point Reactor Kinetics Equations

To describe the reactor kinetics, the number of neutrons and the number of delayed neutron precursors that change with time are considered. The following ordinary differential equations can be gotten if the space dependence of these variables is ignored and the neutron energy is handled in one group. This is called the point reactor kinetics model. Actually, the reactor is not treated as a single point but the assumption is made that the space distribution of parameters does not change with time. When a slow disturbance is treated in spatial asymmetry, the point reactor kinetics model can be used by weighting the reactivity feedback amount determined with the importance function. Generally, the point reactor approximation can be used to approximate a slow change of the space distribution of parameters. It can be applied to many transient events that contain the disturbance to be handled by reactor control. In contrast, when handling a local and fast reactivity disturbance, the space-dependent kinetics model must be used. An example is the accident that may occur if asymmetric control rods are quickly withdrawn.

As the delayed neutron precursors are treated in six groups, the point reactor kinetics equations are as follows:

$$\frac{\mathrm{d}n(t)}{\mathrm{d}t} = [k_{\mathrm{eff}}(1-\beta)-1]\frac{n(t)}{l} + \sum_{i=1}^{6}\lambda_i C_i(t) \quad \left(i = 1 \sim 6, \quad \beta = \sum_{i=1}^{6}\beta_i\right) \tag{2.1}$$

$$\frac{\mathrm{d}C\ (t)}{\mathrm{d}t} = k_{\mathrm{eff}}\,\beta_i\frac{n(t)}{l} - \lambda_i C_i(t) \tag{2.2}$$

where n is the number of neutrons in the reactor, C_i is the number of delayed neutron precursors in group i, β_i is the delayed neutron fraction of group i, λ_i is the

Y. Oka and K. Suzuki (eds.), *Nuclear Reactor Kinetics and Plant Control*,
An Advanced Course in Nuclear Engineering,
DOI 10.1007/978-4-431-54195-0_2,

decay constant for the delayed neutron precursors of group i, k_{eff} is the effective multiplication factor, and ℓ is the prompt neutron lifetime. Equation (2.1) shows balance of the number of neutrons. The left side shows the number of neutrons that changes with time. The first term of the right side shows the difference between the production rate of the prompt neutrons and the loss rate of neutrons. The second term shows the production rate of delayed neutrons. Equation (2.2) shows balance of the number of delayed neutron precursors of group i. The left side shows the time change, and the first term of the right side shows the production rate and the second term shows the loss rate. This is a seven-element simultaneous normal differential equation.

Using reactivity ρ and prompt neutron generation time Λ, we can also write the point reactor kinetics equations as follows:

$$\frac{\mathrm{d}n(t)}{\mathrm{d}t} = \frac{\rho - \beta}{\Lambda} n(t) + \sum_{i=1}^{6} \lambda_i C_i(t) \quad \left(i = 1 \sim 6, \quad \beta = \sum_{i=1}^{6} \beta_i \right) \tag{2.3}$$

$$\frac{\mathrm{d}C_i(t)}{\mathrm{d}t} = \frac{\beta_i}{\Lambda} n(t) - \lambda_i C_i(t) \tag{2.4}$$

Here, the reactor power is proportional to the neutron number (n) in the core. This n is also used as the power.

2.2 Solution for Prompt Supercritical Reactor

If the reactor exceeds the prompt critical which is called the "prompt supercritical state," the effect of delayed neutrons can be ignored. The kinetics equation can be written as follows.

$$\frac{\mathrm{d}n(t)}{\mathrm{d}t} = (k_{\mathrm{eff}} - 1)\frac{n(t)}{l} \tag{2.5}$$

If $k_{\mathrm{eff}} - 1$ is constant and the initial value of neutron number is "n_0," the following equation is obtained.

$$n(t) = n_0 \exp\left(\frac{k_{\mathrm{eff}} - 1}{l} t\right) \tag{2.6}$$

The value ℓ is on the order of 10^{-4}–10^{-3} s for the thermal neutron reactor, and it is on the order of 10^{-8}–10^{-6} s for the fast neutron reactor. When the reactor exceeds the prompt critical state, the number of neutrons rapidly increases with time.

2.3 The Inhour Equation

Here, the reactor kinetics equations with delayed neutrons are solved.

$$n(t) = n_0 e^{\omega t}, \quad C_i(t) = C_{i0} e^{\omega t} \tag{2.7}$$

Assuming the solution as above, and substituting it into Eq. (2.2) gives Eq. (2.8).

$$C_i = \frac{k_{\mathrm{eff}}\,\beta_i}{l} n \frac{1}{\omega + \lambda_i} \tag{2.8}$$

Then, substituting this equation and Eq. (2.7) into Eqs. (2.1) and (2.2) gives Eq. (2.9).

$$k_{\mathrm{eff}} - 1 = \omega l + k_{\mathrm{eff}} \sum_{i=1}^{6} \frac{\omega \beta_i}{\omega + \lambda_i} \tag{2.9}$$

When reactivity ρ is used, Eq. (2.9) is expressed as Eq. (2.10) and from that Eq. (2.11) is obtained.

$$\rho = \frac{\omega l}{k_{\mathrm{eff}}} + \sum_{i=1}^{6} \frac{\omega \beta_i}{\omega + \lambda_i} \tag{2.10}$$

$$\rho = \frac{\omega l}{\omega l + 1} + \frac{1}{\omega l + 1} \sum_{i=1}^{6} \frac{\omega \beta_i}{\omega + \lambda_i} \tag{2.11}$$

Equation (2.11) is the seventh order algebraic equation of ω, and it is called the inhour equation. There are seven roots of the equation for the value of ρ $(-\infty \leq \rho \leq 1)$. The roots are shown in Fig. 2.1.

When $\rho > 0$, ω has one positive root and six negative roots. When $\rho < 0$, ω has seven negative roots. Thus, the solution of the reactor kinetics equations with delayed neutrons is as follows.

$$n(t) = A_0 e^{\omega_0 t} + A_1 e^{\omega_1 t} + A_2 e^{\omega_2 t} + A_3 e^{\omega_3 t} + A_4 e^{\omega_4 t} + A_5 e^{\omega_5 t} + A_6 e^{\omega_6 t} \tag{2.12}$$

If $\rho > 0$,

$$\omega 0 > 0, \ \omega 1, \omega 2, \omega 3, \omega 4, \omega 5, \omega 6 < 0$$

If $\rho < 0$

$$\omega 0, \omega 1, \omega 2, \omega 3, \omega 4, \omega 5, \omega 6 < 0$$

Fig. 2.1 Seven roots of the equation, (2.11)

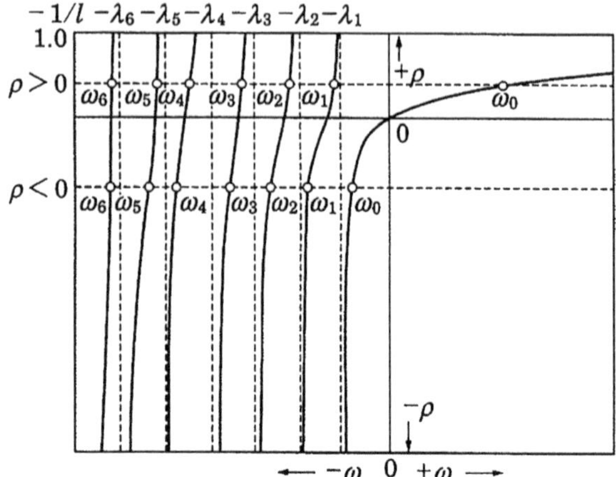

If $\rho > 0$ and only a short time has passed, the second and subsequent terms can be ignored and Eq. (2.13) is obtained.

$$n(t) = A_0 e^{\omega_0 t} \tag{2.13}$$

$T_0 = 1/\omega_0$ is called the reactor period. The reactor period is a period of time that is required to increase the neutron number and the power of reactor by "e"-fold.

If $\rho = 0$, that is, if the reactor is critical, $\omega_0 = 0$. If $\rho \to 1$ limit, $\omega_0 \to \infty$ occurs. It corresponds to the prompt supercritical state. If $\rho \to -\infty$ occurs, $\omega_0 \to \lambda_1$ occurs.

Next, the approximate reactor period is determined when the value of ρ is positive and very small and when it is positive and much larger than β.

If a very small positive reactivity of $\rho \ll \beta$ is added, ω_0 is sufficiently small for ℓ^{-1} and λ_i and value ω_0 is negligible. Thus, the inhour Eq. (2.11) can be approximated as Eq. (2.14).

$$\rho = \omega_0 l + \omega_0 \sum_{i=1}^{6} \frac{\beta_i}{\lambda_i} \tag{2.14}$$

Consequently, the reactor period can be expressed by

$$T_0 = \frac{1}{\omega_0} = \frac{1}{\rho}\left[l + \sum_{i=1}^{6} \frac{\beta_i}{\lambda_i} \right] \cong \frac{\langle l \rangle}{\rho} \cong \frac{\langle l \rangle}{k_{\text{eff}} - 1} \tag{2.15}$$

where $\rho \cong k_{\text{eff}} - 1$, and $\langle l \rangle$ is the mean lifetime of neutrons including the delayed neutrons. From Eq. (2.15), it is clear that the reactor period is determined by this mean neutron lifetime $\langle l \rangle$ when very small positive reactivity is added.

When a large positive reactivity of $\rho \gg \beta$ is added, ω_0 becomes much larger than λ_i, and the inhour Eq. (2.11) can be approximated as Eq. (2.16).

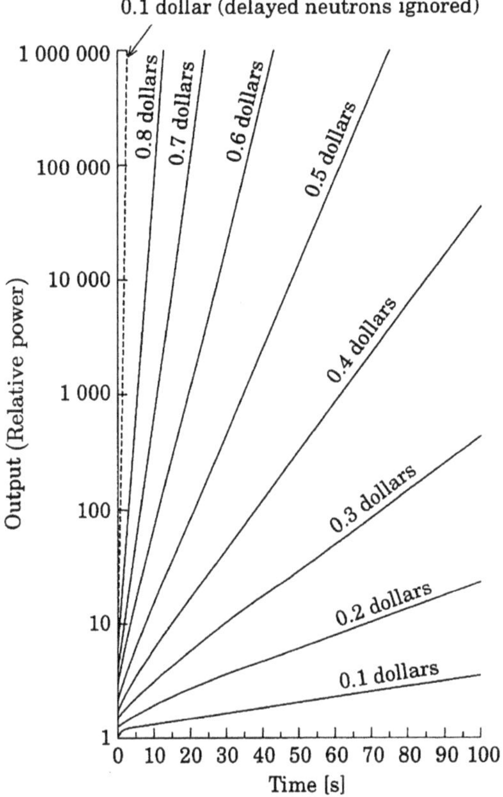

Fig. 2.2 Power variation relative to positive stepwise reactivity insertion

$$\rho \cong \frac{\omega_0 l + \beta}{\omega_0 l + 1} \tag{2.16}$$

The reactor period is defined by the next equation.

$$T_0 = \frac{1}{\omega_0} \cong \frac{l}{k_{\mathrm{eff}}(\rho - \beta)} \cong \frac{l}{k_{\mathrm{eff}} - 1} \tag{2.17}$$

Here, three approximations are used: $\rho - \beta = \rho$, $\rho = k_{\mathrm{eff}} - 1$, and $k_{\mathrm{eff}} = 1$. Equation (2.17) matches the reactor period of Eq. (2.6) which ignores the delayed neutrons. That is, if a large positive reactivity is added, the reactor period is determined by the prompt neutron lifetime $\langle l \rangle$.

The power change of the reactor is shown in Figs. 2.2 and 2.3 if a positive reactivity and a negative reactivity are inserted instantaneously in the critical reactor for time $t = 0$ (stepwise), respectively. The power change is shown with dotted lines if the delayed neutrons are ignored. It is clear that the delayed neutrons significantly affect the power change of the reactor.

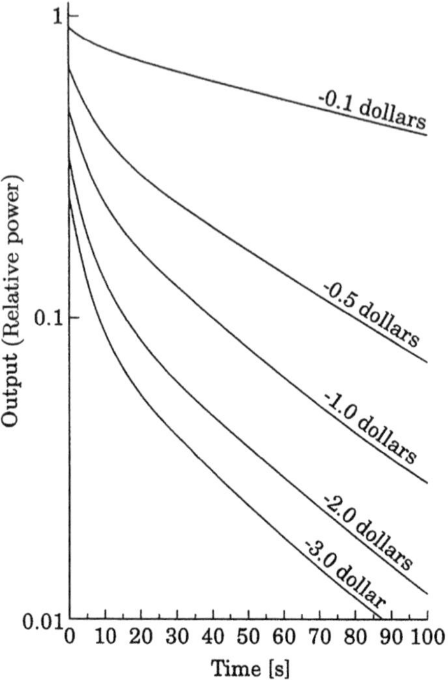

Fig. 2.3 Power variation relative to negative stepwise reactivity insertion

2.4 One Group of Delayed Neutron Approximation

If all the delayed neutrons are approximated in one group, the point kinetics equations can be simplified into a two-dimensional simultaneous equation.

$$\frac{\mathrm{d}n(t)}{\mathrm{d}t} = [k_{\mathrm{eff}}(1-\beta)-1]\frac{n(t)}{l} + \lambda C(t) \tag{2.18}$$

$$\frac{\mathrm{d}C(t)}{\mathrm{d}t} = k_{\mathrm{eff}}\,\beta\frac{n(t)}{l} - \lambda C(t) \tag{2.19}$$

Here, the following approximation is used for the average decay constant.

$$\frac{\beta}{\lambda} = \sum_{i=1}^{6}\frac{\beta_i}{\lambda_i} \tag{2.20}$$

Equation (2.10) is a type of the inhour equation, and it can be simplified as follows.

$$\rho = \frac{\omega l}{k_{\mathrm{eff}}} + \frac{\omega\beta}{\omega+\lambda} \tag{2.21}$$

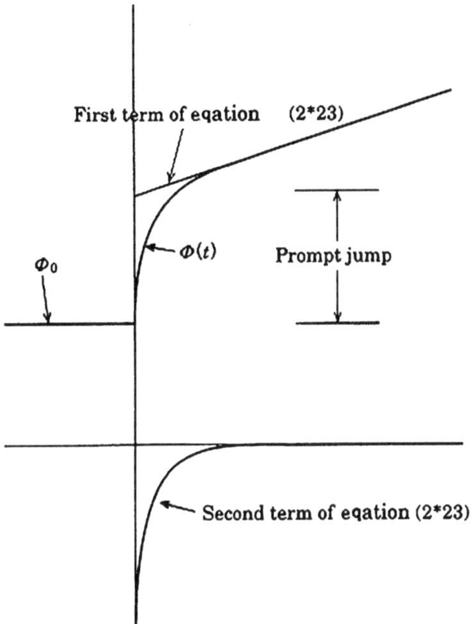

Fig. 2.4 Power change immediately after positive reactivity insertion

The solutions ω_1 and ω_2 of Eq. (2.21) are derived as follows if very small positive reactivity ρ_0 for $\rho \gg \beta$ is added.

In deriving Eq. (2.21) by approximating with $k_{\mathrm{eff}} = 1$ and $\rho_0 \ll \beta$,

$$\omega_1 \cong \frac{\lambda\rho_0}{\beta - \rho_0}, \quad \omega_2 \cong -\frac{\beta - \rho_0}{l}$$

are obtained.

Consequently, the general solution for n is expressed as Eq. (2.22).

$$n(t) = A_1 \exp\left(\frac{\lambda\rho_0}{\beta - \rho_0} t\right) + A_2 \exp\left(-\frac{\beta - \rho_0}{l} t\right) \tag{2.22}$$

Equation (2.19) is obtained by using the initial condition $n(0) = n_0$ and Eqs. (2.18) and (2.19).

$$n(t) = n_0\left[\frac{\beta}{\beta - \rho_0} \exp\left(\frac{\lambda\rho_0}{\beta - \rho_0} t\right) + \frac{\rho_0}{\beta - \rho_0} \exp\left(-\frac{\beta - \rho_0}{l} t\right)\right] \tag{2.23}$$

As prompt neutron lifetime ℓ is very short, the second term of the right side of Eq. (2.23) rapidly decreases immediately after reactivity insertion. The power increases gradually due to the first term. This is shown in Fig. 2.4.

The quick change from n_0 to $\frac{\beta}{\beta-\rho_0}n_0$ immediately after the reactivity insertion is called the "prompt jump." The one-group model of delayed neutrons has an appropriate approximation for the increase of power. When the power decreases, the delayed neutron precursor with the longest halftime is dominant and this approximation becomes inappropriate.

2.5 Approximation with the Constant Delayed Neutron Generation Rate

Now, the case is considered when the reactivity changes stepwise from 0 to ρ_0. It can be assumed that the delayed neutron production rate is constant immediately after a change of reactivity (within approximately 1 s).

$$C_i(t) = C_i(0) \tag{2.24}$$

With this approximation, the point reactor kinetics equations can be written as follows.

$$\frac{\mathrm{d}n(t)}{\mathrm{d}t} = \frac{\rho_0 - \beta}{\Lambda}n(t) + \sum_{i=1}^{6}\lambda_i C_i(0) \quad \left(i = 1 \sim 6,\ \beta = \sum_{i=1}^{6}\beta_i\right) \tag{2.25}$$

$$0 = \frac{\beta_i}{\Lambda}n_0 - \lambda_i C_i(0) \quad (n_0 = n(0)) \tag{2.26}$$

If $\lambda_i C_i(0)$ is eliminated from this equation, the following first-order ordinary differential equation can be obtained.

$$\frac{\mathrm{d}n(t)}{\mathrm{d}t} = \frac{\rho_0 - \beta}{\Lambda}n(t) + \frac{\beta}{\Lambda}n_0 \tag{2.27}$$

Equation (2.28) is the solution to Eq. (2.27).

$$n(t) = n_0 \exp\left(-\frac{\beta - \rho_0}{\Lambda}t\right) + \frac{\beta}{\beta - \rho_0}n_0\left[1 - \exp\left(-\frac{\beta - \rho_0}{\Lambda}t\right)\right] \tag{2.28}$$

As prompt neutron generation time Λ is very short, the term of $\exp\left(-\frac{\beta-\rho_0}{\Lambda}t\right)$ is quickly attenuated with time, and $n(t)$ closes to $\frac{\beta}{\beta-\rho_0}n_0$ with asymptotic behavior. As mentioned above, the constant delayed neutron production rate approximation is suitable to describe the behavior of rapid power change (prompt jump) immediately after reactivity insertion.

2.6 Prompt Jump Approximation

When the time has passed after input of reactivity, most of the output change is affected by the delayed neutron precursors and it can be $\frac{\mathrm{d}n(t)}{\mathrm{d}t} = 0$.

With this approximation, the point reactor kinetics equations are as follows.

$$0 = \frac{\rho(t) - \beta}{\Lambda} n(t) + \sum_{i=1}^{6} \lambda_i C_i(t) \quad \left(i = 1 \sim 6,\ \beta = \sum_{i=1}^{6} \beta_i \right) \tag{2.29}$$

$$\frac{\mathrm{d}C_i(t)}{\mathrm{d}t} = \frac{\beta_i}{\Lambda} n(t) - \lambda_i C_i(t) \tag{2.30}$$

If delayed neutrons are assumed as one group and if C_i is eliminated, the result is:

$$(\rho(t) - \beta)\frac{\mathrm{d}n(t)}{\mathrm{d}t} + \left[\frac{\mathrm{d}\rho(t)}{\mathrm{d}t} + \lambda \rho(t) \right] n(t) = 0 \tag{2.31}$$

and its solution is

$$n(t) = n_0 \exp A(t), \quad \text{where} \quad A(t) \equiv \int_0^t \mathrm{d}\tau \frac{\dot{\rho}(\tau) + \lambda \rho(\tau)}{\beta - \rho(\tau)} \tag{2.32}$$

If $\rho(t) = \rho_0$,

$$n(t) = n_0 \exp\left(\frac{\lambda \rho_0}{\beta - \rho_0} t \right) \tag{2.33}$$

As shown by the constant delayed neutron production rate approximation, the power quickly changes from n_0 to $\frac{\beta}{\beta - \rho_0} n_0$ by the prompt jump immediately after reactivity insertion. If n_0 of Eq. (2.33) is substituted as $n_0 \frac{\beta}{\beta - \rho_0}$, it results in

$$n(t) = n_0 \frac{\beta}{\beta - \rho_0} \exp\left(\frac{\lambda \rho_0}{\beta - \rho_0} t \right)$$

and that matches the first term of Eq. (2.23) which was obtained without the approximations. The quick power change due to the prompt jump can generally be expressed as follows even if the reactor is subcritical or even if negative reactivity is inserted:

$$\frac{n_1}{n_0} = \frac{\beta - \rho_0}{\beta - \rho_1} \tag{2.34}$$

where suffixes 0 and 1 show the respective state before and after reactivity insertion.

2.7 Kinetics in Subcritical State

As the fission chain reaction attenuates when the reactor is subcritical, no steady neutron flux is formed if the neutron source does not exist.

If the neutron source exists, the neutron numbers are kept constant in the subcritical reactor according to the neutron source strength and the amount of subcritical reactivity.

This is explained conceptually as follows using the effective multiplication coefficient k_{eff}, numbers of neutrons n, and neutron source strength S. When fission occurs because of S, the number of produced neutrons is proportional to $k_{eff}S$ in the first generation. It becomes k_{eff}^2S in the second generation, and it becomes k_{eff}^3S in the third generation. In the n-th generation, it becomes k_{eff}^nS. If the effective multiplication coefficients are summed until n becomes infinity, the result is Eq. (2.35).

$$n \propto S + k_{\mathrm{eff}}\,S + {k_{\mathrm{eff}}}^2\,S + \cdots\cdots {k_{\mathrm{eff}}}^n\,S \cdots\cdots = \frac{S}{1 - k_{\mathrm{eff}}} \tag{2.35}$$

In other words, the number of neutrons in the subcritical reactor is uniquely determined by neutron source strength S and subcritical reactivity $1 - k_{eff}$. While the reactor is critical, the neutron numbers of the reactor are mathematically indefinite and can take any value. It can be said that the power of the subcritical reactor is definite but that of the critical reactor is indefinite. In an actual critical reactor, the upper limit of the power is determined by the heat removal rate and other factors. The kinetics equation of the subcritical reactor is expressed as follows when the neutron source strength is S.

$$\begin{cases} \dfrac{\mathrm{d}n(t)}{\mathrm{d}t} = \dfrac{\rho(t) - \beta}{\Lambda} n(t) + \sum \lambda_i C_i + S \\ \dfrac{\mathrm{d}C_i(t)}{\mathrm{d}t} = \dfrac{\beta_i}{\Lambda} n(t) - \lambda_i C_i \end{cases} \tag{2.36}$$

Here, S is constant and neutron absorption of the neutron source itself is ignored in deriving the equation.

Next, the case is considered in which the reactor is still subcritical even if the reactivity is added.

When the reactivity changes stepwise at $t = 0$ from $\rho(t) = \rho_0$ to $\rho_0 + \delta\rho$, the delayed neutron production is considered to be constant for a short time. With the constant delayed neutron production rate approximation, it becomes Eq. (2.37)

$$\lambda_i C_i(t) = \frac{\beta}{\Lambda} n_0 \tag{2.37}$$

and Eq. (2.36) becomes the following.

$$\frac{dn}{dt} = \frac{\rho_0 + \delta\rho - \beta}{\Lambda} n(t) + \frac{\beta}{\Lambda} n_0 + S \tag{2.38}$$

At $t = 0$, the reactor is in the steady state and $n = n_0$. Thus,

$$\frac{\rho_0 - \beta}{\Lambda} n_0 + \frac{\beta}{\Lambda} n_0 + S = 0, -\frac{\rho_0}{\Lambda} n_0 = S$$

Equation (2.38) is solved using these values.

$$n(t) = n_0 e^{\frac{\rho_0 + \delta\rho - \beta}{\Lambda} t} + \frac{\beta - \rho_0}{\beta - \rho_0 - \delta\rho} n_0 \left(1 - e^{\frac{\rho_0 + \delta\rho - \beta}{\Lambda} t}\right) \tag{2.39}$$

Here, Eq. (2.39) can be written as Eq. (2.40), where the neutron number at large "t" is denoted as $n(0)$.

$$n(0) = \frac{\beta - \rho_0}{\beta - \rho_0 - \delta\rho} n_0 \tag{2.40}$$

The prompt jump amount is Eq. (2.41).

$$\delta n_p = n(0) - n_0 = \frac{\delta\rho}{\beta - \rho_0 - \delta\rho} n_0 \tag{2.41}$$

It can be seen that the prompt jump amount increases as subcritical reactivity $\rho_0 = -|\rho_0|$ decreases for the same value of $\delta\rho$.

The reactor finally enters the steady state even for this new subcritical reactivity $\rho_0 + \delta\rho$. The number of steady state neutrons n_1 can be derived using $\rho_0 + \delta\rho$ for Eq. (2.36) and

$$\frac{dn}{dt} = 0, \quad \frac{dC_i}{dt} = 0$$

$$\frac{\rho_0 + \delta\rho - \beta}{\Lambda} n_1 + \frac{\beta}{\Lambda} n_1 + S = 0 \tag{2.42}$$

In contrast, when the reactor is in the steady state for n_0 before the reactivity is inserted, the following equation is derived.

$$\frac{\rho_0 - \beta}{\Lambda} n_0 + \frac{\beta}{\Lambda} n_0 + S = 0 \tag{2.43}$$

From these equations

$$n_1 = \frac{\rho_0}{\rho_0 + \delta\rho} n_0 \tag{2.44}$$

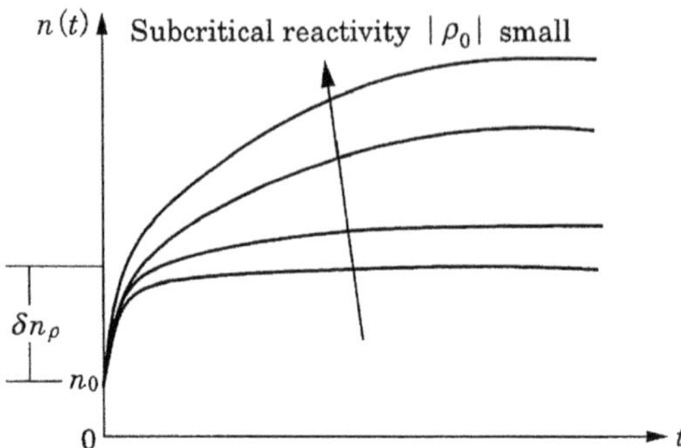

Fig. 2.5 Power variation relative to stepwise reactivity input in subcritical state

$$\delta n_t = n_1 - n_0 = \frac{-\delta\rho}{\rho_0 + \delta\rho} n_0 \tag{2.45}$$

In other words, if $\delta\rho$ is the same, the change to this new equilibrium is large as subcritical reactivity $|\rho_0|$ is small.

The behavior of "*n*" is shown in Fig. 2.5 after the change of reactivity when the reactor is subcritical.

The reactor enters the status when the value first increased for δn_ρ.

Eqs. (2.41) and (2.45) lead to the following.

$$\frac{\delta n_\rho}{\delta n_t} = \frac{-\rho_0 - \delta_\rho}{\beta - \rho_0 - \delta_\rho}$$

Consequently, the entire change of δn_t is large when subcriticality $|\rho_0|$ is small for the same reactivity insertion, δn_{t}.

The ratio of the prompt component δn_ρ to the entire change of δn_{t} is small as subcriticality $|\rho_0|$ is small.

Chapter 2 Exercises

1. Draw a diagram of the power change when 0.1 % positive reactivity is inserted stepwise for 30 s in the critical reactor having steady 1-watt power.
   (Question 3-2 from the seventh examination for license of chief reactor engineers of Japan)
2. Prove that the total amount of released energy, *E*, until time "*t*" seconds, is expressed approximately by the following equation when large reactivity is inserted stepwise in a critical reactor.

$$E = [P(t) - P(0)] \frac{l}{k_{\mathrm{eff}}(\rho - \beta)}$$

ℓ is the prompt neutron lifetime.
ρ is the reactivity insertion rate ($\rho > \beta$).
k_{eff} is the effective neutron multiplication factor.
β is the delayed neutron fraction.
$P(t)$ is the reactor power at time "t" seconds.
The structural change of core materials and the reactivity feedbacks are ignored.
(Question 1-5 from 14th examination for license of chief reactor engineers of Japan)

3. Obtain the transient behavior of the reactor power with neutron source strength S and neutron lifetime ℓ, for the following three cases, and draw them in a figure. Ignore all reactivity feedbacks such as delayed neutron and temperature coefficient. Treat all energies as one group. Use the point-wise reactor approximation for the solution.

Case (1) When neutron source strength S is inserted quickly in the subcritical core (the effective multiplication coefficient $k_{\text{eff}} < 1$ and the neutron density $n = 0$) at time $t = 0$.
Case (2) When neutron source strength S is removed quickly from the subcritical core (the effective multiplication coefficient $k_{\text{eff}} < 1$ and the neutron density $n = n0$) at time $t = 0$.
Case (3) When neutron source strength S remains in the core (the neutron density $n = n_0$) of the critical reactor.

(Question 3-2 of the 17th examination for license of chief reactor engineers of Japan)

4. (a) Prove that the power change is roughly expressed by the following equation when the positive reactivity, δk, is inserted stepwise in the reactor operating at constant power.

$$\frac{dn}{dt} = \frac{\delta k - \beta}{l^*} n + \frac{\beta}{l^*} n_0$$

n is the power (in neutron numbers).
n_0 is the initial steady power (in neutron numbers).
ℓ^* is the average effective neutron lifetime.
β is the delayed neutron fraction.
t is the time.

(b) Solve the equation in (a), derive an approximate expression of the power change, and draw a graph.
(c) Determine the gradient of the power change at time zero.
(Question 3-2 from the 17th examination for license of chief reactor engineers of Japan)

5. When the large reactivity is added stepwise to the subcritical reactor with the neutron source strength S and the reactor becomes supercritical (delayed supercritical), neutron density n rapidly increases at first, and then it increases gradually.

   The rapidly increasing neutron density δn (>0) is small even for the same δk when (1) the initial value of n is small and (2) when S is large. Prove them using equations.

   Note that approximate values can be used for δn and δk as they are small. (Question 3-2 from the 16th examination for license of chief reactor engineers of Japan)

Chapter 3
Temperature Effect of Reactivity

Yoshiaki Oka

3.1 Reactor with Reactivity Feedback

In Chap. 2, it was assumed that effective multiplication coefficient k_{eff} and reactivity ρ do not depend on reactor power n, and the point reactor kinetics equations were solved. Their solutions are applicable to the reactor having almost zero power or zero number of neutrons. It is called the "zero-power reactor" and its reactor temperature does not change. In the actual reactor, however, when its number of neutrons (or its power) changes, the temperature of the reactor changes and therefore, the k_{eff} and ρ values change. These changes affect reactor power. This reactivity change is called the temperature effect of reactivity. The reactivity changes with reactor temperature and moderator density, etc. The reactor power changes with the reactivity. Therefore, this reactor power change is called the reactivity feedback effect.

Reactivity $\rho(t)$ can be expressed as the sum of reactivity $\delta\rho_{\mathrm{ext}}(t)$ which is externally applied to the reactor by control rods and other components, and of feedback reactivity $\delta\rho_{\mathrm{f}}(t)$which is applied to the reactor due to internal factors.

$$\rho(t) = \delta\rho_{\mathrm{ext}}(t) + \delta\rho_{\mathrm{f}}(t) \tag{3.1}$$

Here, $\delta\rho$ is the difference from the equilibrium value.

A block diagram of the reactor experiencing the reactivity feedback effect is shown in Fig. 3.1.

The characteristics of the reactor with reactivity feedback can be represented by combining the upper block which shows the kinetics having no feedback with the lower block which has the feedback mechanism (i.e., the feedback reactivity effect is taken into account).

Y. Oka and K. Suzuki (eds.), *Nuclear Reactor Kinetics and Plant Control*,
An Advanced Course in Nuclear Engineering,
DOI 10.1007/978-4-431-54195-0_3,

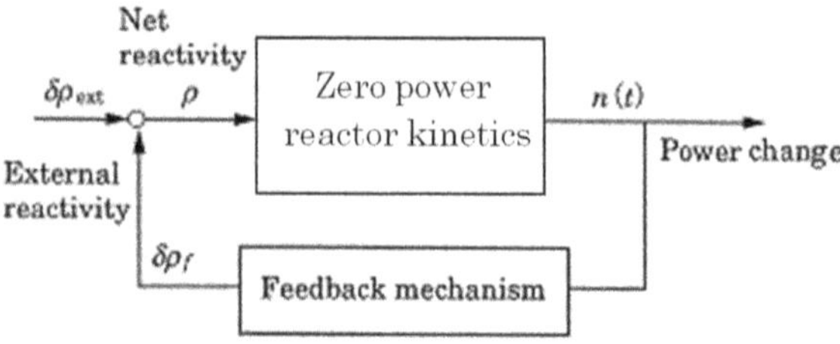

Fig. 3.1 Block diagram of a reactor with feedback

3.2 Reactivity Coefficient

The fuel, coolant, and moderator temperatures change when the reactor power changes. So, the densities change due to the thermal expansion and phase transition. As a result, the macroscopic neutron cross section changes, the neutron reaction rate changes, and the reactivity changes. When the fuel temperature rises, the thermal vibration of the ^{238}U nuclei and other fuel nuclei increases, and the distribution of the relative speed between neutrons and fuel nuclei is broadened by the Doppler effect. As the result, the self-shielding effect of resonance decreases, the resonance absorption rate of ^{238}U increases, and the reactivity decreases.

The major effects due to temperature can be summarized as follows:

<1> Change of resonance absorption due to temperature change of the fuel (Doppler effect).

<2> Change of neutron absorption and neutron spectrum (the neutron energy distribution) due to change of the moderator density and the coolant density.

<3> Other effects such as a buckling of fuel rods due to the differential thermal expansion between the core center part and peripheral part.

The reactivity effect by fuel rod buckling needs to be considered primarily for the fast reactor where the Doppler and the moderator density effects are small.

The effect of temperature change on the reactivity can be expressed by temperature coefficient of reactivity, α_T as

$$\alpha_T = \frac{d\rho}{dT} \tag{3.2}$$

where, ρ is the reactivity, and T is the temperature

$$\rho = \frac{k-1}{k}$$

From this equation, (3.3) is gotten.

$$\alpha_T = \frac{1}{k^2} \cdot \frac{dk}{dT} \tag{3.3}$$

As k is close to 1, it can be approximated using the following equation.

$$\alpha_{\mathrm{T}} = \frac{1}{k} \cdot \frac{\mathrm{d}k}{\mathrm{d}T} \tag{3.4}$$

The temperature coefficient can be divided into temperature coefficient $\alpha_{\mathrm{T}}{}^{j}$ of fuel, moderator, coolant, and others, and the total temperature coefficient is the sum of these individual coefficients.

$$\alpha_{\mathrm{T}} = \sum_{j} \alpha_{\mathrm{T}}{}^{j} \equiv \sum_{j} \frac{\partial \rho}{\partial T_j} \tag{3.5}$$

The time until the reactivity changes depends on the time needed for heat transfer. Although the temperature change of fuel occurs instantaneously due to prompt fission energy deposition on the fuel, the coolant temperature change is delayed due to the time needed for heat transfer from fuel rods to the coolant.

When the power has changed, the amount of reactivity change is indicated by the power coefficient of reactivity. This power coefficient can be defined by reactivity ρ differentiated by power P.

Further, it can be expressed by the temperature coefficient $\alpha_{\mathrm{T}}{}^{j}$ of each core part as follows.

$$\alpha_P \equiv \frac{\mathrm{d}\rho}{\mathrm{d}P} = \sum_{j} \left(\frac{\partial \rho}{\partial T_j}\right)\left(\frac{\partial T_j}{\partial P}\right) = \sum_{j} \alpha_{\mathrm{T}}{}^{j} \left(\frac{\partial T_j}{\partial P}\right) \tag{3.6}$$

If the power coefficient is positive, the reactivity increases when the power increases. So, the power increases further. This is a positive feedback. This is very dangerous situation and the reactor has to be designed to always keep the negative power coefficient in all possible operating conditions. The serious accident at the Chernobyl nuclear power plant was caused in large part by the positive power coefficient. The reactor with a positive power coefficient design cannot be constructed nor licensed now in most countries.

3.3 Fuel Temperature Coefficient (Doppler Coefficient)

In a thermal neutron reactor, most of the fuel temperature coefficient is caused by the Doppler effect. When the fuel temperature rises, the thermal vibration of the fuel nuclei increases and the distribution of relative speed to the target nuclei is broadened even for the same neutron energy. Thus in effect, the sharp peak of resonance cross section becomes mild. The shape is broadened and the peak decreases. An example is given in Fig. 3.2 where the micro neutron absorption cross section is compared between 293 and 1,500 K for resonance of ^{238}U at 36.7 eV.

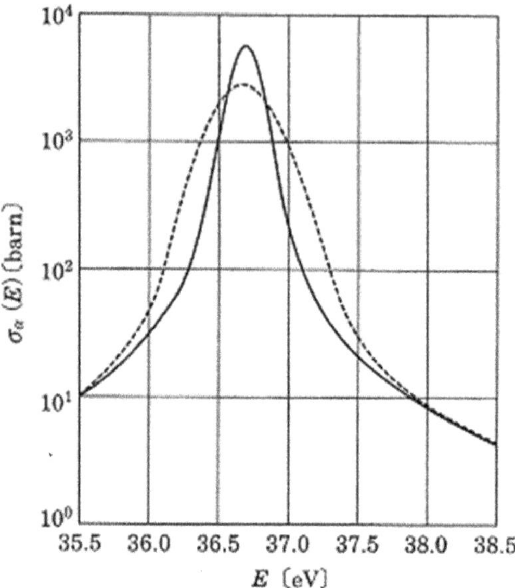

Fig. 3.2 Comparison of resonance absorption cross section of ^{238}U at 293 K (*full lines*) and 1,500 K (*dotted lines*)

Although the neutron spectrum (the distribution of neutron energy) around the resonance decreases due to the resonance absorption (especially for the resonance energy), its level of decrease is small if the temperature is high. This is because the peak of the micro resonance cross section has decreased at high temperature. This is shown in Fig. 3.3.

The resonance absorption can be expressed by the product of the micro cross section and the neutron spectrum. Figure 3.4 shows the difference of absorption (the increment of absorption amount) between 1,500 and 293 K.

The entire absorption rate is larger at 1,500 K than at 293 K. Thus, when the fuel temperature increases, the resonance absorption is increased by the Doppler effect. This is because the self-shielding effect of resonance is decreased.

Because LWRs (light water reactors) use the low-enrichment fuel containing a lot of ^{238}U, the reactivity decreases when the resonance absorption of ^{238}U increases. Thus, the reactivity coefficient becomes a negative value. The Doppler reactivity coefficient of LWRs is in a range of -5 to -1×10^{-5}/K. Figure 3.5 shows the Doppler coefficient of a BWRs (boiling water reactors).

Because the ^{240}Pu is accumulated at the end of cycle, the Doppler coefficient becomes a larger negative value. During the zero-power critical state at low temperature (20 °C and 0 % void fraction) or during the zero-power critical state at high temperature (287 °C and 0 % void fraction), the modulator density is high and the neutron spectrum is softer (there are many thermal neutrons). Because the numbers of resonance neutrons decrease and the resonance absorption rate decreases, the Doppler coefficient is small at zero power. The sodium cooled fast reactor that uses U-Pu mixed oxide fuel also has a negative value of Doppler reactivity coefficient. Because the Doppler reactivity coefficient has an immediate negative reactivity effect, it is important for reactor safety.

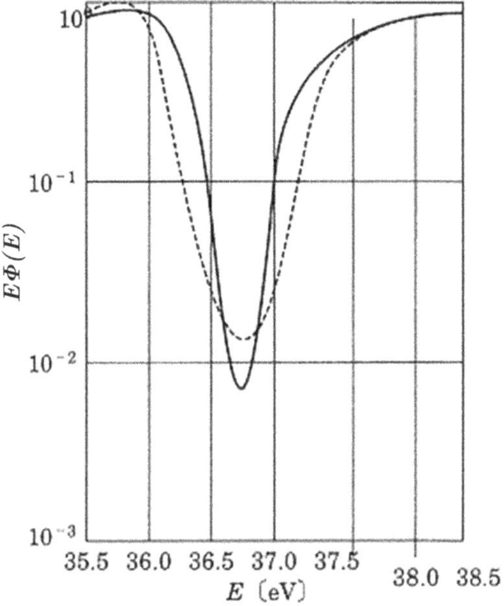

Fig 3.3 Comparison of neutron energy distribution of ^{238}U around the resonance at 293 K and 1,500 K (*dotted lines*)

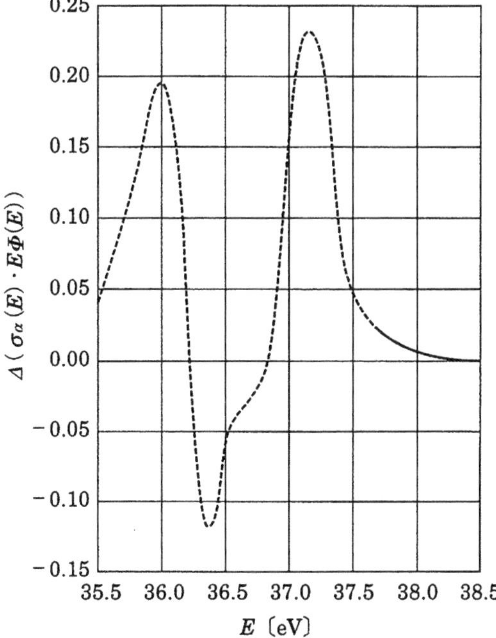

Fig. 3.4 Increment of resonance absorption (difference between 1,500 and 293 K)

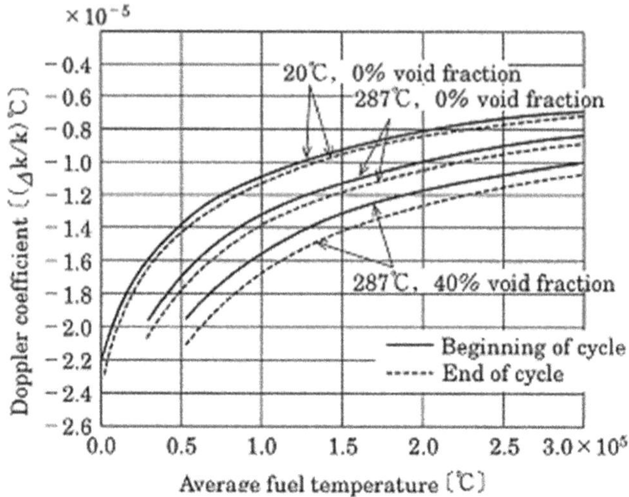

Fig. 3.5 Doppler coefficient of boiling water reactor. From the "Application for permission for installation modification of Chubu Electric Power Company's Hamaoka No. 5 nuclear power plant (4 September 1997)"

3.4 Reactivity Coefficient of Moderator and Coolant

The reactivity coefficients of moderator and coolant are given as change of reactivities divided by the change of temperature, density, pressure, void fraction, and other parameters. When the density of the moderator changes, the following two types of reactivity effect occur.

<1> The neutron absorption rate decreases (the absorption cross section decreases when the energy of neutrons increases) and the neutron leakage increases when the moderator temperature increases because the average energy of neutrons in the thermal neutron reactor is determined by this temperature. In the solid moderated reactor such as a graphite-moderated high temperature gas-cooled reactor, the primary moderator temperature effect is caused by hardening of the thermal neutron spectrum (i.e., shifting to higher energy distribution).

<2> The reactivity changes according to the degree of moderation. The sign of this effect changes depending on the moderator-to-fuel ratio of the core. Figure 3.6 shows the change of neutron multiplication factor with the moderator-to-fuel volume ratio. The neutron multiplication factor reaches the maximum at a certain volume ratio which is called the "optimum moderation state." The moderation is insufficient on the left side of this state in Fig. 3.6, but the moderation is excessive on the right side. In the insufficient moderation area, when the moderator density decreases, the neutron multiplication factor decreases and the reactivity decreases. The reactivity

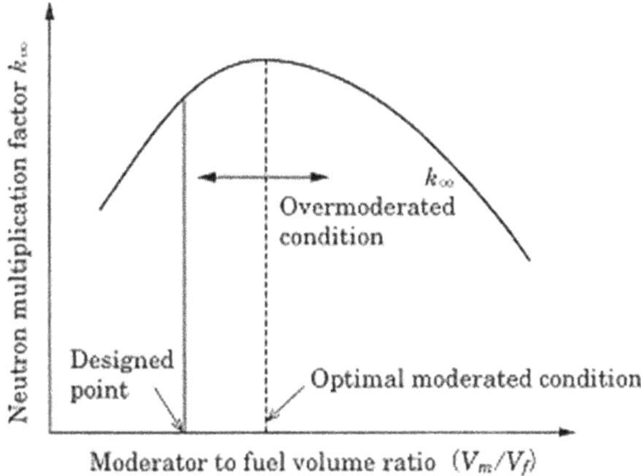

Fig. 3.6 Change of neutron multiplication factor with moderator to fuel volume ratio (Vm/Vf)

coefficient becomes a negative value. On the other hand, in the excessive moderation area, when the temperature increases, the moderator density decreases due to the thermal expansion and boiling. The neutron multiplication factor increases and the reactivity coefficient becomes a positive value. In the actual reactor, the core is designed so that the pitch between fuel rods is determined to have a slightly insufficient moderation in order to keep a negative reactivity coefficient.

In the LWR, the coolant also functions as the moderator. In the BWR, however, the core is cooled by the boiling two-phase flow and the moderator density changes when the volume of vapor bubbles (voids) changes. The (moderator) void coefficient of BWRs is shown in Fig. 3.7.

In the PWR (pressurized water reactor), boric acid is added to the primary cooling water. The boric acid concentration is decreased and it is used to compensate for the decrease in reactivity due to burnup of the fuel. This is called "chemical shim control." In this case, when the moderator or coolant temperature increases and the coolant density decreases, the density of the boric acid also decreases. The neutron absorption rate decreases and the moderator temperature coefficient may become a positive value when the boric acid concentration is high.

It should be noted that a large negative reactivity coefficient is not necessarily good from the viewpoint of safety. For example, if the pump of a shut-down loop is started erroneously, the cold coolant is fed into the core and positive reactivity is added. This is called a cold coolant ingress accident. If the negative reactivity coefficient is large, the inserted positive reactivity increases. In BWRs, if the steam turbine load is lost due to a failure of the power transmission line, the turbine control valve, installed on the main steam line from the reactor to the turbine, is quickly closed in order to protect the turbine. This increases the reactor pressure and decreases the volume of voids in the core coolant. Positive reactivity is inserted. Although some measures can be taken, it is most important to design the core to have the appropriate negative value of void reactivity coefficient. Measures which

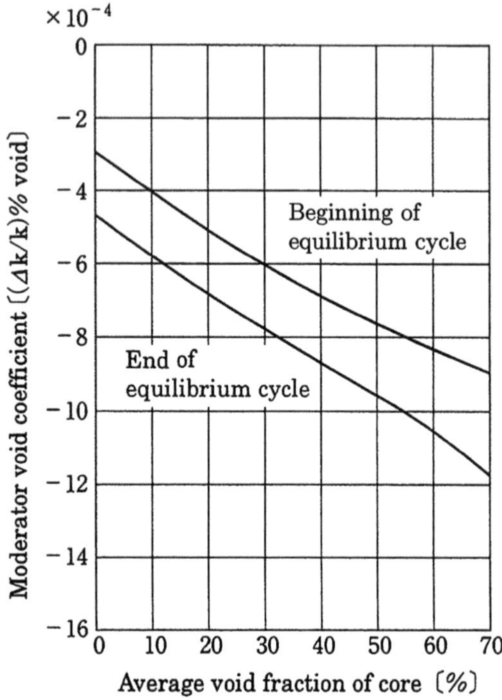

Fig. 3.7 Void coefficient of moderator in boiling water reactor. From the "Application for permission for installation modification of Chubu Electric Power Company's Hamaoka No. 5 nuclear Power plant (April of 1997)"

can be taken include an automatic decrease in reactor power (called set back), the release of steam to the condenser due to opening of the turbine bypass valve, the increase of the number of voids by reducing the recirculation pump speed and reducing core coolant flow.

Table 3.1 summarizes PWR reactivity coefficients.

The Canadian deuterium uranium reactor (CANDU) and the graphite moderated boiling water cooled pressurized tube reactor (RBMK) use moderator and coolant which are separated by pressure tubes. In these reactors, the decrease of coolant density does not cause the decrease of moderator density or hardening of the neutron spectrum. When the coolant density decreases, the neutron absorption rate of the coolant decreases and the reactivity increases. The coolant temperature coefficient may become a positive value. Table 3.2 compares the thermal neutron cross sections and reactor physics parameters of light water, heavy water, and graphite. Although light water has a large moderating ratio and it can moderate fast neutrons effectively with a small volume, but its thermal neutron cross section is large when compared with those of heavy water and graphite. Thus, the coolant void coefficient can often become a positive value in the pressure tube graphite-moderator reactor (for example, the RBMK) which uses light water coolant. It has caused a positive power coefficient and resulted in a nuclear power excursion.

Table 3.1 Reactivity coefficients of PWRs

Moderator temperature coefficient	$(+0.8 \sim -9.4) \times 10^{-4}(\Delta k/k)/°C$
Doppler coefficient	$(-2.1 \sim -5.2) \times 10^{-5}(\Delta k/k)/°C$
Void coefficient	$(+0.7 \sim -3.1) \times 10^{-3}(\Delta k/k)/\%$Void fraction
Pressure coefficient	$(+8.0 \sim -0.5) \times 10^{-5}(\Delta k/k)/(kg/cm^2)$
Moderator density coefficient	$(+0.51 \sim 0)\ (\Delta k/k)/(g/cm^3)$

The moderator temperature coefficient becomes a negative value at high temperature and high power

Table 3.2 Comparison of the thermal neutron cross section of light water, heavy water, and graphite

	Density	Micro absorption cross section (barn)	Macro absorption cross section (10^{-5}, cm^{-1})	Moderating ratio
H_2O	1.0	0.66 (hydrogen)	2,200	1.36
D_2O	1.1	0.001 (deuterium)	3.3	0.18
Graphite	1.6	0.0034 (carbon)	27	0.16

3.5 Reactivity Coefficients of the Fast Reactor

Because the fast reactor has a hard neutron spectrum, its Doppler coefficient is smaller than that of thermal neutron reactor. Also, the fast reactor does not use any moderator. Its reactivity temperature coefficient is generally small. In the fast reactor, the reactivity effect may occur due to bowing of fuel rods due to differential thermal expansion. Thermal expansion is large at the center of the core where the temperature is high. The fuel rods are bent toward the center of the core, when they are restricted to move horizontally both at the top and at the bottom of the core. This increases the reactivity and it was a problem in the early small fast reactors. This problem can be solved by determining the appropriate position and method to hold and support the fuel rods and core.

In a large liquid sodium-cooled fast reactor, the sodium void coefficient tends to be positive in the center of the core. It is a problem treated in hypothetical accidents when the sodium coolant boils. The effect of the sodium void coefficient is a competition between the positive reactivity effect where nuclear fission increases due to the hardened spectrum, and the negative reactivity effect due to the increased leakage of neutrons. As the neutron leakage effect is small at the center of a large core, the local sodium void coefficient becomes a positive value. Countermeasures can be taken: (1) to increase the neutron leakage by designing a flat core; (2) to decrease the void coefficient by softening the neutron spectrum in the core and to increase the Doppler reactivity; or (3) to form a thin hydride layer between the blanket and seed fuel in order to moderate the fast neutrons (produced in the seed at voiding) in this layer and to absorb the moderated neutrons by the blanket fuel (^{238}U).

3.6 Power Defect of Reactivity

Because the reactor is always designed to have a negative value of power reactivity coefficient, the negative reactivity applies to the reactor when the output is increased from the zero-output critical state to the output operation state. This is the power defect of reactivity. After the reactor has been set to the critical state with zero output, we need to add an external reactivity by removing control rods or others so that the reactor output is increased.

When the high power operation starts, fission products such as xenon and samarium having the high neutron absorption are also accumulated rapidly and it is necessary to compensate this negative reactivity. The power defect by reactivity feedback has the time constant of several seconds to minutes, but the accumulation of xenon and samarium has a time constant of several hours to several days.

3.7 Temperature Feedback Model

To calculate the dynamics of the reactor plant having reactivity feedback, the point-reactor kinetics equations described in Chap. 2 are solved together with the equations of reactor thermal hydraulics, and the temperature and the reactivity feedback are determined. In the actual plant calculation, the space-dependent temperature and power distribution are determined and weighted appropriately, and then the reactivity feedback amount to be used for point reactor kinetics equations is calculated. Although numerical analysis is required for the actual plant dynamics calculation, the following typical temperature feedback models can be used for exercises.

<1> Newton cooling model

$$\frac{\mathrm{d}T_{\mathrm{F}}}{\mathrm{d}t} = KP(t) - \gamma(T_{\mathrm{F}} - T_{\mathrm{C}}) \tag{3.7}$$

<2> Adiabatic model

$$\frac{\mathrm{d}T_{\mathrm{F}}}{\mathrm{d}t} = KP(t) \tag{3.8}$$

<3> Constant heat removal model

$$\frac{\mathrm{d}T_{\mathrm{F}}}{\mathrm{d}t} = K(P - P_0) \tag{3.9}$$

Here, K and γ are thermal constants of the core, P is the power, T_{F} is the fuel temperature, and T_{C} is the coolant temperature.

Chapter 3 Exercises

1. The reactor with a negative temperature coefficient $-\alpha[\Delta k/k/\ ^\circ\mathrm{C}]$ and a large heat capacity C[kcal/ °C] is operated in a steady state. Add the reactivity $\delta k[\Delta k/k]$ stepwise and complete the following tasks.

   (a) Calculate the temperature increase until the neutron flux reaches the maximum level.
   (b) Calculate the neutron flux at the maximum level.
   (c) Draw the transient time change of neutron flux φ and temperature (T °C).

Assume the following:

<1> Delayed neutrons can be ignored.
<2> The heat generated after the stepwise reactivity addition is not removed to the outside, but it is used to increase the core temperature only.
<3> The transient change is small and occurs around $k = 1$.
<4> The initial neutron flux is $\varphi_{0,}$ the neutron lifetime is $l(s)$, the conversion factor between the neutron flux and heat is A (kcal/neutron flux * s).
(Question 3.3 from the 11th examination for license of chief reactor engineers of Japan)

Chapter 4
Kinetics Parameters and Reactivity Measurement Experiments

Yoshiaki Oka

Introduction To analyze reactor kinetics, we need to know the kinetics parameters of the reactor. Also, we build the reactor, operate it in the critical state, and determine the reactivity of each control rod. This chapter explains the primary steps of these operations.

4.1 Critical Approach Experiment (Inverse Multiplication Coefficient Method)

The operation to increase the reactivity of subcritical reactor and to set the reactor to the critical state is called the critical approach. The first critical approach of the reactor just constructed is called the "initial critical approach."

The critical approach has the following steps:

<1> Add the fuel gradually
<2> Reduce the soluble toxicant concentration of the moderator
<3> Increase the moderator or reflector

If the subcritical reactor has no neutron source, its fission chain reaction is dropped quickly. If the reactor has the neutron source, the neutron flux is constantly formed according to the intensity of the source and the subcritical reactivity as shown by Eq. (2.35). The inverse number of subcritical reactivity $1 - k$ is written by M as follows:

$$M = \frac{1}{1 - k} \tag{4.1}$$

M^{-1} is called the inverse multiplication factor. When it closes to the critical state, value k closes to 1 and, therefore, M^{-1} closes to zero.

In the critical approach procedure, we place multiple measurement systems in each of the reactor and add the fuel gradually around the neutron source. As the

Y. Oka and K. Suzuki (eds.), *Nuclear Reactor Kinetics and Plant Control*,
An Advanced Course in Nuclear Engineering,
DOI 10.1007/978-4-431-54195-0_4, © Springer Japan 2013

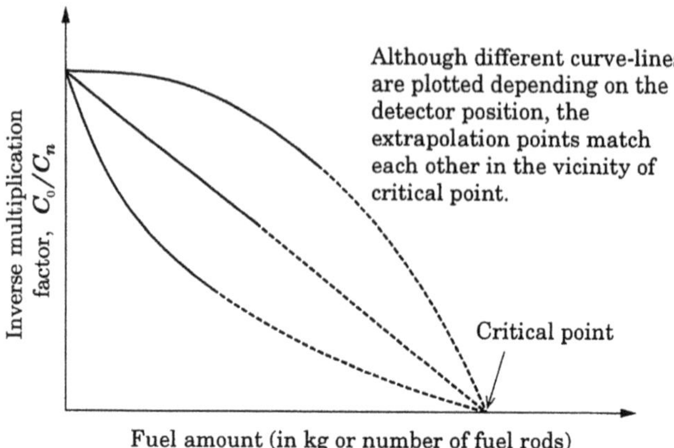

Fig. 4.1 Concept of critical approach with inverse multiplication factor

ratio of counting rate C_i of each step to the initial counting rate C_0 is an inverse multiplication coefficient, we plot the inverse multiplication factor in the figure by setting the fuel amount on the horizontal axis. We extrapolate it and determine a predictive value of fuel amount to have the zero C_0/C_i ratio. In the next step, we assume the amount of loaded fuel is half of the difference from the predicted value and we repeat this operation. The concept of this procedure is shown in Fig. 4.1.

Because we use multiple neutron detectors, we can obtain the predictive value from each plot. We use the minimum one. The extrapolation points match closely to the critical point.

In the initial critical approach, we do not know the fuel amount that causes the reactor to enter the critical state. Therefore, we need to perform the above procedure carefully. If the subcritical reactor does not have a neutron source, the neutron flux is not formed and neutron counting by the fission chain reaction is not obtained. Although different curve-lines are plotted depending on the detector position, the extrapolation points match each other in the vicinity of critical point. If the reactor is in a subcritical state, we need to insert the neutron source into the core and make the critical approach. This is required not only for the initial critical approach but also when we start the reactor and set it to the critical state. If we do not use the neutron source or if we cannot obtain a sufficient counting rate for the counters, and if we erroneously load a large amount of fuel into the core to trigger the supercritical state, the output can rapidly increase when neutrons are suddenly supplied by the cosmic rays or other source, creating a very dangerous state. The neutron sources are californium-252 (252Cf) radiation source, plutonium–beryllium (Pu–Be) radiation source, and antimony–beryllium (Sb–Be) radiation source. The neutron emission mechanism can be the spontaneous fission for the californium-252 radiation source, the reactivity of beryllium (α, n) generated by alpha radiation of plutonium for Pu–Be radiation source and the reactivity of beryllium (γ, n) generated by high-energy gamma radiation from the antimony for Sb–Be radiation source.

The californium-252 and Pu–Be are used as the primary radiation source to be used during startup of initial operation. When the reactor is started, the antimony (Sb) continues to be irradiated in the reactor and the gamma rays are radiated. It is used as the secondary radiation source. When we use the primary radiation source, deterioration due to the neutron irradiation occurs. In a research reactor, therefore, we often pull out the radiation source from the core after startup of reactor operation and insert it each time when we start the reactor. We can determine that the reactor has reached the critical state by confirming that output is being maintained at a certain level when we remove the neutron source. If the critical reactor has an external neutron source (a neutron source not by fission chain reaction), logically a linear output increase occurs in proportion to the intensity of neutron source.

4.2 Calibration of Control Rods

Here, we will discuss the entire reactivity of a single control rod, or how to measure the equivalent reactivity per unit length. The period technique, the control rod drop technique, and the comparison technique are often used.

4.2.1 Period Method

The period method is used to determine the equivalent reactivity per unit length of control rod. When we pull control rods slightly out of the critical reactor and when we suddenly input (stepwise) a small positive reactivity, the reactor enters the transient state and its output increases by following the single exponential function as shown by Eq. (2.13). $T_0 = 1/\omega_0$ is called the reactor period. By measuring the T_0, we can determine the reactivity using the following equation that is conducted with the small ℓ from Eq. (2.10).

$$\rho = \sum_{i=1}^{6} \frac{\beta_i}{1 + \lambda_i T_0} \tag{4.2}$$

This reactivity is the reactivity equivalent to the control rods that we have pulled up. In the actual measurement, we measure the time when the output is doubled, divide it by 0.693, and determine the T_0. We do not measure it immediately after the reactor has reached the critical state. We need to measure it after the reactor has maintained the critical state for more than a few minutes, that is, after sufficient time has passed until the delayed neutron precursors can reach the equilibrium concentration.

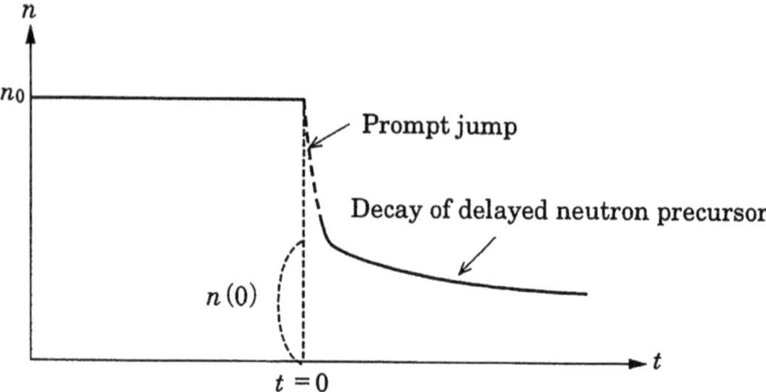

Fig. 4.2 Control rod drop method

4.2.2 Control Rod Drop Method

The control rod drop method is used to determine the entire reactivity of a single control rod or to determine the reactivity of several control rods. When the reactor is operating with a constant output and if we suddenly add the negative reactivity $-\delta\rho$ by inserting control rods, the reactor output immediately drops by the prompt jump. Then, the output gradually drops by the decay of delayed neutron precursors (Fig. 4.2).

If the initial output is n_0 and the output that is dropped by prompt jump is $n(0)$, the following is obtained as shown by Eq. (2.34).

$$\frac{n(0)}{n_0} = \frac{\beta}{\beta + \delta\rho} \tag{4.3}$$

We can calculate the $n(0)/n_0$ by extrapolating the time change of counting to $t = 0$. From this result, we can determine the initially input reactivity $\delta\rho$ using Eq. (4.3).

To measure the reactivity by the rod drop technique, we can measure the drop of reactor output after dropping of control rods over the time using the multichannel neutron counter, determine its integral value, and determine the reactivity using the following Eq. (4.4).

$$\delta\rho = \frac{n_0}{\int_0^\infty n(t)\mathrm{d}t}\left[l + \sum_{i=1}^{6}\frac{\beta_i}{\lambda_i}\right] \tag{4.4}$$

The accuracy of this technique is higher than the technique that determines the $n(0)$ by extrapolation. We need to keep the reactor in the critical state before insertion of control rods until the delayed neutron precursors reach the equilibrium state, and this is the same as for the period technique. Figure 4.3 shows the equivalent reactivity per unit length of control rods, and the change of equivalent reactivity due to the pull-out amount of control rods.

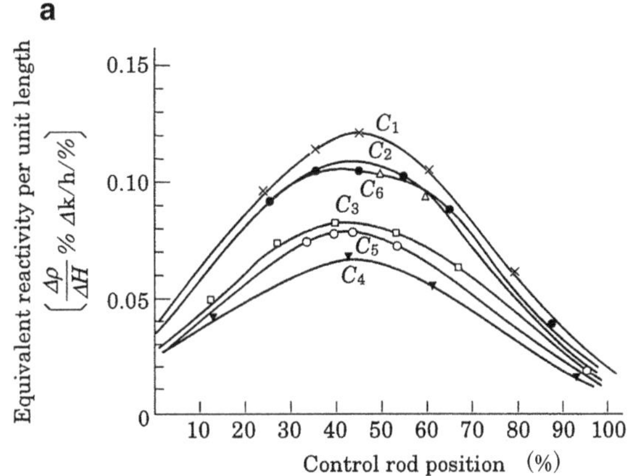

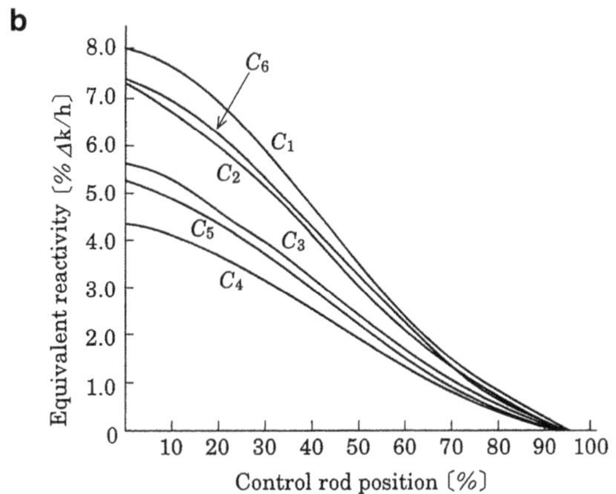

Fig. 4.3 Equivalent reactivity of control rod. (**a**) Equivalent reactivity per unit length (differential curve). (**b**) Variation in reactivity by the amount that control rod is pulled out (integral curve)

The equivalent reactivity per unit length becomes the maximum value when the control rods exist at the center of the core. When it is integrated, the integral curve of (b) is obtained.

4.2.3 Comparison Method

Assume that we have calibrated one control rod in the period method, and we insert this control rod and a control rod to be calibrated in the critical reactor. If we move one control rod and make the reactor critical again, we can determine the equivalent reactivity of the control rod to be calibrated by comparing it with the equivalent reactivity of the already calibrated control rod. If we repeat this operation in each stroke of rods, we can calibrate the entire reactivity. If the control rods are closely mounted in the reactor, we need to take care not to interfere with the rods with each other when calibrating the rods in this technique.

4.3 Neutron Source Multiplication Method

In the subcritical reactor, the neutron source strength S and neutron number n have the relation of Eq. (2.35).

$$n \propto \frac{S}{1 - k_{\text{eff}}}$$

If neutron n^* is measured with the known subcriticality $1 - k_{\text{eff}}{}^*$, we can determine k_{eff} by measuring n for the known $1 - k_{\text{eff}}$.

$$1 - k_{\text{eff}} = (1 - k_{\text{eff}}*)\frac{n^*}{n} \tag{4.5}$$

It can be used to measure the light subcritical reactor (that is, the subcritical reactor close to the critical state).

4.4 Neutron Source Pull-Out Method

When the subcritical reactor maintains neutron number (output) n_0 using the neutron source having the strength S, quickly pull out the neutron source from the reactor. Measure the neutron numbers n_1 after the prompt jump, and determine the subcriticality ρ_0.

If we use the approximation of delayed neutron constant generation rate for the point reactor kinetics equation (2.36) of the subcritical reactor, we can obtain the following equation:

$$n_1 = \frac{\beta}{\beta - \rho_0} n_0 \tag{4.6}$$

By deforming it, we can obtain the following solution.

$$\frac{\rho_0}{\beta} = 1 - \frac{n_0}{n_1} \tag{4.7}$$

We can determine ρ_0 by measuring n_1. Here, we ignore the effect on the reactivity of neutron absorption effect of the neutron source itself.

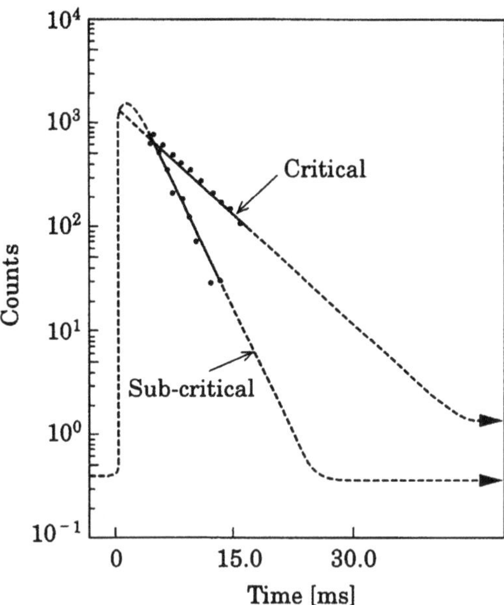

Fig. 4.4 Damping of pulsed neutron in critical system and subcritical system

4.5 Pulsed Neutron Technique

Using an accelerator, generate the pulsed neutrons in the critical or subcritical reactor, and measure the attenuation of prompt neutrons.

The attenuation of neutrons after pulse generation can be expressed by the following equation.

$$n(t) \propto \mathrm{e}^{-\alpha t} \tag{4.8}$$

where α is the attenuation constant of prompt neutron, and

$$\alpha = \frac{\beta - \rho_0}{\Lambda} = \frac{1 - k_{\mathrm{eff}}(1 - \beta)}{l} \tag{4.9}$$

it is $\alpha_c = \beta/\Lambda$ during critical state. If we measure the α_c, we can obtain the following using the ratio to α:

$$\frac{\alpha}{\alpha_c} = \frac{\beta - \rho_0}{\beta} = 1 - \frac{\rho_0}{\beta} \tag{4.10}$$

Thus, we can determine the subcriticality ρ_0 in dollars. A measurement example is given in Fig. 4.4.

Although the accelerator is required, we can measure the deep subcriticality using the pulsed neutron technique.

4.6 Control Rod Oscillator Technique

If very small cyclic disturbance $\delta\rho$ is given by using the control rods or absorbing material of the critical reactor, the reactor output oscillates cyclically. Using this oscillation, we can determine the reactor transfer function (the frequency response of reactor) (for details, see Part II).

The disturbance, zero-output transfer function $G(j\omega)$, and oscillation δn of reactor output have the following relationship.

$$\frac{\delta n}{n_0 \delta\rho} = G(j\omega) \tag{4.11}$$

In a high frequency, the transfer function can be:

$$G(j\omega) \sim \frac{1}{\omega\Lambda} \tag{4.12}$$

Therefore,

$$\delta\rho = \omega\Lambda \frac{\delta n}{n_0} \tag{4.13}$$

From this equation, if value Λ is known, we can determine value $\delta\rho$ by measuring δn.

4.7 Reactor Noise Analysis Technique

The neutron numbers in a reactor fluctuate around the average even in the steady state. When the absorption and fission occur stochastically, this variation should show the Poisson distribution. Actually, however, two or more neutrons are generated by fission. As its chain has the correlation and a displacement occurs from the Poisson distribution, we analyze this displacement.

4.7.1 Feynman-α Method

If we measure the counting distribution of constant time T and the average rate by changing value T, we can determine prompt neutron attenuation constant α.

4.7.2 Rossi-α Method

If we measure the distribution of time period for continuous pulses by using a pulse from the neutron detector as the time reference (trigger), we can determine prompt neutron attenuation constant α.

The Feynman-α method is suitable for the measurement of the thermal neutron reactor system, and the Rossi-α method is suitable for the measurement of the fast reactor system.

If we use the detector that can measure the neutron as a continuous amount of current (but not the pulses), we can determine the kinetics parameters from the correlation function of current fluctuation or the power spectrum density of its Fourier transformation. The correlation functions and the definition of power spectrum density are given below.

- Auto-correlation function

$$\varphi_{xx}(\tau) = \lim_{T\to\infty} \frac{1}{2T} \int_{-T}^{T} x(t)\, x(t+\tau)\mathrm{d}t \tag{4.14}$$

- Cross-correlation function

$$\varphi_{xy}(\tau) = \lim_{T\to\infty} \frac{1}{2T} \int_{-T}^{T} x(t)\, y(t+\tau)\mathrm{d}t \tag{4.15}$$

Because the power spectrum density is the Fourier transformation of correlation function,

- Auto-power spectrum density

$$\Phi_{xx}(\omega) \equiv \frac{1}{\pi} \int_{-\infty}^{\infty} d\tau e^{-j\omega\tau} \varphi_{xx}(\tau) \tag{4.16}$$

- Cross-power spectrum density

$$\Phi_{xy}(j\omega) \equiv \frac{1}{\pi} \int_{-\infty}^{\infty} d\tau e^{-j\omega\tau} \varphi_{xy}(\tau) \tag{4.17}$$

Because the auto-correlation function is an even function of $\Phi_{xx}(-\tau) = \Phi_{xx}(\tau)$, the auto-power spectrum density is a real number. The cross-power spectrum density is a complex number. The cross-correlation function has the following relationship $\Phi_{xy}(-\tau) = \Phi_{xy}(\tau)$.

The auto-power spectrum density $\Phi_{ii}(\omega)$ contains a square of gain of reactor transfer function $G(j\omega)$ as shown in Eq. (4.18).

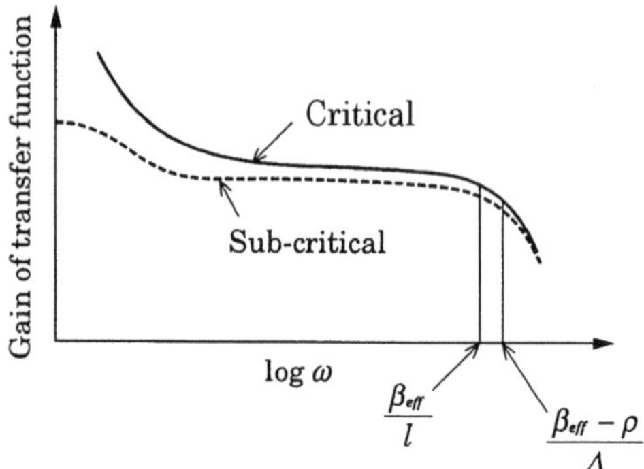

Fig. 4.5 Angular frequency dependence of gain in reactor transfer function

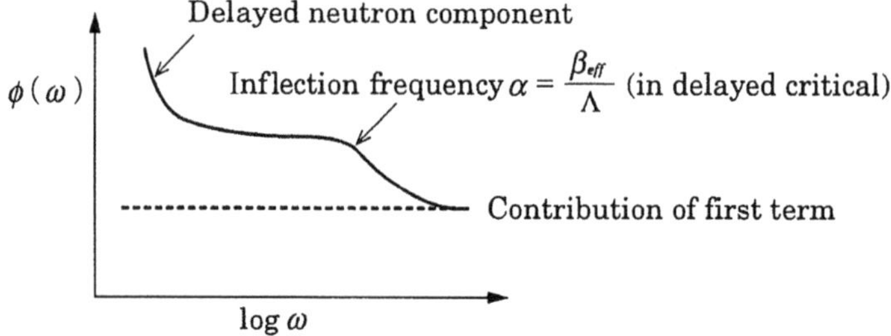

Fig. 4.6 Angular frequency dependence of auto-power spectrum density

$$\Phi_{ii}(\omega) = \varepsilon P_0 + \varepsilon^2 \frac{\overline{\nu(\nu-1)}}{\nu^{-2}} P_0 |G(j\omega)|^2 \tag{4.18}$$

The first term shows the noise of detector, and the second term shows the contribution of the nuclear fission branching process. ε is the efficiency of detector, ν is the number of prompt neutrons per fission, and P_0 is the nuclear fission rate.

As shown in Fig. 4.5, a high-order breaking point of gain for the nuclear transfer function, that is, the inflection point at the higher side of angular frequency locates at $\alpha_c = \beta/\Lambda$ for the critical reactor and at $\alpha = (\beta - \rho_0)/\Lambda$ for subcritical reactor.

If the auto-power spectrum density is drawn for the angular frequency, it has the shape as shown in Fig. 4.6.

If we measure the auto-power spectrum density and determine the high-order inflection point, it corresponds to the high-order inflection point of the reactor transfer function. So, we can determine value α_c for the critical reactor and value α for the subcritical reactor. If β/λ is known, we can determine subcriticality ρ_0.

The first term of Eq. (4.18) is the noise of detector. If the angular frequency is high or if the efficiency of detector is poor and if the first term of Eq. (4.18) is greater than the second term, we can measure and determine the reactor output (the fission rate) from the auto-power spectrum density. This is called the Campbell's method.

The reactor noise technique has advantageous measurement techniques that do not require the need for manual addition of disturbance to the reactor. The phase information of the transfer function cannot be obtained.

Chapter 4 Exercises

1. The reactor has control rods at approximately 50 % position and is being maintained at the low-output critical state. When we have pulled the control rods out by 15 cm, the output has increased with period $T = 30$ s. In this case, how much reactivity is there per 1 cm of control rods? Here, we assume that the reactor kinetics is affected by a group of delayed neutrons. Use the following equations and constants.

Reactivity	$\delta k = l/T + \beta/(1 + \lambda T)$
Neutron average lifetime	$l = 10\text{–}3$ s
Ratio of delayed neutrons	$\beta = 0.0064$
Decay constant of delayed neutron	$\lambda = 0.077\ \text{s}^{-1}$

   (The 4th test for license of chief reactor engineers of Japan, question 3-3)

2. After the reactor has reached the critical state, we inserted the control rods from the 100 % pullout position into the 80 % position in order to calibrate the control rods. When we measured the neutrons using the neutron counter, it was 38,728 cpm. The equivalent reactivity for insertion of control rods was −0.38 %. After that, when we inserted the control rods into the 60 %, 40 %, 20 %, and 0 % position, the respective counting values were 7,889, 3,602, 2,240, and 1,848 cpm. Determine the equivalent reactivity of control rods at each position. (The 4th test for license of chief reactor engineers of Japan, question 3-4)

Part II
Actual Nuclear Reactor Plant Control

Chapter 5
Control System Basics and PID Control

Katsuo Suzuki

5.1 Basic Configuration of Control System

The first automatic control system is said to be the governor (a speed regulator) for the steam engine, invented by James Watt. The speed regulator enabled the steam engine to be used as a practical power source, starting the industrial revolution. "Control" is defined as "adding required operation to an object so that it can be adapted for a certain purpose" (JIS automatic control terms). Control can be divided into two main categories: automatic and manual. Automatic control is divided into feedback control, feed-forward control, sequential control, and others. Automatic control can be implemented by control systems among which single-variable control systems (single-input, single-output systems) are one type.

This chapter deals with these single-variable control systems. Feedback control is defined as "a control that compares between a controlled variable and a desired value (command signal), and performs a corrective operation to match between them" (JIS automatic control terms). Figure 5.1 shows the basic configuration of feedback control using main control system components. Arrows in the figure indicate the flow (transfer) of control signals, not energy flow.

This is how the block diagrams for control systems differ from those for electrical circuits. The way deviding system to blocks is not uniquely determined. However, this devision must be done to transfer control signals unidirectionally. A signal transfer path from the point detecting a controlled variable to the summation point is referred to as the backward path, and a signal transfer path from the output at the summation point to a controlled variable is the forward path.

This block diagram indicates three following features of the feedback control system. <1> The control signal transfer path is closed (a closed loop). <2> Control signals are transferred through the closed loop unidirectionally. <3> There is a desired value to which a controlled variable should be matched.

Control systems can be divided into the following three control types according to time-dependent characteristics of the desired value $v(t)$. A control with the $v(t)$ value that is unchanged with time is called constant value control or fixed command

Y. Oka and K. Suzuki (eds.), *Nuclear Reactor Kinetics and Plant Control*,
An Advanced Course in Nuclear Engineering,
DOI 10.1007/978-4-431-54195-0_5, © Springer Japan 2013

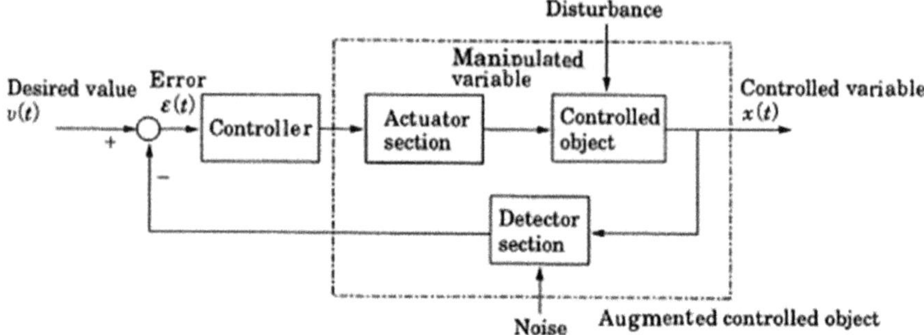

Fig. 5.1 Basic configuration of feedback control system

control. A control with the $v(t)$ given by a known function $f(t)$ is called program control, and a control with the $v(t)$ given by an unknown function is calledfollow-up control. Those three types differ from each other only in how the desired value $v(t)$ is given. The technique to minimize the error $\varepsilon(t)$ between the controlled variable $x(t)$ and the desired value $v(t)$, or to diminish the error caused by disturbances is common to all three types.

5.2 Transfer Function

5.2.1 *Component Transfer Function*

The transfer function is useful for theoretical control studies. We use the electrical circuit of Fig. 5.2 to understand the transfer function. When the switch S is closed at $t = 0$, we obtain a transfer function from the power supply $e(t)$ to the circuit current $i(t)$.

Dynamics characteristics of the current $i(t)$ flowing through this diagram can be described by the following equation:

$$L\frac{\mathrm{d}i(t)}{\mathrm{d}t} + Ri(t) + \frac{1}{C}\int_0^t i(\tau)\,\mathrm{d}\tau = e(t) \tag{5.1}$$

Applying Laplace transform to both side yields the following equation:

$$L\{sI(s) - i(0+)\} + RI(s) + \frac{1}{Cs}I(s) = E(s)$$

where $I(s)$ and $E(s)$ are Laplace transforms, respectively, of $i(t)$ and $e(t)$.

$$I(s) = \int_0^\infty i(t)\,\mathrm{e}^{-st}\,\mathrm{d}t, \qquad E(s) = \int_0^\infty e(t)\,\mathrm{e}^{-st}\,\mathrm{d}t$$

Fig. 5.2 Electrical circuit

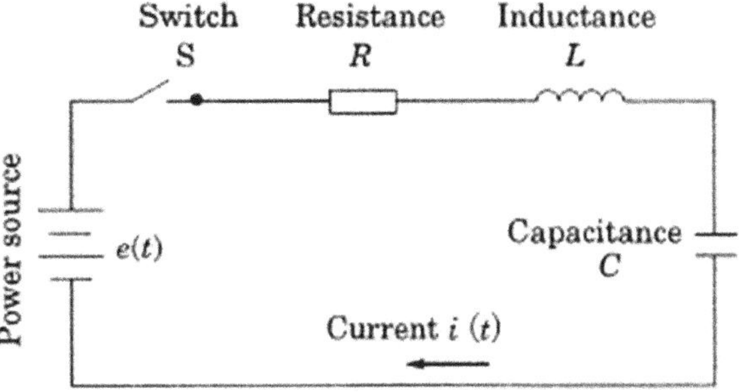

The initial value of circuit current is $i(0+) = 0$.

Thus, when we regard $e(t)$ as an input signal and $i(t)$ as an output signal, and if we define the circuit transfer function as the ratio between $I(s)$ and $E(s)$, then we obtain the following expression.

$$G(s) \equiv \frac{I(s)}{E(s)} = \left(\frac{1}{R}\right)\frac{T_2 s}{T_1 T_2 s^2 + T_2 s + 1}, \quad T_1 = \frac{L}{R}, \; T_2 = RC \tag{5.2}$$

As this example shows, the transfer function can be defined as "the ratio between Laplace transform of an output signal and that of an input signal when all initial values are set to zero." We should note that the transfer function is defined in components where a signal travels unidirectionally, e.g., voltage $e(t) \rightarrow$ current $i(t)$. (Here, "unidirection" denotes a relation between $i(t)$ and $e(t)$ for which any change on $i(t)$ does not cause $e(t)$ to change.) This unidirectionality of signal transfer permits us to handle several components connected in cascade as one transfer component, or to apply equivalent transform for a set of transfer components in a block diagram.

The component transfer function of Eq. (5.2) has been derived from a differential equation that represents the input/output relation. In practice, however, there are many cases where the input/output relations of the components or the system are too complex to be represented by a simple differential equation. The following describes how to obtain transfer functions in such cases.

If we measure the input signal $x(t)$ and output signal $y(t)$ of an object component or system, and let $X(s)$ and $Y(s)$ denote the respective Laplace transformed functions, then we can obtain a transfer function from the following expression.

$$G(s) = \frac{Y(s)}{X(s)} \tag{5.3}$$

If the input $x(t)$ is a unit size impulse-like function, then $X(s) = 1$, and we obtain

$$G(s) = Y(s) \tag{5.4}$$

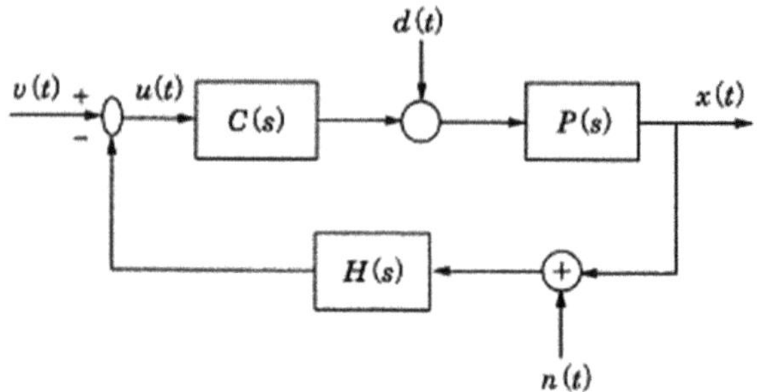

Fig. 5.3 Block diagram of a feedback control system

This means that if we input an impulse-like $x(t)$, which is actually not possible, then Laplace transform of the output simply yields the transfer function of the objective component or system. If we use the input $x(t)$ of the unit step function $u(t)$, then we get the following expression.

$$X(s) = \frac{1}{s}$$

Thus, we obtain the component transfer function

$$G(s) = sY(s) \tag{5.5}$$

Even if a differential equation representing the input/output relation of the components is not known, we can obtain a transfer function by this principle-based approach. In order to measure actual input/output signals, however, we must solve problems concerning measurement experiments. These problems include generation of input signals, elimination of measurement noise, range of linearity/nonlinearity, and restrictions of operation. To get the correct solution, a method has been proposed to measure transfer functions using other types of input instead of impulse-like or step-like input.

5.2.2 *The Transfer Function in Feedback Control Systems*

Figure 5.3 illustrates the connection of transfer components in general feedback control systems.

In this figure, $C(s)$ is the transfer function of a controller, $P(s)$ is that of a controlled object, and $H(s)$ is that of a feedback component. Additionally, $x(t)$ denotes a controlled variable and $v(t)$ its desired value. $d(t)$ denotes a disturbance entering the controlled object which disturbs the controlled variables, and $n(t)$ is noise associated with detector.

Based on this figure, we can obtain three types of transfer functions in a feedback control system, i.e., from the desired value $v(t)$, disturbance $d(t)$, and noise $n(t)$, respectively, to the controlled variable $x(t)$, provided that $G(s) = C(s)P(s)$.

$$G_{xv}(s) = \frac{G(s)}{1 + G(s)H(s)}, \qquad G_{xd}(s) = \frac{P(s)}{1 + G(s)H(s)},$$
$$G_{xn}(s) = -\frac{G(s)H(s)}{1 + G(s)H(s)} \tag{5.6}$$

As shown above, for negative feedback systems, $1 + G(s)H(s)$ appears in the denominator of all transfer functions. The $G(s)H(s)$ is the product of all component transfer functions existing on a closed loop and it is referred to as the open-loop transfer function. As described later, the stability of a feedback control system is discussed on the basis of the open-loop transfer function.

When we combine the desired value $V(s)$, disturbance $D(s)$ and noise $N(s)$, fluctuation of the controlled variable $X(s)$ becomes Eq. (5.7).

$$X(s) = G_{xv}(s)V(s) + G_{xd}(s)D(s) + G_{xn}(s)N(s) \tag{5.7}$$

Therefore, $\varepsilon(t) = v(t) - x(t)$, the deviation of the controlled variable from the desired value is expressed as the following expression:

$$\begin{aligned} E(s) &= V(s) - X(s) \\ &= \left(1 - \frac{G(s)}{1 + G(s)H(s)}\right)V(s) - \frac{P(s)}{1 + G(s)H(s)}D(s) + \frac{G(s)H(s)}{1 + G(s)H(s)}N(s) \end{aligned} \tag{5.8}$$

In general, a control system design should minimize Eqs. (5.6) and (5.8), as well as assure its own stability. Because those two requirements contradict each other, they present certain design difficulties.

5.3 Stability and Performance

A desired value varies in actual control systems, and the systems are affected by disturbances or noise. The most important feature required for a control system is its own stability. This means that even if the system state is disturbed by hindrances, the error $\varepsilon(t)$ must return to zero after an appropriate time. A control system is said to be unstable if a controlled variable deviates from a desired value and diverges, or oscillates persistently. The second most important feature is the characteristic that determines the speed and accuracy return to zero when an error occurs. The characteristic is called control performance. Increasing stability decreases performance, and vice versa. The two features cannot be determined independently. Adjusting them to an optimum status is the most important challenge for control system design. We describe stability first, and then performance.

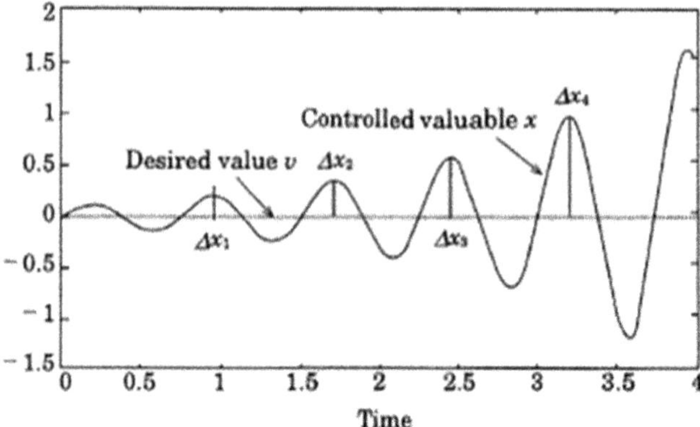

Fig. 5.4 Behavior of unstable control system

5.3.1 Evaluation of Stability

[1] Stable control system

We suppose that the desired value $v(t)$ varies $\Delta\theta$ in a step-like manner. Then the control system attempts to match the controlled variable $x(t)$ to the desired value and, if it is stable, the controlled variable eventually settles at a certain constant value (which may be the identical to the desired value, or slightly different). If the control system is unstable, however, its attempt at matching the controlled variable to the desired value cannot be adequate, causing the controlled variable to repeat oscillating and increase in amplitude. Eventually the controlled variable oscillates persistently at the ultimate amplitude brought about by nonlinearity of the system, resulting in partial damage to the control system.

Figure 5.4 illustrates conceptual behavior of an unstable control system.

As indicated in the figure, the difference (Δx) between the value of the controlled variable x and the desired value v becomes $\Delta x_1 < \Delta x_2 < \Delta x_3 <$ as time passes. The value diverges and never settles at the desired value.

On the other hand, the stable control system behaves as shown in Fig. 5.5. The error of the controlled variable is gradually decreased as $\Delta x_1 > \Delta x_2 > \Delta x_3 > \ldots$, and the variable eventually matches the desired value.

[2] Stability criteria

It is necessary to consider how we can determine whether or not a control system is stable. The fluctuation and error of a controlled variable caused by disturbance in a control system are expressed respectively as Eqs. (5.7) and (5.8). If the disturbance is bounded, therefore, the expansion theorem of Heviside enables us to determine stability or instability according to whether real parts of the poles of $G_{xv}(s)$, $G_{xd}(s)$, and $G_{xn}(s)$ are positive or negative. If all real parts of the poles are negative, the system is stable. (Even if poles with positive real parts exist, the system is stable if all the poles are canceled out by the zero point of the numerator.)

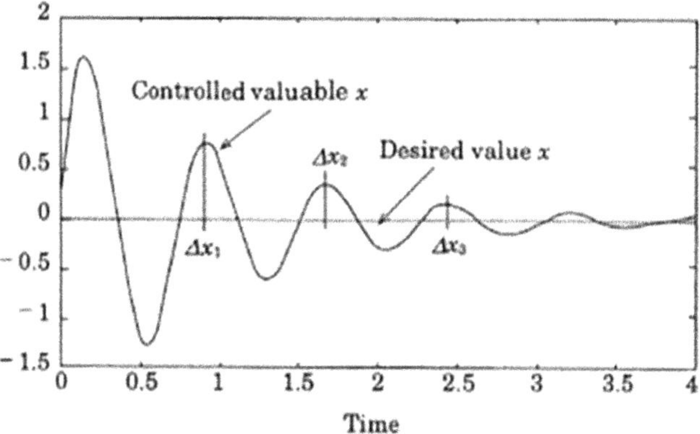

Fig. 5.5 Behavior of stable control system

Finally, stability of a control system can be determined by the root of Eq. (5.9) which is referred to as a characteristic equation.

$$1 + G(s)H(s) = 0 \tag{5.9}$$

The transfer function is usually denoted by a rational function of the Laplace transform variable s, so the characteristic equation can be an algebraic equation as shown in the following.

$$a_0 s^n + a_1 s^{n-1} + a_2 s^{n-2} + \cdots + a_{n-1} s + a_n = 0 \tag{5.10}$$

In the days when computers were not as developed as they are today, it was extremely difficult to find the root of the Eq. (5.10). For this reason, various approaches were taken to determine whether the real part of a root is positive or negative. The Routh–Hurwitz criterion is particularly well known; it determines the existence of an unstable root through the use of a sequence obtained by applying simple operations to coefficients $a_0, a_1, a_2, \ldots$. Because high-speed digital computers and computational algorithms are available today, we can obtain the root of an equation of degree n with a high level of accuracy to determine the stability in a straightforward manner. Both the direct determination (using the root of an equation of degree n) and the Routh–Hurwitz method are mathematical approaches and have the drawback that they cannot visually grasp the stability of a control system.

The Nyquist criterion is a method to eliminate the drawback. It determines the existence of an unstable root by generating the vector locus of the open-loop transfer function $G(s)H(s)$ of a control system. Usually in a nuclear reactor plant or a chemical plant, $G(s)H(s)$ does not have a pole on the right half region of the plane s, and it converges to zero or a certain value when $s \to \infty$. For those types of plants, the Nyquist method can be described as having three steps.

Step <1> In the open-loop transfer function $G(s)H(s)$, let $s = j\omega$, and generate a vector locus for $\omega = -\infty$ to $+\infty$.

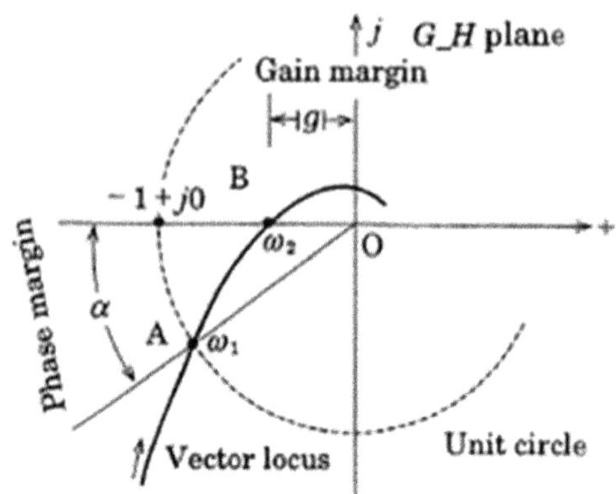

Fig. 5.6 Conceptual diagram of Nyquist locus and stability margins

Step <2> When ω is changed from $-\infty$ to $+\infty$, assume that a vector uniting the point $(-1, +j0)$ on the real axis and a point on this vector locus rotates around the point $(-1, +j0)$ counterclockwise on R.

Step <3> Then the control system is stable only if $R = 0$.

For easier use in actual designing, we can restate these steps as "when changed in the range of $\omega = 0$ to ∞, the system is stable if the vector locus of the open-loop transfer function $G(s)H(s)$ sees the point $(-1, +j0)$ on the left, and unstable if the function sees it on the right. Figure 5.6 illustrates a conceptual Nyquist diagram of a stable system.

The intersection of the vector locus and the circle with radius 1 centered at the zero point is denoted as A (its angular frequency is ω_1), and the intersection of the vector locus and the negative real axis is denoted as B (its angular frequency is ω_2). The angle α formed by OA and the negative axis is a phase margin, and the inverse of OB denoted in the units of dB is called a gain margin. Those margins show how far the Nyquist locus is located from the point $(-1, j0)$, or how much margin is left for the system to be unstable. Thus, we can grasp the stability visually.

Controlled objects such as automatically operationed aircraft and rockets are unstable by themselves, and so have poles on the right half region. For this case, we can extend step <3> in the Nyquist method.

Step <3-i> Investigate the number of poles (unstable poles) existing on the right half of the G_H plane. Assume that the number of poles is P.

Step <3-ii> Then the control system is stable only if $R = P$.

[3] Example of use of criteria

An example determination of the stability of a control system is shown in Fig. 5.7; it consists of a position follow-up control system using an amplidyne and an electric motor.

In the electronic device section, K_α denotes the constant of amplifier gain and K_θ is the constant of position-to-voltage conversion. In the electric motor section, $x(t)$ denotes a rotation angle, R_α is an armature resistance, K_T is a torque conversion coefficient, J is inertia, and K_e is an armature reaction coefficient. Frictional resistance of the electric motor is neglected.

If we denote the output voltage for the input current $i(t)$ by $e(t)$, the transfer function of the amplidyne is expressed as given below:

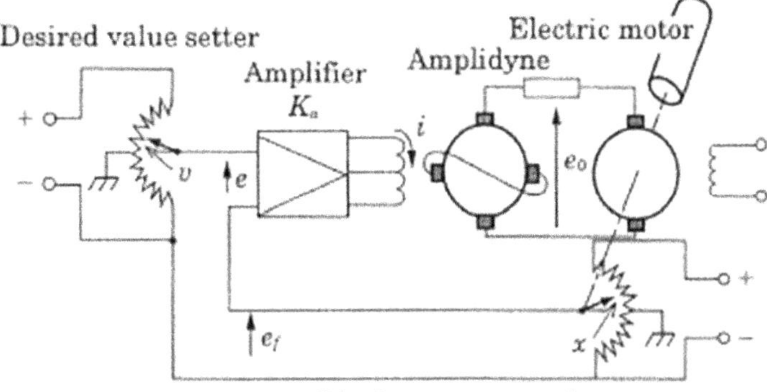

Fig. 5.7 Desired value (position) follow-up control system with amplidyne and electric motor (Izawa 1954)

$$\frac{E(s)}{I(s)} = \frac{K_\alpha}{T_f s + 1}$$

where the time constant T_f is assumed to be 0.2 s.

The transfer function of the electric motor between the applied voltage $e(t)$ and rotation angle $x(t)$ is given by the following expression:

$$\frac{X(s)}{E(s)} = \frac{\left(\frac{1}{R_a}\right)(K_T)\left(\frac{1}{Js^2}\right)}{1 + \left(\frac{1}{R_a}\right)(K_T)\left(\frac{1}{Js^2}\right)(K_e s)} = \frac{\frac{1}{K_e}}{s(T_m s + 1)}$$

where $T_m = \frac{JR_a}{K_e K_T}$ and the time constant T_m is assumed to be 0.5 s.

Figure 5.8 shows a block diagram of the desired value follow-up control system based on this argument.

The open-loop transfer function $G(s)H(s)$ of this block diagram is as given below.

$$G(s)H(s) = \frac{K}{s(0.2s + 1)(0.5s + 1)}, \quad K = \frac{K_\alpha K_g K_\theta}{K_e}$$

The characteristic equation of the control system is given by the next expression.

$$1 + G(s)H(s) = 1 + \frac{K}{s(0.2s + 1)(0.5s + 1)} = 0$$

This indicates that stability of the control system is dependent on the value of the constant K. In practice, if $K = 2.0$ in this algebraic equation of the third degree, then we get the root -5.88, $-0.561 \pm 1.76i$. The values show straightforwardly that the

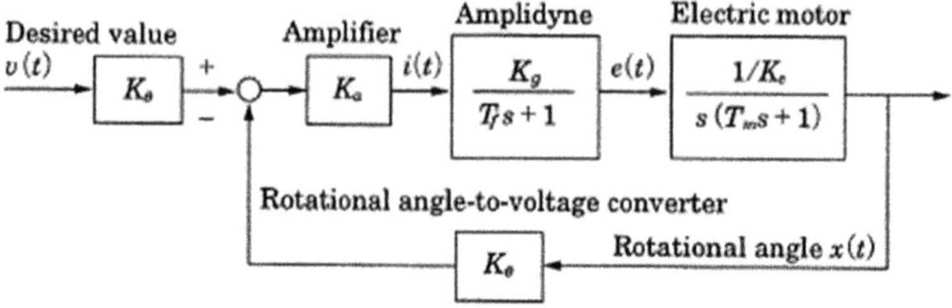

Fig. 5.8 Block diagram of desired value follow-up control system

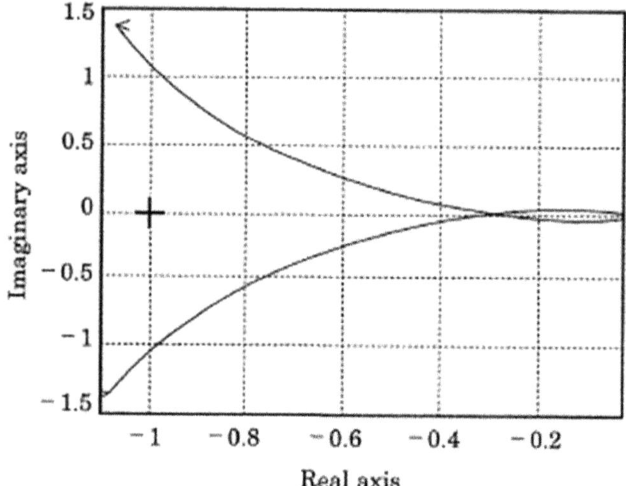

Fig. 5.9 Nyquist locus with $K = 2.0$ (stable)

control system is stable. If $K = 10.0$, then we get the root -7.66, $0.229 \pm 3.65i$, indicating the system is unstable. The gains of the amplifier, amplidyne, and voltage converter (K_α, K_g, K_θ) as well as electric motor constants are defined by the design to obtain the stable K value.

Next, we apply the Nyquist method to determine stability. Figure 5.9 shows the Nyquit locus of $G(j\omega)H(j\omega)$ when $K = 2.0$.

Because the vector locus of $G(j\omega)H(j\omega)$ does not contain the point $(-1, j0)$ inside (the locus sees the point on the left), the control system is stable. On the other hand, Fig. 5.10 shows the vector locus when $K = 10$. Because it contains the point $(-1, j0)$ inside, the control system is unstable.

5.3.2 *Evaluation of Control Performance*

Stability is the most important feature for a control system, and good control response (performance) is the second most important. Requirements for control response cannot be completely matched in all control systems used in diversified industrial fields, but

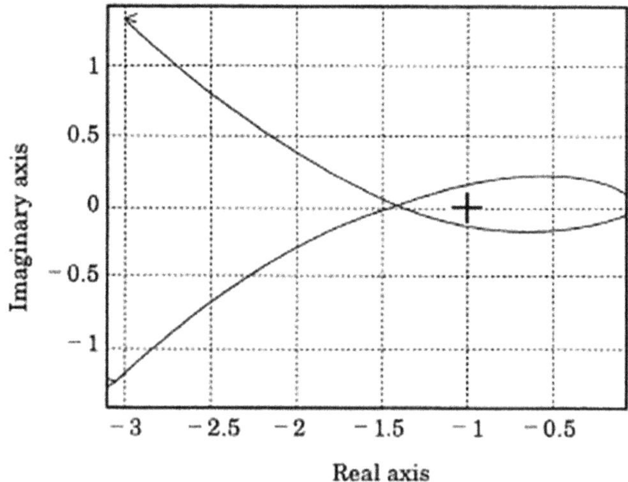

Fig. 5.10 Nyquist locus with $K = 10$ (unstable)

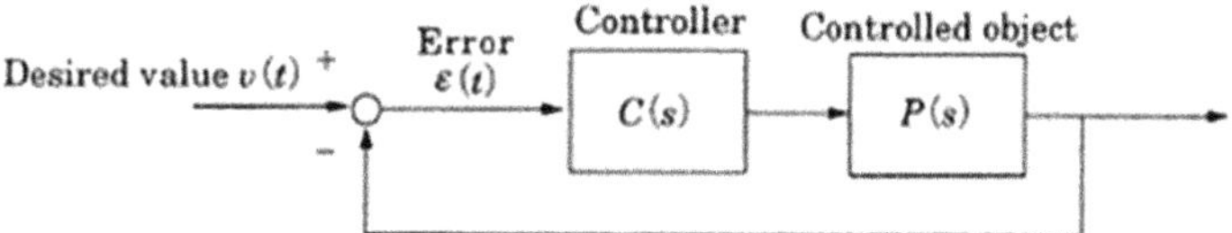

Fig. 5.11 Unity feedback control system

there can be some desirable control performances common to those systems. To determine whether a control response is desirable or not, usually three response characteristics are evaluated: <1> steady-state response, <2> transient response, and <3> frequency response.

[1] Evaluation of steady-state response

For simplification, we consider the unity feedback control system as shown in Fig. 5.11. $C(s)$ in the figure denotes a controller and $P(s)$ is a controlled object.

This assumes $d(t) = 0$, $n(t) = 0$, and $H(s) = 1$ as shown in Fig. 5.3. From the expression (5.8), therefore, error $\varepsilon_v(t)$ between the desired value variation and controlled variable is defined as the following.

$$E_{\upsilon}(s) = \frac{1}{1 + C(s)P(s)} V(s) \tag{5.11}$$

This expression implies that the desired value variation $V(s)$ appears in the error as magnified by the function $S(s)$ defined by the following and called the sensitivity function.

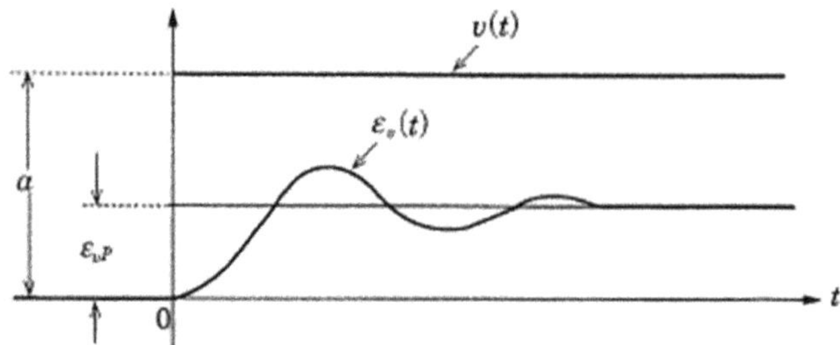

Fig. 5.12 Conceptual diagram of steady-state deviation

$$S(s) = \frac{1}{1 + C(s)P(s)} \tag{5.12}$$

If the disturbance $d(t)$ or noise $n(t)$ as shown in Fig. 5.3 exists, its error is similarly multiplied by S.

This indicates for control system designing that, the smaller $S(s)$ is, the smaller the error is for the desired value variation or disturbances. For example, if we change the desired value step-like by $\alpha[v(t) = \alpha u(t)$, where $u(t)$ is a unit step function], we obtain the error after a sufficient time (called a steady-state position error or steady-state offset), using the final-value theorem of the Laplace transform.

$$\varepsilon_{vP} = \lim_{t\to\infty} \left[\varepsilon_v(t)\right]_{v(t)=au(t)} = \lim_{s\to 0} s \cdot \left[\frac{1}{1 + C(s)P(s)} \cdot \frac{a}{s}\right] = \frac{a}{1 + C(0)P(0)}$$

This indicates that if $C(0)P(0)$ is a finite value, the error proportional to the variation width α of the desired value remains in the controlled variable. Figure 5.12 illustrates the concept of this state.

A smaller ε_{vP} value is desirable, which should be approximately 0.02–0.2 of the variation width α for usual design.

So far we have discussed the steady-state error of the unity feedback control system shown in Fig. 5.11. A similar approach can be applied to the case where a transfer component is contained in the backward path as shown in Fig. 5.3. Also a steady-state error for a desired value varying at a certain speed can similarly be evaluated. Based on this understanding, a design method has been developed that introduces a compensating circuit to boost the component gain in the low frequency band.

[2] Evaluation of transient response

The previous section discussed the steady-state error associated with control accuracy, i.e. one of the important control quality factors. Even if we have succeeded in minimizing the steady-state error, it cannot be said that we have achieved good response if we took a long time to mitigate the steady-state error. The following describes the transient characteristic, another control quality.

A frequently used method for investigating the transient characteristic is to change input signals (such as a desired value or disturbances) step-like and evaluate the transient response quality of a controlled variable at that time. The following two methods can be used to evaluate the quality. The first focuses on damping characteristics pertaining to the transient response of an error, and we can discuss it based on the characteristic equation ($1 + G(s)\,H(s) = 0$).

The second focuses on the amount of a certain time integration value pertaining to the transient response $\varepsilon(t)$ of the error, and we must specifically obtain the $\varepsilon(t)$ to perform its time integration. The latter requires a longer computation time and a more complicated computational algorithm than the former.

Method <1> Method based on damping characteristics

The control system shown in Fig. 5.3 matches the controlled variable $x(t)$ to the desired value $v(t)$, where $d(t)$ denotes a disturbance entering the forward path and $n(t)$ is noise added to the backward path. In addition, $\varepsilon(t) = v(t) - x(t)$ is expressed in a Laplace transform as

$$\begin{aligned} E(s) &= V(s) - X(s) \\ &= \left(1 - \frac{G(s)}{1 + G(s)H(s)}\right)V(s) - \frac{G_d(s)}{1 + G(s)H(s)}D(s) + \frac{G(s)H_n(s)}{1 + G(s)H(s)}N(s) \end{aligned} \tag{5.13}$$

where $G(s) = C(s)P(s)$.

The first term is an error generated by variation of a desired value, the second term is caused by disturbances, and the third is a component generated by noise. Considering that $\varepsilon(t)$ is obtained from an inverse Laplace transform, therefore, the $\varepsilon(t)$ time response is determined by the poles of the characteristic equation $1 + G(s)H(s) = 0$ for any of the $v(t)$, $d(t)$, or $n(t)$ input. This expression is an algebraic equation with real number coefficients, and it can be converted into:

$$\prod_j (s + a_j) \prod_i \left(s^2 + 2\zeta_i \omega_{ni} s + {\omega_{ni}}^2\right) = 0$$

Thus the expansion theorem of Heviside using the root of above expression gives time response of the deviation $\varepsilon(t)$ as follows:

$$\varepsilon(t) = \sum_j A_j \mathrm{e}^{-a_j t} + \sum_i B_i \mathrm{e}^{-\zeta_i \omega_{ni} t} \sin\left\{\sqrt{1 - \zeta_i^2}\,\omega_{ni} t + \phi_i\right\} + C \tag{5.14}$$

where A_j, B_i, Φ_i, C are constants determined by initial conditions of the control system.

The expression (5.14) implies that $-a_j$ and $-\zeta_i \omega_{ni}$ must be negative real numbers in a stable system because the root of the characteristic equation exists on the left half region of the plane s. The expression also implies that the $\varepsilon(t)$ value is damped as time passes, and the damping speed is governed by the smallest value of a_j and $\zeta_i \omega_{ni}$.

Because of this, the minimum absolute value of the real part of the root is called the damping rate, and it is used as a factor to evaluate the quality of control response. Depending on the controlled object, damping of only an oscillating response component pertaining to the second term may be used to evaluate the control response. In this case, the smallest value (ζ) of ζ_i from the expression (5.14) is used and it is called the damping coefficient. The following are recommended for damping coefficient values.

<1> For control with fixed set-point: $\zeta = 0.2$–0.4.
<2> For follow-up control: $\zeta = 0.6$–0.8.

Method <2> Method based on time integration of error

This method considers the following S_1 and S_2 time integrations (control area) for the error of the step response $\varepsilon(t)$ when changing a desired value step-like and uses them as evaluation functions to evaluate the transient characteristic quality of a control system.

$$S_1 = \int_0^{\infty} |\varepsilon(t) - \varepsilon(\infty)|\, \mathrm{d}t$$

$$S_2 = \int_0^{\infty} \{\varepsilon(t) - \varepsilon(\infty)\}^2\, \mathrm{d}t$$

In general, computing S_1 and S_2 is more difficult than dealing with damping coefficients described in the previous method <1>. Some other evaluation functions have been proposed in addition to them.

[3] Evaluation of frequency characteristic

Substitution of $s = j\omega$ (ω: angular frequency) into the expression (5.7) yields the following expression.

$$X(j\omega) = G_{xv}(j\omega)V(j\omega) + G_{xd}(j\omega)D(j\omega) + G_{xn}(j\omega)N(j\omega) \qquad (5.15)$$

It is the frequency response of a control system for frequency components of the input signals $v(t)$, $d(t)$, and $n(t)$. The first term denotes the transient characteristic in which a controlled variable matches a desired value that has changed, the second term is the variance characteristic of a controlled variable affected by external disturbance, and the third term is the variance characteristic affected by noise. This leads to the following consequences.

<1> Within the range of ω where $V(j\omega)$ exists, $|G_{xv}(j\omega)| = 1$ is desirable.
<2> Within the range of ω where $D(j\omega)$ exists, $|G_{xd}(j\omega)| = 0$ is desirable.
<3> Within the range of ω where $N(j\omega)$ exists, $|G_{xn}(j\omega)| = 0$ is desirable.

Next, we describe a simplified method to get the response of a controlled variable for step-like variation of a desired value when the frequency response

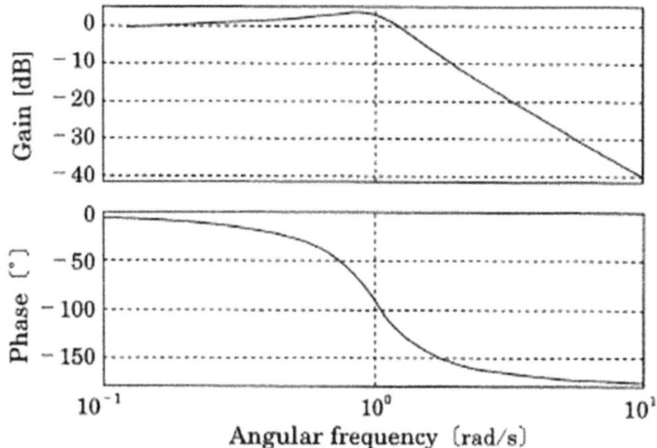

Fig. 5.13 Example of Bode diagram of Eq. (5.16)

function $G_{xv}(j\omega)$ is given as a function of a control system between the desired value and the controlled variable.

We consider an example in which the transfer function of a control system between the desired value and the controlled variable is given as Eq. (5.16).

$$G_{xv}(s) = \frac{1}{s^2 + 1.2s + 1} \tag{5.16}$$

Figure 5.13 shows the Bode diagram.

In general, the transfer function of an ideal filter is given by the following expression which contains two constants, ω_0 and ϕ_0.

$$W_i(s) = \mathrm{e}^{-\left(\frac{\phi_0}{\omega_0}\right)s} \tag{5.17}$$

Therefore, the frequency transfer function of an ideal filter is given below.

$$(1) \quad 0 < \omega < \omega_0 : W_i(j\omega) = \mathrm{e}^{-j\left(\frac{\phi_0}{\omega_0}\right)\omega} \tag{5.18}$$

$$(2) \quad \omega_0 < \omega : W_i(j\omega) = 0 \tag{5.19}$$

If the desired value of a control system changes step-like, the time (t_H) required for the controlled variable $x(t)$ to reach 0.5 of $x(\infty)$ and the time (t_1) required for it to reach the first peak can be computed by the following expression using the constants ω_0 and ϕ_0.

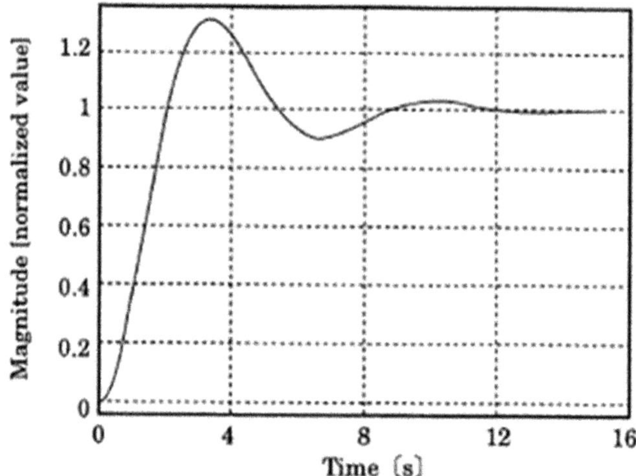

Fig. 5.14 Step response of controlled variable by numerical simulation

$$t_{\mathrm{H}} = \frac{\phi_0}{\omega_0}, \qquad t_1 = \frac{\phi_0}{\omega_0} + \frac{\pi}{\omega_0} \tag{5.20}$$

Then, ω_0 is defined as the frequency 1.5 rad/s to make the gain in the Bode diagram –3db, so that the Bode diagram of Fig. 5.13 is approximated by an ideal filter. The phase ϕ_0 corresponding to this frequency is approximated as 130° (2.3 rad). By substituting those constants into the expression (5.20), the response characteristic times of the controlled variable for the step change of the desired value in the control system are found to be $t_{\mathrm{H}} = 1.5$ s and $t_1 = 3.6$ s.

The step response of the control system given by the transfer function $G_{xu}(s)$ of the expression (5.16) is simulated numerically and Fig. 5.14 shows the results.

The t_{H} and t_1 are approximately 1.7 s and 3.5 s, respectively, which are close to those obtained above for the ideal filter.

5.4 Design Method of Control Systems

5.4.1 Design Procedure

Design work for automatic control systems depends on the system to be designed. The challenge is how to meet requirements for sophisticated performance. It is certainly difficult to develop a design method applicable to all systems, but the following design procedure may be commonly applied to all control systems.

Step <1> Planning performance requirements and basic design policy for a control system.

This step is intended to provide full understanding of the functions of a controlled object, control performance such as required accuracy or response speed, and types and sizes of disturbances affecting the controlled object. In addition, this step is intended to plan the basic policy that determines operating environments of the control system, allowable expenses, and methods to procure control devices and energy sources within the determined conditions.

Step <2> Grasping static and dynamic characteristics of the objective system

When designing a control system for an existing plant, we need to measure the main process variables (such as temperature, pressure, or flow rate) by performing static and dynamic experiments. If it is difficult or impossible to perform experiments using an actual operational plant or a plant that is not yet built, so the static and dynamic characteristics should be considered using a simulation model of the plant.

Step <3> Studying required control performance based on control theories

This step is intended to study whether the required performances of <1> above are possible. Through the study, for example, we can search for the allowable range of a steady-state error, the amount of overshoot or delay in tracking response for a step variation of the desired value, or the damping characteristic of controlled variable fluctuated by disturbances.

Step <4> Creating block diagrams for the control system, inserting compensators, determining undefined constants, etc.

This step expresses the control system by block diagrams in order to achieve required control performances. The diagrams evolve and are rewritten as the design work progresses because compensators or other factors will be inserted. Undefined constants contained in the diagrams will also be corrected if so required.

Step <5> Verifying control characteristics by experiments using an actual plant or numerical simulation this step is intended to verify whether the control performance satisfies the requirement. The verification process is performed through static and dynamic experiments using an actual plant into which the designed controller is incorporated, or using numerical simulation models.

Step <6> Consideration for producing a control system

This step is intended to transform the block diagrams of the designed control system into equivalent ones that are convenient for the producing and installing. This saves cost and reduces faults in production or installation, which are extremely important for practical purposes.

5.4.2 Evaluating Characteristics of the Control System

The design of a control system is intended to configure the control system having desirable characteristics. We consider next what kinds of characteristics are desirable, and how we can evaluate them. The following considerations are based on the feedback control system illustrated in Fig. 5.3.

Sections 5.1–5.3 in this chapter have concluded that a control system is stable if the system finally settles at a certain value or zero without a controlled variable or its error diverging when a desired value varies step-like. The sections have also shown that a control system is stable if all roots of its characteristic equations exist on the left half region of the s plane. Furthermore, evaluation of control characteristics has already been expounded using indices related to the steady-state characteristics, transient characteristics, and frequency characteristics.

We need to consider whether these efforts are sufficient for characteristic evaluation of a control system. No assessment technique is applicable to all control systems because different types of control systems are used in different industrial fields, so the control performances required for the systems are diverse. However, we can cite five important evaluation items common to all control systems.

Item <1> Desired value follow-up characteristic

This characteristic is evaluated in terms of steady-state position error, steady-state velocity error, rise time, damping ratio, etc. of the controlled variables. Those terms can be determined by the transfer function from a desired value to a controlled variable [$G_{xv}(s)$ of the expression (5.6)]. Therefore, the gain of the following transfer function should be 1 in the frequency band of a desired value.

$$G_{xv}(s) = \frac{G(s)}{1 + G(s)H(s)} = \frac{C(s)P(s)}{1 + C(s)P(s)H(s)} \tag{5.21}$$

Item <2> Disturbance suppression characteristic

This can be measured by the ratio between the transfer function from a disturbance to a controlled variable in a feedback control system [$G_{xd}(s)$ in Eq. (5.6)], and the transfer function from a disturbance to a controlled variable in an open-loop system ($P(s)$). The ratio is expressed as (5.22).

$$S(s) \equiv \frac{G_{xd}(s)}{P(s)} = \frac{1}{1 + C(s)P(s)H(s)} \tag{5.22}$$

To improve the disturbance suppression, therefore, this $S(s)$ should be reduced in the frequency band of a disturbance. This $S(s)$ is the same sensitivity function that we mentioned earlier (Sect. 5.3.2).

Item <3> Sensitivity related to characteristic variation of a controlled object:

Designing a control system involves using the model of a controlled object. This approach is fraught with unavoidable modeling error (system parameter accuracy, neglect of high-frequency or nonlinear components pertaining to actual system dynamic characteristics, etc.). The problem is how far the errors affect a controlled variable. It is also known that the smaller the sensitivity function $S(s)$ is, the smaller is the characteristic variation of a control system caused by modeling errors. Therefore, our design should reduce $S(s)$ in an area with as wide a frequency as possible.

Item <4> Stability margin

The smaller the complementary sensitivity function is that is defined by the expression,

$$T(s) = \frac{C(s)P(s)H(s)}{1 + C(s)P(s)H(s)} \tag{5.23}$$

Table 5.1 Design methods of a control system

Evaluation item	Design policy for frequency response of control system	Applied control method
(a) Desired value follow-up characteristic	$G_{xv}(s)$ should be gain of 1 in the frequency area of a desired value	Two-degrees-of-freedom control
(b) Disturbance suppression characteristic (c) Sensitivity	Sensitivity function $S(s)$ should be minimized in the low frequency band	Feedback control system
(d) Stability margin (e) Noise effect	Sensitivity function $T(s)$ should be minimized in the high-frequency band	Feedback control system

the larger is the stability margin. This is a generalization of a gain margin or phase margin in the conventional Nyquist criterion.

Item <5> Effect suppression of detector noise

Effect of detector noise on a controlled variable is determined by the transfer function $G_{xn}(s)$ of the expression (5.6), which is equivalent to the complementary sensitivity function $T(s)$ defined above. Therefore, diminishing $T(s)$ reduces the effect of detection noise.

Among these five items, <1> is referred to as a desired value characteristic, and <2> to <5> are feedback characteristics. A feedback control system is intended to keep the feedback characteristic at a favorable condition. In general, we cannot satisfy both the feedback characteristic and desired value characteristic at the same time. To satisfy both types of the required specifications, we must rely on two-degrees-of-freedom control that generates the control signal (u) using the desired value signal (v) and control signal (x) independently.

Table 5.1 summarizes the design policy for three transfer functions of Eqs. (5.21)–(5.23).

5.5 PID Control and Parameter Tuning Technique

5.5.1 PID Control

PID control is used in 84 % of the control systems currently used for plants and facilities in various industries. If advanced variants are included, this percentage rises to more than 90 % (Furuta and Tomita 1990).

Figure 5.15 illustrates the basic configuration of unity feedback PID control.

In this figure, the disturbance $d(t)$ affecting a controlled object is added to a manipulated variable, and noise $n(t)$ is to a controlled variable. A large error shows that a controlled variable deviates widely from a desired value and, to correct the error, a large value of the manipulated variable is required. On the other hand, it is proper to add a smaller value of a manipulated variable in the case of a small error. This is the generation rule for the manipulated variable proportional to an error, and it is referred to as a P action (proportional action).

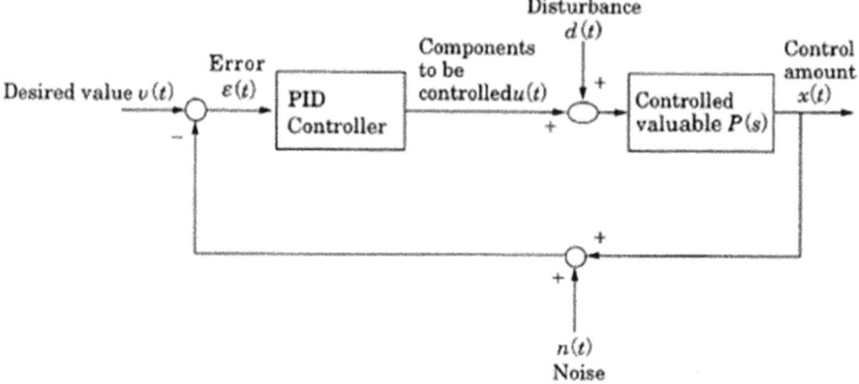

Fig. 5.15 Basic configuration diagram of PID control system

This P action itself is not capable of eliminating a steady-state error of the controlled variable for a step-like change of desired value and disturbances, i.e. the controlled variable settles at a value apart from the desired value. To eliminate the error, an I action (integral action) is required that generates a manipulated variable proportional to the integral of the error. This action resets the error to zero, so it may also be referred to as a reset action. Furthermore, a differential of the error can be used to generate the manipulated variable. It is called a D action (derivative action). This action is proportional to the variation rate of the error, so it may also be referred to as a rate action.

PID control is a control using those three actions as the control rule. If control includes only one or two of the actions, it may be called P control, PI control, or PD control. The PID control rule can be expressed as

$$u(t) = K_{\mathrm{P}}\varepsilon(t) + K_{\mathrm{I}}\int_0^t \varepsilon(\tau)\,\mathrm{d}\tau + K_{\mathrm{D}}\frac{\mathrm{d}}{\mathrm{d}t}\varepsilon(t)$$

It is conventionally given by (5.24):

$$u(t) = K_{\mathrm{P}}\left\{\varepsilon(t) + \frac{1}{T_{\mathrm{I}}}\int_0^t \varepsilon(\tau)\,\mathrm{d}\tau + T_{\mathrm{D}}\frac{\mathrm{d}}{\mathrm{d}t}\varepsilon(t)\right\} \tag{5.24}$$

where K_{P} is a constant called proportional gain, the units of which are obtained by dividing the units of a manipulated variable by those of an error. T_{I} and T_{D} are also constants and called integral time and derivative time, respectively.

Equation (5.24) gives the transfer function of a controller as the next equation.

$$C(s) = K_{\mathrm{P}}\left(1 + \frac{1}{T_{\mathrm{I}}s} + T_{\mathrm{D}}s\right) \tag{5.25}$$

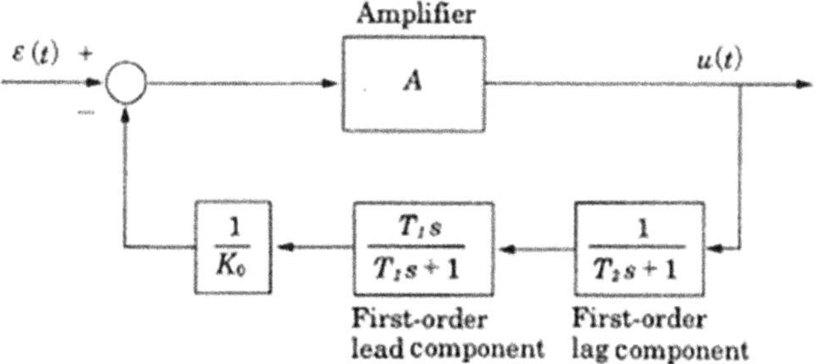

Fig. 5.16 Configuration example of PID controller with power amplifier (Suda et al. 1992)

However, a derivative action of the third term cannot be completely realized. Therefore, the actual design work substitutes an electrical circuit performing approximate derivative for it.

It is no exaggeration to say that PID control is extremely versatile because the proportional, integral, and derivative actions match simple human intuition. Proportional actions are calculated according to the size of the present error, integral actions are based on the accumulation of past errors, and derivative actions are based on the trend of future error. That is why we can intuitively understand the rationality of those overall actions. The practical know-how of PID control is based on accumulated experiences. Even today when various new control theories and techniques have been proposed and come into practical use, still a lot can be learned from PID control systems because they have many advantages including being versatile, simple, and practical.

To take actual control actions, a power source is required that can drive the actuator of a controlled object. For practical easiness, PID controllers have been designed which consist of a power amplifier and the PID control rule. Pneumatic and electric types of PID controllers are commercially available today. Both types adopt a configuration that gives feedback to a high gain-power amplifier via first-order lag or first-order lead components. Figure 5.16 illustrates an example.

That the configuration approximately denotes the PID controller transfer function of Eq. (5.25) can easily be seen from a simple computation because the amplifier gain A is extremely large. The PID controller $C(s)$ of Eq. (5.25) shows that the three constants (K_P, T_I and T_D) are independent. The configuration example of Fig. 5.16 shows, however, that the three constants interfere with each other. The reciprocal interference coefficient is $(1 + T_2/T_1)$.

5.5.2 *Tuning Methods of PID Control*

Constants for the PID control rule (K_P, T_I and T_D) must be set to appropriate values. The following describes two constant tuning methods, the Ziegler–Nichols ultimate sensitivity method and the step response method.

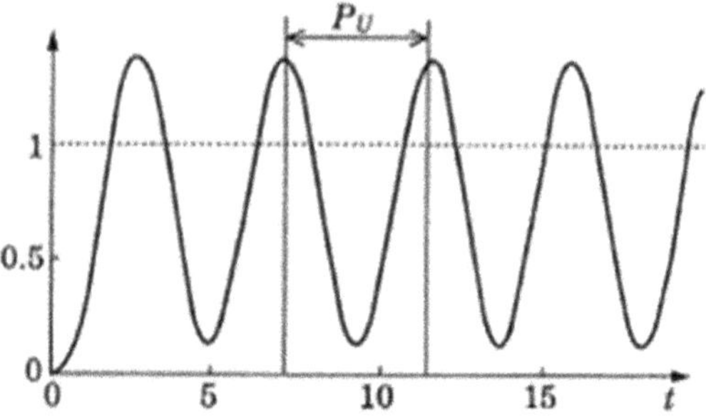

Fig. 5.17 Step responses (stable ultimate oscillation for tuning K_P)

Calibrating constants using the ultimate sensitivity method is done in the following manner. When we limit the PID controller to the proportional gain K_P, and increase it gradually, the step response begins persistent oscillation as shown in Fig. 5.17. This is called stable ultimate oscillation.

The proportional gain that yields a stable ultimate oscillation is denoted as K_{PU}, and the period of persistent oscillation is denoted as P_U. Based on the values, PID parameters of Eq. (5.25) are defined as follows:

<1> For only proportional action ($T_1 = \infty$, $T_D = 0$): $K_P = 0.5\ K_{PU}$
<2> For proportional + integral actions ($T_D = 0$): $K_P = 0.45\ K_{PU}$, $T_I = 0.83\ P_U$
<3> For proportional + integral + derivative actions: $K_P = 0.6\ K_{PU}$, $T_I = 0.5\ P_U$, $T_D = 0.125 P_U$

The ultimate sensitivity method is thus based on the proportional gain (K_{PU}) and the oscillation period (P_U) during stable ultimate oscillation to set PID parameters.

The step response method is the second tuning method and it is done as follows. First the step response of a controlled object is measured. One of the following transfer functions $G(s)$ is fitted to the measurement data and the values of parameters K, L, and T are estimated.

$$G(s) = \frac{1}{Ts}e^{-Ls} \quad \text{or} \quad G(s) = \frac{K}{Ts+1}e^{-Ls} \tag{5.26}$$

The step response method uses the estimated values and defines PID constants as follows: Let $R = K/T$.

<1> For only proportional action ($T_1 = \infty$, $T_D = 0$): $K_P = 1/RL$
<2> For proportional + integral actions ($T_D = 0$): $K_P = 0.9/RL$, $T_1 = L/0.3$
<3> For proportional + integral + derivative actions: $K_P = 1.2/RL$, $T_1 = 2\ L$, $T_D = 0.5\ L$

Based on the thus configured PID constants, the step response of a control system is damped by 25 % (0.25 of the damping rate per one basic oscillation period of the controlled variable response). The 25 % damping has been recommended as a desirable damping characteristic in many control experiments.

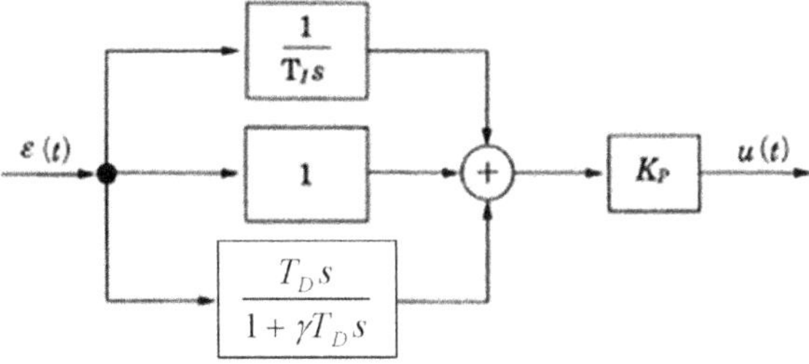

Fig. 5.18 Implementation of the position type PID controller

However, values obtained by those methods are no more than candidates for the PID constants. The values are actually used as initial values to reconfigure the PID constants by trial-and-error and to find the optimum values.

5.5.3 *Implementing the PID Control Rule*

[1] Analog implementation

1) Derivative action
   As already mentioned, the derivative action represented by $T_D s$ of Eq. (5.25) cannot completely be realized. Therefore, some approximated derivative must be used as an alternative for implementing the PID controller. Usually for the approximated derivative, the following is used in combination with a first-order lag filter.

$$T_D s \cong \frac{T_D s}{1 + \gamma T_D s} \tag{5.27}$$

   Here, $1/\gamma$ is called derivative gain, and its sufficiently larger value. Then $T_D s$ shows that the approximation accuracy is favorable. Most commercially available PID controllers have approximately $\gamma = 0.1$.
2) Position type and velocity type
   Implementation of the position type PID controller uses Eq. (5.25) as it is. Figure 5.18 shows the implementation. For a derivative action, we use an approximate derivative of the expression (5.27).

On the other hand, implementation of the velocity type controller is by Eq. (5.28) derived from Eq. (5.25).

$$C(s) = K_P \left(\frac{1}{T_I} + s + T_D s^2 \right) \left(\frac{1}{s} \right) \tag{5.28}$$

For actual implementation,

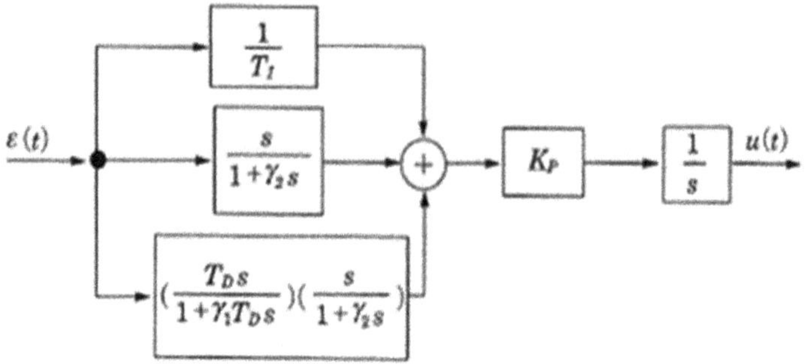

Fig. 5.19 Implementation of the velocity type PID controller

$$C(s) = K_{\mathrm{P}}\left(\frac{1}{T_{\mathrm{I}}} + \frac{s}{1+\gamma_2 s} + \frac{T_{\mathrm{D}}s}{1+\gamma_1 T_{\mathrm{D}}s}\cdot\frac{s}{1+\gamma_2 s}\right)\left(\frac{1}{s}\right) \tag{5.29}$$

is adopted which uses an approximated derivative. If the actuator has an integral characteristic, the integrator component in the last term of Eq. (5.29) is naturally replaced by the characteristic. Figure 5.19 shows the block diagram.

[2] Digital implementation

Compared with analog implementations, digital implementations can make more flexible computations to deal with nonlinear operation and logical determination.

With the recent advancements in microprocessors, digital PID controllers have come to be widely adopted. In digital control, the manipulated variable $u(m\Delta t)$ is derived from sampling values $\{\varepsilon(k\Delta t), k = 0, 1, 2, \ldots\}$ of an error per the interval Δt. For an integral or derivative action, a proper method is adopted from different kinds of numerical integration or differentiation operations. Usually the integral is approximated simply by a finite sum, and the derivative is approximated by backward difference because Δt is sufficiently short compared with the changing speed of a signal. The digital control rules for the position type and the velocity type controllers are as follows.

1) Position type

$$u(m\Delta t) = K_{\mathrm{P}}\left(\varepsilon(m\Delta t) + \frac{\Delta t}{T_{\mathrm{I}}}\sum_{k=0}^{m}\varepsilon(k\Delta t) + \frac{T_{\mathrm{D}}}{\Delta t}\left\{\varepsilon(m\Delta t) - \varepsilon\left(\overline{m-1}\Delta t\right)\right\}\right) \tag{5.30}$$

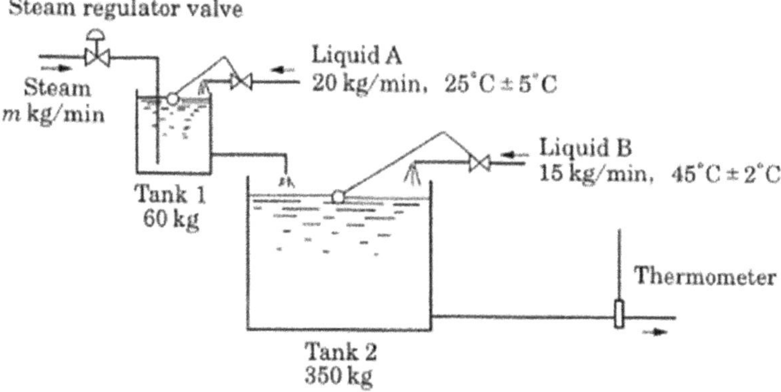

Fig. 5.20 Controlled objective process (Izawa 1954)

2) Velocity type

$$\begin{aligned}\Delta u(m\Delta t) &:= u(m\Delta t) - u((m-1)\Delta t)\\ &= K_{\mathrm{P}}\left(\varepsilon(m\Delta t) - \varepsilon(\overline{m-1}\Delta t) + \frac{\Delta t}{T_{\mathrm{I}}}\varepsilon(m\Delta t)\right.\\ &\quad\left. + \frac{T_{\mathrm{D}}}{\Delta t}\left\{\varepsilon(m\Delta t) - 2\varepsilon(\overline{m-1}\Delta t) + \varepsilon(\overline{m-2}\Delta t)\right\}\right)\end{aligned} \tag{5.31}$$

They generate a manipulated variable by adding $u(m\Delta t) = \sum_{k=0}^{m} \Delta u(k\Delta t)$.

The above approximates the derivative action of a PID control rule by backward difference, but approximate differentiation of (5.27) is also often used. Digital implementations have more superior features than analog ones with respect to the flexibility of logical determination and computation, downsizing the controller, easiness of maintenance and inspection, etc. and they are expected to be widely used in the future.

5.6 Design Examples

5.6.1 Design Specifications

This section provides a design example for a PID control system for a chemical process consisting of two tank systems that mix Liquid A and Liquid B to produce Liquid C. Figure 5.20 illustrates the process.

Liquid A is heated by steam in Tank A, and then it flows into Tank 2 where it is mixed with Liquid B to produce Liquid C product. We consider the design issues for

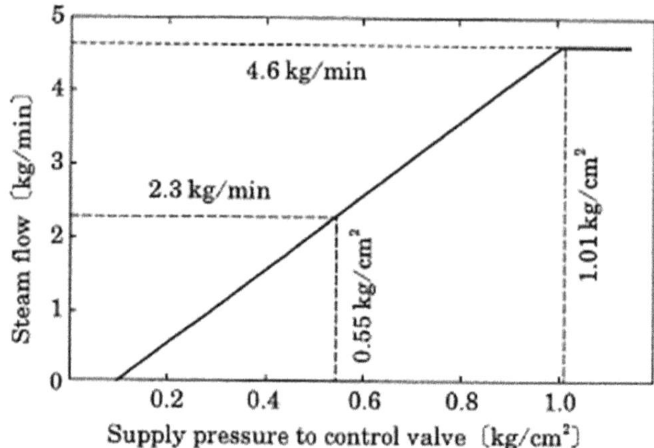

Fig. 5.21 Characteristics diagram of diaphragm valve

a steam flow control system that maintains the temperature of the Liquid C product at 65 °C. The seven design specifications are as follows:

Spec <1> Liquid A flows into Tank 1 at the constant rate of 20 kg/min. Its temperature fluctuates in the range of 25 °C ± 5 °C with a fluctuation period of approximately 5 min. It is assumed that the specific heat of Liquid A is 1 kcal/kg.

Spec <2> Liquid B flows into Tank 2 at the constant rate of 15 kg/min. Its temperature fluctuates in the range of 45 °C ± 2 °C with a long fluctuation period of a few hours or more. It is assumed that the specific heat of Liquid B is 1 kcal/kg.

Spec <3> Liquid B changes in quality if it is heated to 84 °C or higher.

Spec <4> The Liquid C temperature is controlled at 65 °C ± 0.5 °C.

Spec <5> The capacity of Tank 1 is 60 kg and that of Tank 2 is 350 kg.

Spec <6> The Liquid C temperature is detected at the piping outlet of Tank 2 as shown in the figure. Time constant of the thermocouple is neglected. The quantity of Liquid C inside piping between Tank C and the detection point is 37.5 kg.

Spec <7> A pneumatic diaphragm valve is used for steam regulation. Figure 5.21 shows the relation between the supplied air pressure and steam flow.

5.6.2 *Static Characteristics*

First, we need to know the temperature and the steam feed rate in Tank 1 from static characteristics of the controlled object. We obtain the steam feed rate (m kg/min) that raises the Liquid C temperature to 65 °C when the Liquid A and Liquid B temperatures are at their respective nominal values. The latent heat of flow steam is assumed to be 550 kcal/kg.

Incoming calories into Tank 1 Q_{i1}: $Q_{i1} = m \times 550 + 25 \times 20 \times 1$

Outgoing calories from Tank 1 Q_{o1} are given as: $Q_{o1} = T_1 \times (20 + m) \times 1$, where T_1 is the temperature of liquid flowing from Tank 1. In the steady state, $Q_{i1} = Q_{o1}$, so we obtain Eq. (5.32).

$$550m + 500 = 20T_1 + mT_1 \tag{5.32}$$

In the similar way, balancing between in- and out-calories in the steady-state Tank 2 yields Eq. (5.33).

$$20T_1 + mT_1 + 15 \times 45 = (20 + m + 15) \times 65 \tag{5.33}$$

$T_1 = 79$ °C and $m = 2.3$ kg/min are found from those equations.

5.6.3 Dynamic Characteristics

The fluctuation from the steady-state steam flow rate is denoted by z kg/min and fluctuations of Liquid A and Liquid B temperatures are shown by d_1 and d_2, respectively. Fluctuations of the outgoing liquid temperatures from Tank 1 and Tank 2 are denoted by c_1 and c_2, respectively. Fluctuation of the detection point temperature is indicated by x °C. If fluctuation z of steam flow is negligibly small compared with steady-state flow rate, then the following equation for the Tank 1 liquid temperature holds:

$$60\frac{\mathrm{d}c_1(t)}{\mathrm{d}t} = (550z(t) + 20d_1(t)) - 20c_1(t)$$

By applying Laplace transform to obtain $C_1(s)$, we get

$$C_1(s) = \frac{55/2}{3s+1}Z(s) + \frac{1}{3s+1}D_1(s) \tag{5.34}$$

Similarly, we can obtain the following expression for the Tank 2 liquid temperature $C_2(s)$.

$$C_2(s) = \frac{4/7}{10s+1}C_1(s) + \frac{3/7}{10s+1}D_2(s) \tag{5.35}$$

The liquid temperature flowing out of Tank 2 is detected at the position in the piping shown in Fig. 5.20. Because of transportation inside the pipe, the Tank 2 outlet temperature $c_2(t)$ is detected as the temperature $x(t)$ with the delay time of 37.5 kg ÷ 37.3 kg/min = 1 min This relationship is expressed as

$$X(s) = \mathrm{e}^{-1s}C_2(s) \tag{5.36}$$

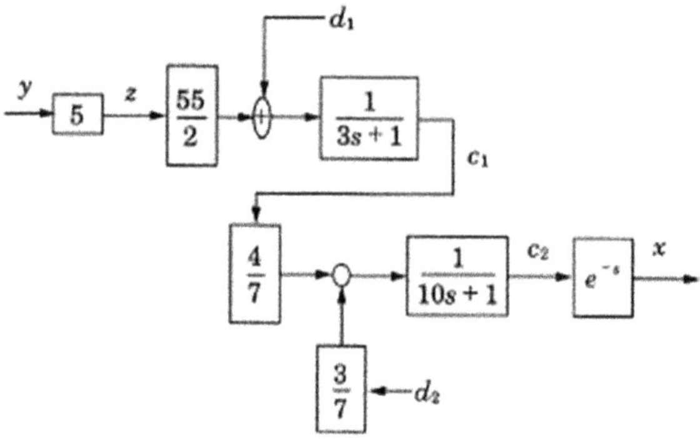

Fig. 5.22 Block diagram of the controlled objective

The pressure supplied to the steam regulation valve and the steam flow rate show a symmetrical linear characteristic concerning steady-state points as shown in Fig. 5.21. The linearity enables us to approximate the relation between the fluctuation $z(t)$ from steady-state steam flow (2.3 kg/min) and the fluctuation $y(t)$ from steady-state operating pressure (0.55 kg/cm) as per the following expression:

$$Z(s) = \frac{4.6 - 0}{1.01 - 0.09} Y(s) = 5Y(s) \tag{5.37}$$

5.6.4 *Control System Designing and Stability Margin*

By combining expressions (5.34)–(5.37), the controlled object $P(s)$ with input as the steam flow rate $y(t)$ and output as the sensor temperature $x(t)$ is expressed as a block diagram shown in Fig. 5.22.

The transfer function of the controlled object derived from Fig. 5.22 becomes

$$P(s) = \frac{X(s)}{Y(s)} = \frac{(550/7)\,e^{-s}}{(3s+1)(10s+1)} \tag{5.38}$$

Figure 5.23 shows the Bode diagram representing this expression.

When a vector locus cuts the actual axis (phase: $-180°$), the angular frequency is $\omega_u = 0.64$ rad/min and the size of its segment is $|P(j\omega_u)| = 14.9$ dB $=$ 5.56 cm^2 °C/kg.

Now, we turn our attention to designing a control system arranged as shown in Fig. 5.24.

The control system can be expressed as a block diagram shown in Fig. 5.25.

Restricting the controller to only P action and changing its proportional gain K_P [kg/cm^2 °C] gradually to larger values, we find the whole control system begins

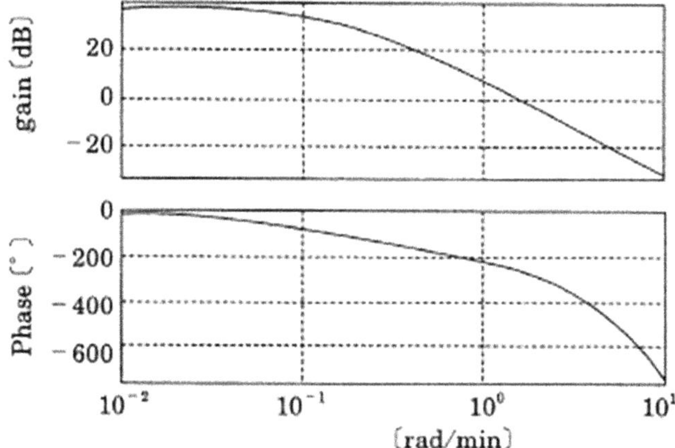

Fig. 5.23 Bode diagram of the controlled objective

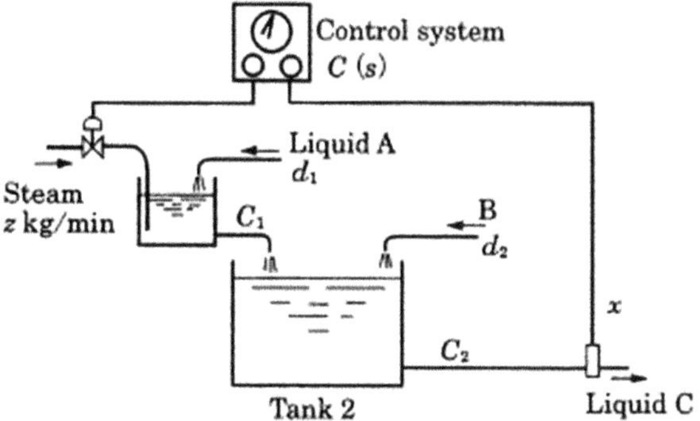

Fig. 5.24 Illustration of control system (Izawa 1954)

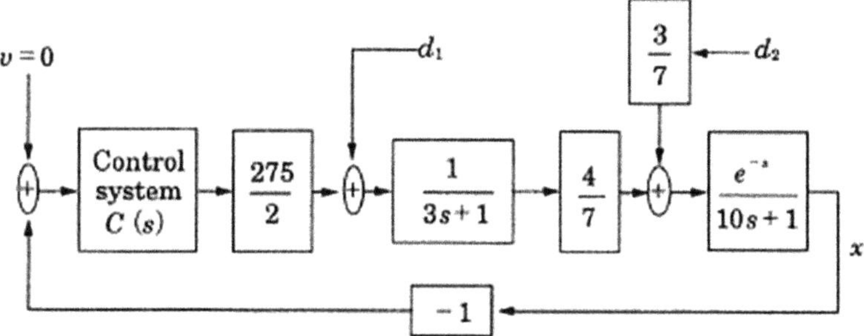

Fig. 5.25 Block diagram of the controlled system

to experience persistent oscillation at a certain value K_{PU}. This K_{PU} is the ultimate sensitivity of a P action controller. Then, using Eq. (5.9), the following holds true.

$$1 + K_{\mathrm{PU}}P(j\omega_u) = 0$$

Table 5.2 PID controller constants tuned using ultimate sensitivity method

	K_P [kg/(cm$^2\cdot$°C)]	T_I (min)	T_D (min)
P control	0.09	–	–
PI control	0.08	8.2	–
PID control	0.11	4.9	1.2

This expression yields

$$K_{PU}|P(j\omega_u)| = 1$$
$$\angle P(j\omega_u) = -180°$$

From $\omega_u = 0.64$ rad/min that we already obtained, the period P_u of persistent oscillation is derived as

$$P_U = \frac{2\pi}{\omega_u} = \frac{2\pi}{0.64} = 9.82\text{min}$$

As $|P(j\omega_u)| = 14.9\,\text{dB} = 5.56\,\text{cm}^2\cdot°\text{C/kg}$, we obtain the following for K_{PU}:

$$K_{PU} = \frac{1}{5.56} = 0.18\ (\text{kg}/(\text{cm}^2\cdot°\text{C}))$$

Table 5.2 summaries constants of PID controller, resulted from applying the Ziegler–Nichols ultimate sensitivity method for tuning to the obtained K_{PU} and the persistent oscillation period P_U.

The following shows the open-loop transfer functions of a control system consisting of the controller with constants shown in Table 5.2.

<1> P control: $C(s)P(s) = \frac{7.07\,e^{-s}}{(3s+1)(10s+1)}$

<2> PI control: $C(s)P(s) = \frac{0.767(8.2s+1)}{s(3s+1)(10s+1)}\,e^{-s}$

<3> PID control: $C(s)P(s) = \frac{1.76(5.88s^2+4.9s+1)}{s(3s+1)(10s+1)}\,e^{-s}$

Figures 5.26, 5.27, and 5.28 show Bode diagrams for the respective open-loop transfer functions.

Table 5.3 shows the phase margin and gain margin values of each control system read from the Bode diagrams. Values of the gain margin are all the same level, while values of the phase margin are different, decreasing in the order of PID control, P control, and PI control.

5.6.5 *Evaluation of Control Characteristics*

Next, we evaluate the disturbance suppression characteristic of the previously designed PID control system. For the disturbance $d_1 = \pm 5$ °C (with a fluctuation

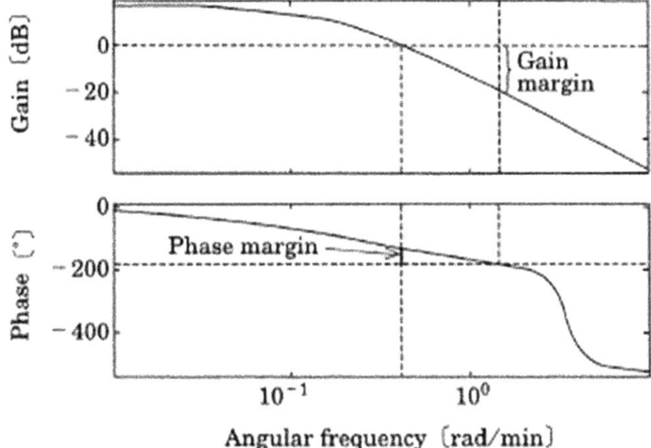

Fig. 5.26 Bode diagram of open-loop transfer function of P control system

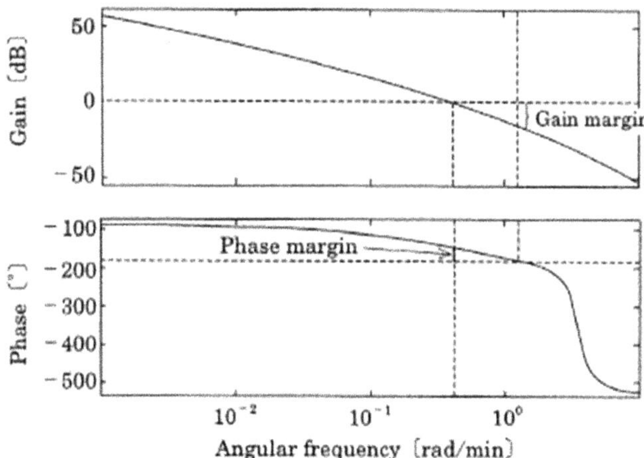

Fig. 5.27 Bode diagram of open-loop transfer function of PI control system

period of 5 min) and for the disturbance $d_2 = \pm 2\ ^\circ\mathrm{C}$ (with a fluctuation period > 6 h), we evaluate two items. <1> Whether the Tank 1 temperature can be maintained at 84 °C or lower. The outlet temperature of Tank 1 is 79 °C in the steady state, so c_1 in Fig. 5.24 should fall within the range of ±5 °C. <2> Whether temperature x of the Liquid C product can be controlled at 65 ± 0.5 °C. The measured temperature x in Fig. 5.24 should fall within the range of ±0.5 °C.

Table 5.2 gives the following PID controller $C(s)$.

$$C(s) = 0.11\left(1 + \frac{1}{4.9s} + 1.2s\right) \tag{5.39}$$

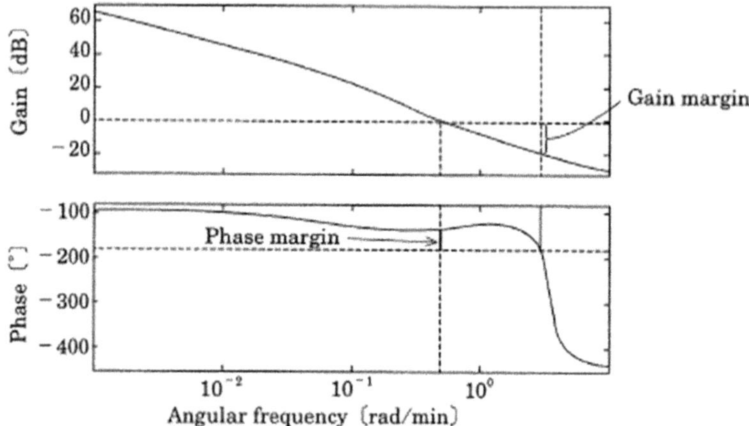

Fig. 5.28 Bode diagram of open-loop transfer function of PID control system

Table 5.3 Stability margins of each control system (values in parentheses are angular frequencies)

	Phase margin (°)	Gain margin (dB)
P control	47 (0.4 rad/min)	19 (1.5 rad/min)
PI control	33 (0.4 rad/min)	18 (1.3 rad/min)
PID control	51 (0.5 rad/min)	18 (2.9 rad/min)

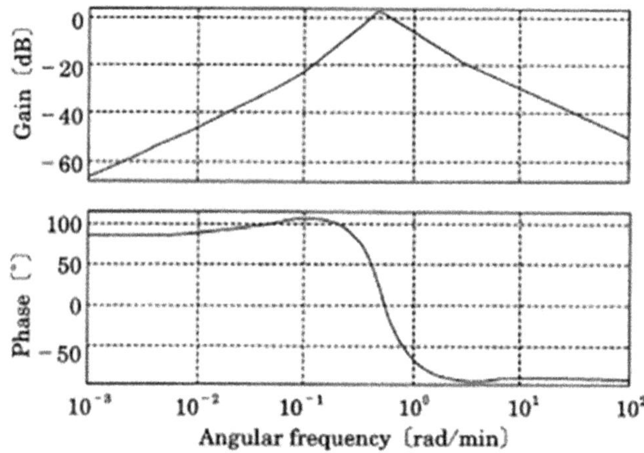

Fig. 5.29 Frequency responses of control system concerning the influence on c_1 from disturbance d_1

[1] The disturbance suppression characteristic for the disturbance d_1

Figure 5.29 shows the frequency response concerning influence of the disturbances d_1 on c_1.

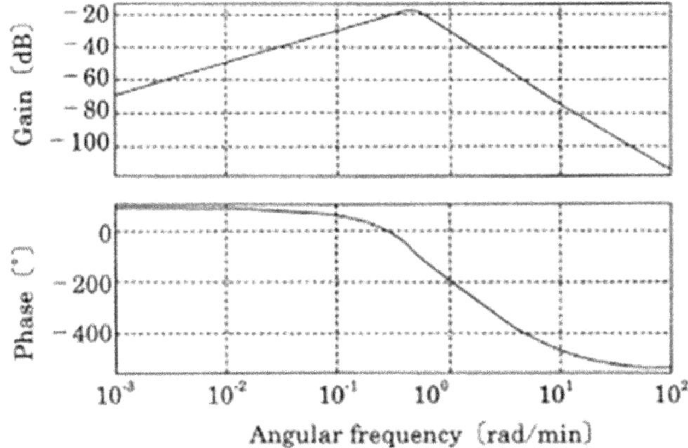

Fig. 5.30 Frequency responses of control system concerning the influence on x from disturbance d_1

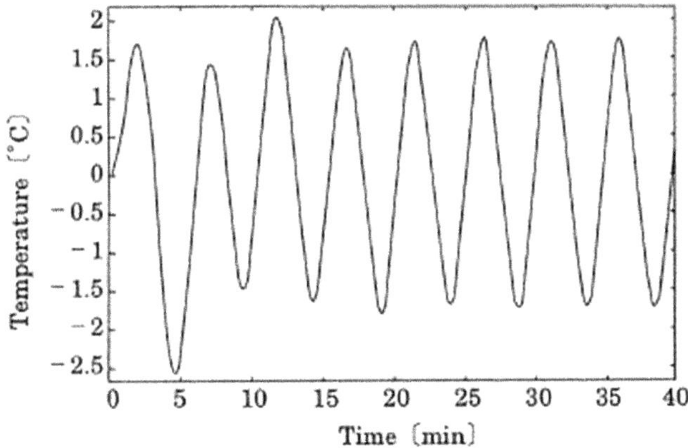

Fig. 5.31 Simulation result of time variation of c_1 by sinusoidal disturbance d_1

The natural frequency of the control system is approximately 0.5 rad/min (natural period of approximately 13 min), and that is approximated to the fluctuating frequency of 1.3 rad/min of the disturbance d_1 (fluctuation period of 5 min). However, the control system gain for the fluctuating frequency of this disturbance is approximately −20 dB. Influence of the disturbance d_1 on c_1 may be suppressed to a low level. The frequency response from the disturbance d_1 to the controlled variable x is shown in Fig. 5.30, indicating that influence of the disturbance d_1 on x is sufficiently suppressed to a low level.

We can confirm those speculations by using the dynamic simulation of the temperature c_1 and the controlled variable x for the sinusoidal disturbance d_1 with amplitude 5 °C and frequency 1.3 rad/min. Figures 5.31 and 5.32 show results of the simulation.

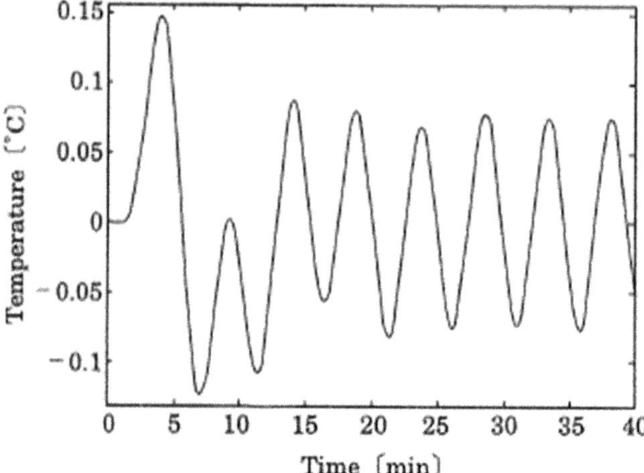

Fig. 5.32 Simulation result of time variation of x by sinusoidal disturbance d_1

Table 5.4 Evaluation of control characteristics against disturbance fluctuation

Disturbance	Fluctuation c_1 of the Tank 1 temperature	Fluctuation x of the Tank 2 temperature
$d_1 = 5$ °C	Approx. 2.0 °C	0.15 °C
$d_2 = 2$ °C	Approx. 1.5 °C	0 °C
Total fluctuation	Approx. 3.5 °C	0.15 °C
Design specifications	5 °C or less	0.5 °C or less

Those results indicate that the 5 °C fluctuation of the disturbance d_1 is suppressed approximately to 2 °C at a maximum as fluctuation of c_1. This satisfies the limit of approximately 5 °C or less, the allowable fluctuation band of c_1. Fluctuation is also suppressed to 0.15 °C at a maximum as fluctuation of x, which satisfies the limit of 0.5 °C, the allowable fluctuation band.

[2] Evaluation of the disturbance suppression characteristic for the disturbance d_2

We investigate into suppression effect of influence of the disturbance d_2 on c_1 and x. The fluctuation band of d_2 is assumed to be 2 °C, and its fluctuation period is 5 h in the following discussion. This period is extremely long compared with 13 min, the previously mentioned natural period (13 min) of the control system. By applying the final-value theorem of the Laplace transform, and multiplying 2 °C of the d_2 fluctuation by a steady-state gain, we obtain the deviation of c_1 as 1.5 °C and that of x as 0 °C.

Table 5.4 summarizes the results. We can confirm that the designed control system satisfies the design specifications.

The commercial control system software MATLAB is used for the frequency response computation and dynamic characteristic simulation. For the computation and simulation, we deal with dead time (transport lag time) contained in the controlled object as the (2, 2) type Pade approximation.

Laplace Transform Formulae

Definitions of Laplace transform

We denote the output as $y(t)$ when the input $x(t)$ is added to a system.

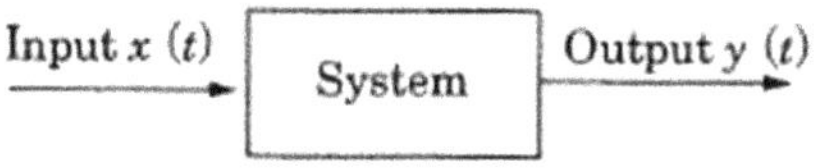

Input $x(t)$ → System → Output $y(t)$

A transfer function connects the input and output by

(Laplace transform of output) = (transfer function)
× (Laplace transform of input)

which is based mathematically on the Laplace transform.

The Laplace transform is an integral transform that maps the time function $f(t)$ defined by $t \geq 0$ to the complex function $F(s)$ using the integral kernel e^{-st} (s is a complex number, and called a Laplace operator). The following defines the Laplace transform. Usually, $f(t) = 0$ if $t < 0$.

$$F(s) = \int_0^{\infty} f(t)\,\mathrm{e}^{-st}\,\mathrm{d}t$$

As the integral range of this expression reaches $[0, \infty)$, it is a concern whether the integral is convergent. We may regard it as convergent for time functions commonly encountered in control engineering.

Next, the operation that obtains the time function $f(t)$ corresponding to the given Laplace transform $F(s)$ is called the inverse Laplace transform, and it is expressed below.

$$f(t) = \frac{1}{2\pi j}\int_{c-j\infty}^{c+j\infty} F(s)\,e^{st}\,\mathrm{d}s$$

Laplace Transform Formulae

Laplace Transform Formula for Time Derivative

$$\int_0^{\infty} f'(t)e^{-st}\,\mathrm{d}t = sF(s) - f(0)$$

$$\int_0^{\infty} f^{(n)}(t)\,\mathrm{e}^{-st}\,\mathrm{d}t = s^n F(s) - s^{n-1}f(0) - s^{n-2}f'(0) - \cdots - f^{(n-1)}(0)$$

where $f(0)$, $f^{(k)}(0)$ denote the initial values of the kth order derivative. When every initial value is zero, the nth order derivative in a time domain is dealt with by multiplying $F(s)$ by s^n in the s domain.

Laplace Transform Formula for Time Integration

The Laplace transform formula for time integration are given by the following expression.

$$\int_0^\infty \left(\int_0^t f(\tau)\mathrm{d}\tau \right) \mathrm{e}^{-st}\mathrm{d}t = \frac{1}{s}F(s)$$

$$\int_0^\infty \left(\int_0^t \left(\int_0^{t_{n-1}} \cdots \left(\int_0^{t_2} \left(\int_0^{t_1} f(\tau)\mathrm{d}\tau \right) \mathrm{d}t_1 \right) \mathrm{d}t_2 \cdots \right) \mathrm{d}t_{n-1} \right) \mathrm{e}^{-st}\,\mathrm{d}t = \frac{1}{s^n}F(s)$$

An nth multiple integral in a time domain is dealt with by dividing $F(s)$ by s^n in the s domain.

Time Transition Formula

The time transition formula is as follows.

$$\int_0^\infty f(t-a)\,\mathrm{e}^{-st}\,\mathrm{d}t = \mathrm{e}^{-as}F(s), \quad a > 0$$

Convolution Integral in a Time Domain

The convolution integral in a time domain is as follows.

$$\int_0^\infty \left(\int_0^t f_1(\tau) f_2(t-\tau)\mathrm{d}\tau \right) \mathrm{e}^{-st}\,\mathrm{d}t = F_1(s)F_2(s)$$

Functions Familiar to Control Engineering

The following functions are commonly used in control engineering (hereafter, the Laplace transform is expressed by the symbol L).

$$(1) \quad L(au(t)) = \frac{a}{s}, \quad u(t): \quad \text{Step function}$$

$$(2) \quad L(a\delta(t)) = a, \quad \delta(t): \quad \text{Impulse function}$$

$$(3)\quad L(e^{at}) = \frac{1}{s-a}$$

$$(4)\quad L\left(\frac{1}{a}e^{-\frac{t}{a}}\right) = \frac{1}{1+as}$$

$$(5)\quad L(t) = \frac{1}{s^2}$$

$$(6)\quad L\left(\frac{1}{(n-1)!}t^{n-1}\right) = \frac{1}{s^n}$$

$$(7)\quad L\left(e^{-\delta t}\sin(\omega t \pm \phi)\right) = \frac{\omega \cos\phi \pm (s+\delta)\sin\phi}{(s+\delta)^2+\omega^2}$$

$$(8)\quad L\left(e^{-\delta t}\cos(\omega t \pm \phi)\right) = \frac{(s+\delta)\cos\phi \mp \omega\sin\phi}{(s+\delta)^2+\omega^2}$$

Expansion Theorem of Heviside

The Laplace transform $F(s)$ is assumed to be given by the following rational function.

$$F(s) = \frac{Ns}{D(s)} = \frac{b_m s^m + b_{m-1}s^{m-1} + \cdots + b_1 s + b_0}{s^n + a_{n-1}s^{n-1} + \cdots + a_1 s + a_0}, \quad (n > m)$$

Then the inverse transform $f(t)$ can be obtained in the following expression:

$$f(t) = k_1 e^{p_1 t} + k_2 e^{p_2 t} + \cdots + k_n e^{p_n t}$$

where

$p_1, p_2, \ldots, p_n$: The root of the denominator polynomial $D(s)$ (no repeated root is assumed)

$k_i = \frac{N(p_i)}{D'(p_i)}, \quad i = 1, 2, \ldots, n, D'(s)$ is the derived function of $D(s)$.

Final-Value Theorem and Initial-Value Theorem

When $sF(s)$ is nonsingular on the right-hand side of the s plane, including the imaginary axis, the following theorems may be applied.

Final-value theorem: $\lim_{t\to\infty} f(t) = \lim_{s\to 0} s\cdot F(s)$

Initial-value theorem: $\lim_{t\to 0+} f(t) = \lim_{s\to\infty} s\cdot F(s)$

Pade Approximation

Pade Approximation Method

It is assumed that $G(s)$ is a given transfer function. We consider a problem to approximate it by a rational function with a polynomial of the nth degree of s as denominator, and a polynomial of the mth degree as numerator. This is called the (n, m) type Pade approximation, which can be obtained by the following procedure.

Step <1>

Expand the given $G(s)$ by Maclaurin's theorem.

The expression is

$$G(s) = g_0 + g_1 s + g_2 s^2 + \cdots + g_k s^k + \cdots + g_{m+n} s^{m+n} + \cdots \tag{5.40}$$

where

$$g_k = \left[\frac{1}{k!} \cdot \frac{\mathrm{d}^k}{\mathrm{d}s^k} G(s) \right]_{s=0}, \quad k = 0, 1, 2, 3, \ldots, \tag{5.41}$$

Step <2>

Give the rational function $H(s)$ that approximates $G(s)$ by the following expression:

Usually $H(s)$ is assumed to be a proper rational function and $m \leqq n$.

$$H(s) = \frac{b_0 + b_1 s + \cdots + b_m s^m}{1 + a_1 s + \cdots + a_n s^n} = h_0 + h_1 s + \cdots + h_{m+n} s^{m+n} + \cdots \tag{5.42}$$

Here, the last expression is a polynomial resulted by performing division calculation of the numerator by the denominator of the rational function.

Step <3>

Find $(m + n + 1)$ numbers of unknown coefficients ($a_1, a_2, \ldots, a_n$, and $b_0, b_1, b_2, \ldots, b_m$) of the expression (5.42) so that the initial $(m + n + 1)$ terms of the expressions (5.40) and (5.42) match each other.

Actually, unknown coefficients a_i and b_j can be obtained by using the following steps. First, the following equation is obtained from Step <3>:

$$(g_0 + g_1 s + \cdots + g_k s^k + \cdots)(1 + a_1 s + \cdots + a_n s^n) = b_0 + b_1 s + \cdots + b_m s^m$$

The left side is expanded to yield the following expression:

$$\sum_{l=1}^{\infty} \left(\sum_{k+i=l} g_k a_i \right) s^l - \sum_{j=0}^{m} b_j s^j = - \sum_{k=0}^{\infty} g_k s^k \tag{5.43}$$

where other than $n \geq i \geq 1$: $a_i = 0$ other than $m \geq j \geq 1\ :\ b_j = 0,\ \ k < 0 : g_k = 0$

Assuming that coefficients from the 0th order term to the $(n + m)$th order term of s on both sides of Eq. (5.43) are equal, then we obtain the following $(n + m + 1)$ equations.

0th order term: $-\,b_0 = -g_0$

1st order term: $(g_0 a_1) - b_1 = -g_1$

2nd order term: $(g_1 a_1 + g_0 a_2) - b_2 = -g_2$

3rd order term: $(g_2 a_1 + g_1 a_2 + g_0 a_3) - b_3 = -g_3$

$\vdots$

mth order term: $(g_{m-1} a_1 + g_{m-2} a_2 + \cdots + g_{m-n} a_n) - b_m = -g_m$

$(m + 1)$th order term $(g_m\, a_1 + g_{m-1}\, a_2 + \cdots + g_{m-(n-1)} a_n) + 0 = -g_{m+1}$

.

.

$(n + m)$th order term $(g_{n+m-1}\, a_1 + g_{n+m-2}\, a_2 + \cdots + g_m a_n) + 0 = -g_{m+n}$

Those $(n + m + 1)$ numbers of equations can be expressed as the following easily viewable equations by using a coefficient matrix, unknown variable vectors, and known vectors.

$$\begin{bmatrix} 0 & 0 & 0 & \cdots & 0 & -1 & \cdots & 0 \\ g_0 & 0 & 0 & \cdots & 0 & 0 & \vdots & 0 \\ g_1 & g_0 & 0 & \cdots & 0 & 0 & \vdots & 0 \\ g_2 & g_1 & g_0 & \cdots & 0 & 0 & \vdots & 0 \\ \vdots & \vdots & \vdots & \vdots & \vdots & 0 & \cdots & 0 \\ g_{m-1} & g_{m-2} & g_{m-3} & \cdots & g_{m-n} & 0 & \cdots & -1 \\ g_m & g_{m-1} & g_{m-2} & \cdots & g_{m-n+1} & 0 & \cdots & 0 \\ \vdots & \vdots & \vdots & \cdots & \vdots & \vdots & \vdots & \vdots \\ g_{n+m-1} & g_{n+m-2} & g_{n+m-3} & \cdots & g_m & 0 & \vdots & 0 \end{bmatrix} \begin{bmatrix} a_1 \\ a_2 \\ a_3 \\ \vdots \\ \vdots \\ a_n \\ b_0 \\ b_1 \\ \vdots \\ b_m \end{bmatrix} = \begin{bmatrix} -g_0 \\ -g_1 \\ -g_2 \\ \vdots \\ -g_m \\ -g_{m+1} \\ -g_{m+2} \\ \vdots \\ -g_{m+n} \end{bmatrix}$$

This matrix and vector representation enables us to obtain values $a_1, a_2, \ldots, a_n$ from the following expression.

$$\begin{bmatrix} g_m & g_{m-1} & g_{m-2} & \cdots & g_{m-n+1} \\ g_{m+1} & g_m & g_{m-1} & \cdots & g_{m-n+2} \\ \vdots & \vdots & \vdots & \cdots & \vdots \\ g_{n+m-1} & g_{n+m-2} & g_{n+m-3} & \cdots & g_m \end{bmatrix} \begin{bmatrix} a_1 \\ a_2 \\ a_3 \\ \vdots \\ a_n \end{bmatrix} = \begin{bmatrix} -g_{m+1} \\ -g_{m+2} \\ \vdots \\ -g_{m+n} \end{bmatrix}$$

It is

$$\begin{bmatrix} a_1 \\ a_2 \\ a_3 \\ \vdots \\ a_n \end{bmatrix} = \begin{bmatrix} g_m & g_{m-1} & g_{m-2} & \cdots & g_{m-n+1} \\ g_{m+1} & g_m & g_{m-1} & \cdots & g_{m-n+2} \\ \vdots & \vdots & \vdots & \cdots & \vdots \\ g_{n+m-1} & g_{n+m-2} & g_{n+m-3} & \cdots & g_m \end{bmatrix}^{-1} \cdot \begin{bmatrix} -g_{m+1} \\ -g_{m+2} \\ \vdots \\ -g_{m+n} \end{bmatrix} \tag{5.44}$$

Similarly, $b_0, b_1, b_2, \ldots, b_m$ can be obtained from the following expression:

$$\begin{bmatrix} b_1 \\ b_2 \\ b_3 \\ \vdots \\ b_m \end{bmatrix} = \begin{bmatrix} 0 & 0 & 0 & \cdots & 0 \\ g_0 & 0 & 0 & \cdots & 0 \\ g_1 & g_0 & 0 & \cdots & 0 \\ \vdots & \vdots & \vdots & \cdots & \vdots \\ g_{m-1} & g_{m-2} & g_{m-3} & & g_{m-n} \end{bmatrix} \cdot \begin{bmatrix} a_1 \\ a_2 \\ a_3 \\ \vdots \\ a_n \end{bmatrix} + \begin{bmatrix} g_0 \\ g_1 \\ g_2 \\ \vdots \\ g_m \end{bmatrix} \tag{5.45}$$

Pade Approximation in a Dead Time System

We obtain the (2, 2) type Pade approximation of the transfer function $G(s) = \mathrm{e}^{-Ls}$ in a dead time system. First, according to Step <1> above, to expand from the first to 4th order terms, Maclaurin's theorem is used.

$$\mathrm{e}^{-Ls} \cong 1 + (-L)s + \left(\frac{L^2}{2}\right)s^2 + \left(-\frac{L^3}{6}\right)s^3 + \left(\frac{L^4}{24}\right)s^4 + \cdots \tag{5.46}$$

Therefore, we obtain the next expression.

$$g_0 = 1, \quad g_1 = -L, \quad g_2 = \frac{L^2}{2}, \quad g_3 = -\frac{L^3}{6}, \quad g_4 = \frac{L^4}{24}, \ldots, g_k = \frac{(-L)^k}{k!}, \ldots$$

Substituting these into the expression (5.44) yields

$$\begin{bmatrix} a_1 \\ a_2 \end{bmatrix} = \begin{bmatrix} \frac{L^2}{2} & -L \\ -\frac{L^3}{6} & \frac{L^2}{2} \end{bmatrix}^{-1} \cdot \begin{bmatrix} \frac{L^3}{6} \\ -\frac{L^4}{24} \end{bmatrix}$$

$$a_1 = -\frac{L}{2}, \qquad a_2 = \frac{L^2}{12}$$

Also by using the expression (5.44), we obtain $b_0 = 1, \quad b_1 = -\frac{L}{2}, \quad b_2 = \frac{L^2}{12}$. Consequently, the (2, 2) type Pade approximation in a dead time system is:

$$G(s) = \mathrm{e}^{-Ls} \cong \frac{1 - \frac{1}{2}Ls + \frac{1}{12}(Ls)^2}{1 + \frac{1}{2}Ls + \frac{1}{12}(Ls)^2} \tag{5.47}$$

Chapter 5 Exercises

1. Obtain Laplace transforms of the following functions
   <1> sin ωt, where ω is positive.
   <2> con ωt, where ω is positive.
   <3> $f(at)$ where a is positive.
2. The impulse input $x(t) = \delta(t)$ has been given to a system with the initial value 0 for all the state variables. Then the response is measured as follows:

$$y(t) = \begin{cases} 0 : & t < 0 \\ \frac{a}{T}t : & 0 \le t \le T \\ 0 : & T < t \end{cases}.$$

   Obtain the system transfer function from this input port to the output.
3. Obtain the Laplace transform of the following function

$$f(t) = \begin{cases} 0 & : \; t < 0 \\ a \sin\left(\frac{2\pi}{T}t\right) & : \; 0 \le t \le \frac{T}{2} \\ 0 & : \; \frac{T}{2} < t \end{cases}.$$

4. Obtain the Laplace transforms of a square wave and a saw-tooth wave which are shown in the following figures.
   <1> Square waveform (with amplitude 1, and cycle $2T$)
   <2> Saw-tooth waveform (with unilateral amplitude 1, and cycle T)

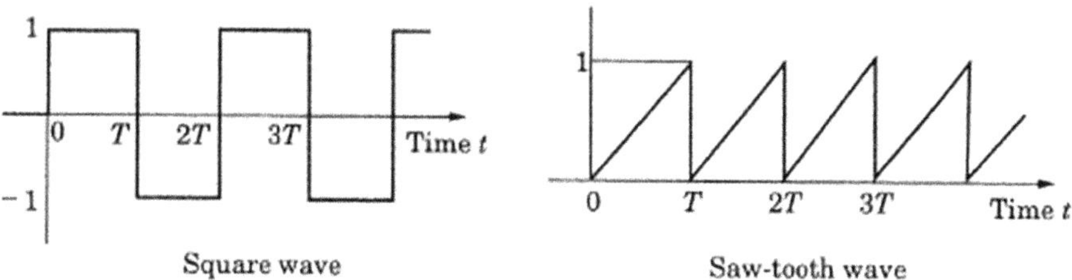

Square wave   Saw-tooth wave

5. Obtain the inverse Laplace transform of the following expression.

$$F(s) = \frac{k}{as(1 + Ts)}$$

(Clue) Use the expansion theorem $L^{-1}\left[\frac{P(s)}{sQ(s)}\right] = \frac{P(0)}{Q(0)} + \sum_{h=2}^{n} \frac{P(s_h)}{s_h Q'(s_h)} e^{s_h t}$.

Assume that $Q(s) = 0$ has $n - 1$ pieces of simple root.

6. Derive the transfer function of the expression (5.6).
7. Consider a system expressed by the following motion equation.

$$m\frac{d^2y}{dt^2} + c\frac{dy}{dt} = kx$$

where x is an input variable, y is an output variable, and m, c, and k are constants of the system.

(1) Obtain the transfer function $G(s)$ of this system.

(2) Suppose that feedback control is incorporated in the system as shown in the figure. Obtain the closed-loop transfer function from the input u to output y (the 39th examination for chief engineer of reactors in Japan)

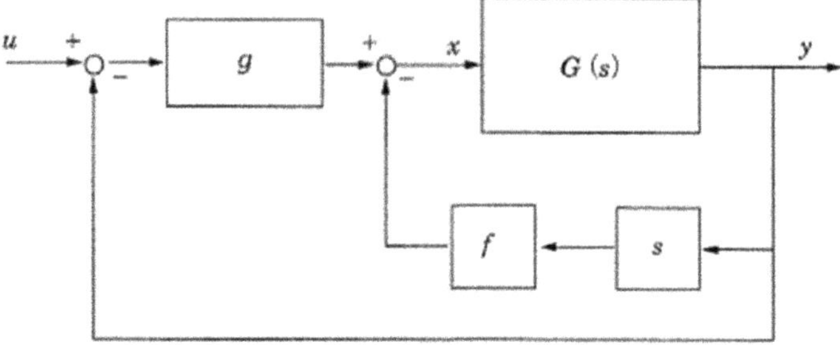

8. Prove that the block diagram shown in Fig. 5.16 is equivalent to that of a PID controller. Also confirm that the mutual interference coefficient is $(1 + T_2/T_1)$.
9. In the design example in Sect. 5.6, we neglected the steam manipulated variable $z(t)$ because it is negligibly small compared with the steady-state flow rate m (=2.3 kg/m), and obtained the expressions (5.34) and (5.35). Obtain the transfer function when the steam flow rate $z(t)$ is not neglected.
10. It is natural that the (2, 2) type approximation (5.47) indicates a different response from that of the dead time system e^{-Ls}. However, to approximate the transfer function of (a dead time system) + (a high order system) by the rational function of s, a Pade approximation is effective. Discuss the reasons.
11. Obtain the Pade approximation of $\frac{5}{s^2+2.5s+5}e^{-Lt}$, a coupled system of dead time system + 2nd order oscillation system.

Bibliography

1. Bennett S (1998) History of control engineering 1800–1930, and history of control engineering 1930–1955. Corona Sha, Tokyo (in Japanese)
2. Kondo B, Fujii K (1972) Control engineering for college course. Ohm Sha, Tokyo (in Japanese)
3. Izawa K (1954) Introduction to automatic control. Ohm Sha, Tokyo (in Japanese)
4. Suda N et al (1992) PID control. Asakura Publishing, Tokyo (in Japanese)
5. Furuta K, Tomita T (1990) Report of current advanced control technology. J Soc Instrum Control Eng 29(10):77–82 (in Japanese)
6. The MathWork MATLAB Ver.5.2, User's Guide (1998)

Chapter 6
Reactor Stability Study

Katsuo Suzuki, Hiroshi Ono, and Shuhei Miyake

6.1 Reactor Transfer Function

6.1.1 Transfer Function of Reactor with No Feedback

[1] One-point core dynamic approximation model

Usually, the one-point core dynamic approximation model can be expressed by the following equation system.

$$\frac{\mathrm{d}N(t)}{\mathrm{d}t} = \frac{\rho(t) - \beta}{\Lambda} N(t) + \sum_{i=1}^{6} \lambda_i C_i(t) + S(t) \tag{6.1}$$

$$\frac{\mathrm{d}C_i(t)}{\mathrm{d}t} = \frac{\beta_i}{\Lambda} N(t) - \lambda_i C_i(t), \quad i = 1, 2, \ldots, 6 \tag{6.2}$$

where

$N(t)$: total number of neutrons in the core
$C_i(t)$: total number of the ith group delayed neutron precursors in the core
$S(t)$: neutron supply rate from the entire neutron sources (numbers per second)

$\rho(t) \equiv (k_{\mathrm{eff}} - 1)/k_{\mathrm{eff}} = \delta k/k_{\mathrm{eff}}$, where k_{eff} is the effective multiplication factor
Λ: neutron generation time (s)
β_i: ratio of the number of delayed neutron precursors of the ith group
β: delayed neutron generation rate for a single fission
λ_i: decay constant of delayed neutron precursors of the ith group (1/s)

Equations (6.1) and (6.2) are approximation by considering the entire reactor core as one point. However, they are useful enough for discussion of reactor control issues. As (6.1) is a nonlinear equation having the product of $\rho(t)$ and $N(t)$, the linearized approximation or another is used according to the reactor control issues.

Y. Oka and K. Suzuki (eds.), *Nuclear Reactor Kinetics and Plant Control*,
An Advanced Course in Nuclear Engineering,
DOI 10.1007/978-4-431-54195-0_6, © Springer Japan 2013

Table 6.1 Nuclear kinetics parameters (an example of BWR core)

Group number, i	Decay constant, λ_i (s^{-1})	Generation ratio, β_i(−)	β_i/λ_i (s)
1	0.0127	0.247×10^{-3}	0.0194
2	0.0327	1.38×10^{-3}	0.0422
3	0.115	1.22×10^{-3}	0.0106
4	0.311	2.64×10^{-3}	0.0085
5	1.40	0.832×10^{-3}	5.94×10^{-4}
6	3.87	0.169×10^{-3}	4.37×10^{-5}
Total		0.0063	0.0813

Coefficient Λ of the one-point core approximate equation is 10^{-3} s order for a graphite reactor and a heavy water reactor, 10^{-4} s order for a light water reactor, and shorter 10^{-7} s order for a fast reactor. The β_i and λ_i values of a general BWR core are shown in Table 6.1.

[2] Equilibrium state of reactor

Equation (6.1), having the product of operation quantity $\rho(t)$ and state quantity $N(t)$, is a nonlinear differential equation, and it is also called a bilinear equation. The six equations of (6.2) are linear differential equations. During power output, the contribution of neutron supply rate, $S(t)$, is small when compared with the one during fission and it can be ignored.

1) Equilibrium state when reactor is critical

Equations (6.1) and (6.2) have two types of equilibrium state. One of them is the critical state ($\rho(t) = 0$) without neutron source. As (6.3) given below satisfies (6.1) and (6.2), it is clear that a single equilibrium state has occurred.

$$\rho(t) = 0, \qquad S(t) = 0, \qquad N(t) = N_0, \qquad C_i(t) = \frac{\beta_i}{\lambda_i \Lambda} N_0 \tag{6.3}$$

At this time, the critical reactor enters the equilibrium state with any output N_0. If the system has no property to stable its control quantity (such as N and C_i) by itself (that is, if the system has no self-equilibrium property), it is called the astatic controlled object.

2) Equilibrium state when reactor is subcritical

The other equilibrium state can occur in the subcritical reactor when the neutron source is inserted into it. In this case, the state quantity is:

$$\rho(t) = \rho_0 < 0, \quad S(t) = S_0, \qquad N(t) = \frac{\Lambda}{-\rho_0} S_0, \qquad C_i(t) = \frac{\beta_i}{-\lambda_i \rho_0} S_0 \tag{6.4}$$

In this case, output $N(t)$ is not any level but is a fixed value that is determined by the neutron generation time (Λ), subcritical reactivity ($-\rho_0$), and neutron source strength (S_0). The subcritical reactor having the neutron source is the static controlled object.

[3] Reactor transfer function
1) Derivation of transfer function

Using the deviation from equilibrium state, let us describe variables $\rho(t)$, $N(t)$, and $C_i(t)$ that change in time as follows:

$$\begin{aligned} \rho(t) &= \rho_0 + \delta\rho(t) \\ N(t) &= N_0 + \delta N(t) \\ C_i(t) &= C_{i0} + \delta C_i(t) \end{aligned} \tag{6.5}$$

Then, assign them to (6.1) and (6.2), and we can get the following equations.

$$\frac{\mathrm{d}\delta N(t)}{\mathrm{d}t} = -\frac{\rho_0 - \beta}{\Lambda}\delta N(t) + \sum_{i=1}^{6} \lambda_i \delta C_i(t) + \frac{\beta}{\Lambda} N_0 \rho(t) + \frac{1}{\Lambda}\delta N(t)\delta\rho(t) \tag{6.6}$$

$$\frac{\mathrm{d}\delta C_i(t)}{\mathrm{d}t} = \frac{\beta_i}{\Lambda}\delta N(t) - \lambda_i \delta C_i(t), \quad i = 1, 2, \ldots, 6 \tag{6.7}$$

Assume that the reactor is almost in the equilibrium state and if we ignore the very small value of product $\delta N(t)$ and $\delta\rho(t)$ in Item 4 of (6.6), we can get the following linear differential equation of $\delta N(t)$ and $\delta C_i(t)$.

$$\frac{\mathrm{d}}{\mathrm{d}t}\delta N(t) = \frac{\rho_0 - \beta}{\Lambda}\delta N(t) + \sum_{i=1}^{6} \lambda_i \delta C_i(t) + \frac{N_0}{\Lambda}\delta\rho(t) \tag{6.8a}$$

$$\frac{\mathrm{d}}{\mathrm{d}t}\delta C_i(t) = \frac{\beta_i}{\Lambda}\delta N(t) - \lambda_i \delta C_i(t) \tag{6.8b}$$

These equations indicate the nuclear kinetics of a zero-output reactor in a state of near equilibrium by ignoring the feedback reactivity caused by the fuel and coolant temperature, pressure, voids and other factors. In this case, the initial conditions are as follows:

$$\delta N(0) = 0, \qquad \delta C_i(0) = 0 \ (i = 1, 2, \ldots, 6), \qquad \delta\rho(0) = 0 \tag{6.9}$$

If (6.8) is converted by Laplace transform under these initial conditions, we can get:

$$s\delta N(s) = \frac{\rho_0 - \beta}{\Lambda}\delta N(s) + \sum_{i=1}^{6} \lambda_i \delta C_i(s) + \frac{N_0}{\Lambda}\delta\rho(s) \tag{6.10a}$$

$$s\delta C_i(s) = \frac{\beta_i}{\Lambda}\delta N(s) - \lambda_i \delta C_i(s), \quad i = 1, 2, \ldots, 6 \tag{6.10b}$$

Therefore, we can determine the transfer function of a zero-output reactor as follows:

$$G_{\rho 0}(s) \equiv \frac{\left(\frac{\delta N(s)}{N_0}\right)}{\left(\frac{\delta \rho(s)}{\beta}\right)} = \frac{1}{s\left(\frac{\Lambda}{\beta} + \sum_{i=1}^{6} \frac{\beta_i/\beta}{s+\lambda_i}\right) - \frac{\rho_0}{\beta}} \tag{6.11}$$

As described in Sect. 5.3 of Chap. 5, if we determine the polarity (the root of the seventh polynomial equation of denominator) of transfer function of (6.11), we can evaluate the stability of reactor. Thanks to the progress of computer technologies, software that can accurately and easily calculate the root of high-order polynomial equations has been developed. Before development of such software, people calculated the root of those equations through various methods. One of them is a solution utilizing diagrams. The solution is still useful for gaining a visual understanding of reactor stability.

2) Frequency response of critical reactor (Bode diagram)

For the zero-output critical reactor, $(\rho_0/\beta) = 0$ in (6.11).

$$G_0(s) \equiv \frac{\left(\frac{\delta N(s)}{N_0}\right)}{\left(\frac{\delta \rho(s)}{\beta}\right)} = \frac{1}{s\left(\frac{\Lambda}{\beta} + \sum_{i=1}^{6} \frac{\beta_i/\beta}{s+\lambda_i}\right)} \tag{6.12}$$

It is clear that this transfer function has zero as one pole. It shows that the critical reactor system is astatic as described earlier.

The frequency response of critical reactor can be approximated according to the level of angular frequency ω as follows:

$$\text{(1)} \quad \text{if } \omega \ll \lambda_i \ll \frac{\beta}{\Lambda}, \quad G_0(j\omega) = \frac{\beta}{(j\omega)\sum_{i=1}^{6} \frac{\beta_i}{\lambda_i}} \tag{6.13a}$$

$$\text{(2)} \quad \text{if } \omega \approx \lambda_i \ll \frac{\beta}{\Lambda}, \quad G_0(j\omega) = \frac{\beta}{(j\omega)\sum_{i=1}^{6} \frac{\beta_i}{j\omega+\lambda_i}} \tag{6.13b}$$

$$\text{(3)} \quad \text{if } \lambda_i \ll \omega \approx \frac{\beta}{\Lambda}, \quad G_0(j\omega) = \frac{\beta}{j\omega\Lambda + \beta} \tag{6.13c}$$

$$\text{(4)} \quad \text{if } \lambda_i < \frac{\beta}{\Lambda} \ll \omega, \quad G_0(j\omega) = \frac{\beta}{j\omega\Lambda} \tag{6.13d}$$

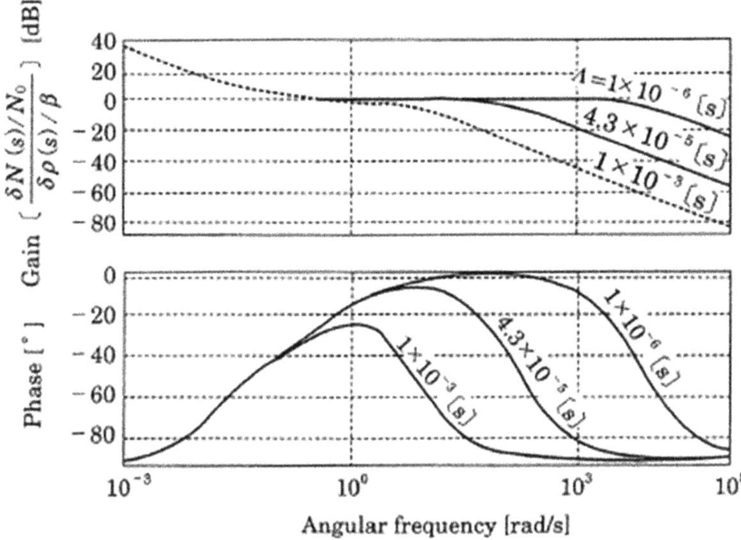

Fig. 6.1 Frequency responses of critical reactor for various neutron generation times

From these approximate expressions, it is clear that the frequency response of critical reactor does not depend on neutron generation time Λ in the low frequency range but depends on Λ in the high-frequency range. This state can be understood well using Fig. 6.1 where a single Bode diagram covering various Λ values (10^{-6}, 4.3×10^{-5}, and 10^{-3} s) is drawn using the kinetics parameters of Table 6.1.

This figure shows that (1) the gain increases limitlessly when frequency ω closes to zero (this is the typical property of astatic system frequency response), and the frequency response does not depend on value Λ when the frequency is low, (2) if value ω is large, it can be approximated in the primary delayed system, (3) the gain is constant (0 dB) in the intermediate frequency range and this frequency range depends on Λ, and (4) the phase depends on value ω and it is advanced from $-90°$ to $0°$, and if value ω is further increased, the phase is delayed for $-90°$ again.

3) Frequency response of subcritical reactor

When the neutron source is inserted in the subcritical reactor ($\rho_0/\beta < 0$), the reactor enters the equilibrium state as expressed by (6.4). If reactivity disturbance $\delta\rho(s)/\beta$ \$ is applied to this equilibrium state, the transfer function to output fluctuation $\delta N(s)/N_0$ (determined by equilibrium state output N_0) can be expressed by (6.11). Figure 6.2 shows the Bode diagram of the frequency response for various values (10^{-6}, 4.3×10^{-5}, 10^{-3} s) of Λ in a subcritical ($\rho_0/\beta = -2\$$) reactor.

This figure shows that (1) the frequency response of subcritical reactor is static as described earlier, (2) the frequency response can be approximated for the primary delayed system, (3) the gain is constant and no phase delays in the low frequency range, but the gain drops inversely proportional to the frequency, and the

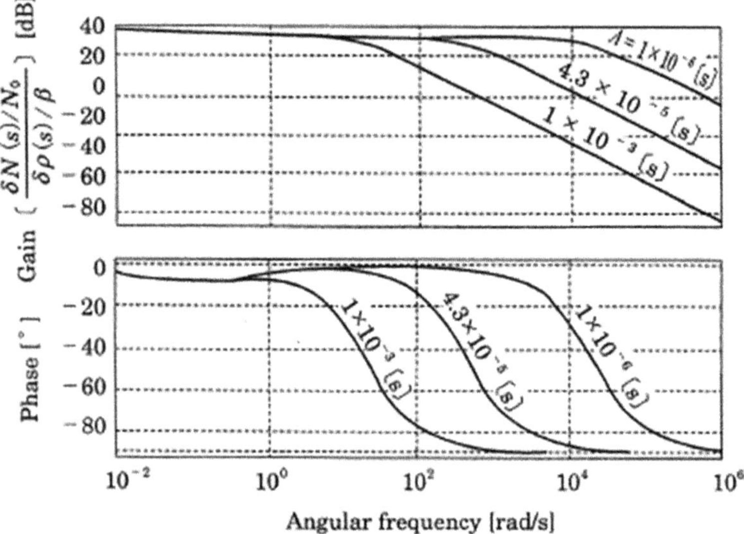

Fig. 6.2 Frequency responses of subcritical reactor for various neutron generation times ($\rho_0/\beta = -2\$$)

phase is finally delayed for 90°, and (4) the frequency response does not depend on Λ in the low frequency range (this is similar to the critical reactor), but it depends on value Λ in the high-frequency range.

6.1.2 Transfer Function of High-Output Reactor Having the Feedback Reactivity

The reactivity feedback caused by the temperature change of fuel and core structure, the temperature change of coolant, and the change of void fraction and others can be ignored if the reactor core output is small. However, when the output is increased, the reactivity feedback must be considered. If these feedback phenomena are processed in the general way, the equation system of reactor kinetics becomes nonlinear and its mathematical operation becomes complicated. However, if various state quantities are close to the stationary state and they are experiencing small changes, the reactivity feedback can be expressed by the linear equation system and the transfer function can be determined. The following describes how to obtain the transfer function of reactor that has the reactivity feedback caused by the changed temperature and coolant voids.

Assume that the kinetics of core temperature $\theta(t)$ and coolant void fraction $V(t)$ can be expressed by the following equations.

$$\frac{\mathrm{d}\theta(t)}{\mathrm{d}t} = k_\theta N(t) - \gamma_\theta(\theta(t) - \theta_c), \quad \text{where } \theta_c \text{ is the temperature of reactor coolant}$$

$$\frac{\mathrm{d}V(t)}{\mathrm{d}t} = k_v N(t) - \gamma_v V(t), \quad \text{where } V(t) \text{ is the void fraction of reactor core}$$

Therefore, if small fluctuation $\delta N(t)$ occurs in the reactor output, temperature deviation $\delta\theta(t)$ and void deviation $\delta V(t)$ from the stationary state can be expressed by the following kinetics equation system. In this case, however, we assume that the coolant temperature (θ_c) does not change for small output fluctuation.

$$\frac{\mathrm{d}\delta\theta(t)}{\mathrm{d}t} = K_\theta \delta N(t) - \gamma_\theta \delta\theta(t), \quad \text{where } \delta\theta(0) = 0 \tag{6.14}$$

$$\frac{\mathrm{d}\delta V(t)}{\mathrm{d}t} = K_v \delta N(t) - \gamma_v \delta V(t), \quad \text{where } \delta V(0) = 0 \tag{6.15}$$

Also, we assume that the reactivity feedback for the temperature and void fluctuation is given by the following equations.

$$\delta\rho_\theta(t) = -\alpha_\theta \delta\theta(t)$$
$$\delta\rho_v(t) = -\alpha_v \delta V(t)$$

If these equations are converted by Laplace transform, the transfer function of temperature and void reactivity feedback for reactor output fluctuation δN can be expressed by the following equations.

<1> Transfer function of temperature:

$$G_\theta(s) = \frac{\delta\theta(s)}{\delta N(s)} = \frac{k_\theta}{1 + T_\theta s} \tag{6.16}$$

<2> Transfer function of void fraction:

$$G_v(s) = \frac{\delta V(s)}{\delta N(s)} = \frac{k_v}{1 + T_v s} \tag{6.17}$$

<3> Transfer function of temperature feedback reactivity ($):

$$\delta\rho_\theta(s) = -\left(\frac{1}{\beta}\right)\alpha_\theta \delta\theta(s) \tag{6.18}$$

<4> Transfer function of void feedback reactivity ($):

$$\delta\rho_v(s) = -\left(\frac{1}{\beta}\right)\alpha_v \delta V(s) \tag{6.19}$$

Table 6.2 Thermal kinetics parameter values (an example of BWR core)

Parameter	Symbol	Value
Temperature time constant	T_θ	2.0 [s]
Temperature gain	k_θ	0.056 [°C/MW]
Temperature coefficient	α_θ	-1.7×10^{-5} [δk/°C]
Void time constant	T_v	5.0 [s]
Void gain	k_v	0.061 [% void fraction/MW]
Void coefficient	α_v	-7.8×10^{-4} [δk/% void fraction]

where

$T_\theta \equiv 1/\gamma_\theta$: time constant for temperature change (s)

$T_v \equiv 1/\gamma_v$: time constant for void fraction change (s)

$k_\theta \equiv K_\theta/\gamma_\theta$: gain for temperature change (°C/MW)

$k_v \equiv K_v/\gamma_v$: gain for void generation (%/MW)

α_θ : temperature feedback coefficient (δk/°C)

α_v : void feedback coefficient (δk/%)

An example of thermal kinetics parameter values of an ordinary BWR reactor core is given in Table 6.2.

If the temperature and void feedback reactivity elements of (6.16)–(6.19) are added to the transfer function $N_0G_0(s)$ of critical reactor having output N_0, the nuclear thermal coupling kinetics can be expressed by the block diagram of Fig. 6.3.

The feedback elements for temperature and void parallel binding of this block diagram can be expressed by single element $G_F(s)$ as the following equation.

$$G_F(s) = \left(\frac{1}{\beta}\right) \frac{(\alpha_\theta k_\theta T_v + \alpha_v k_v T_\theta)s + (\alpha_\theta k_\theta + \alpha_v k_v)}{(T_\theta T_v)s^2 + (T_\theta + T_v)s + 1} \tag{6.20}$$

Therefore, if $G_{xv}(s)$ of equation (5.6) is used, closed-loop transfer function $G_{N0}(s)$, from disturbance reactivity $\delta\rho_{ex}(s)(\$)$ to normalized output $\delta N(s)/N_0$, can be obtained as follows:

$$G_{N_0}(s) \equiv \frac{\left(\frac{\delta N(s)}{N_0}\right)}{\delta\rho_{ex}(s)} = \frac{G_0(s)}{1 + N_0 G(s) G_F(s)} \tag{6.21}$$

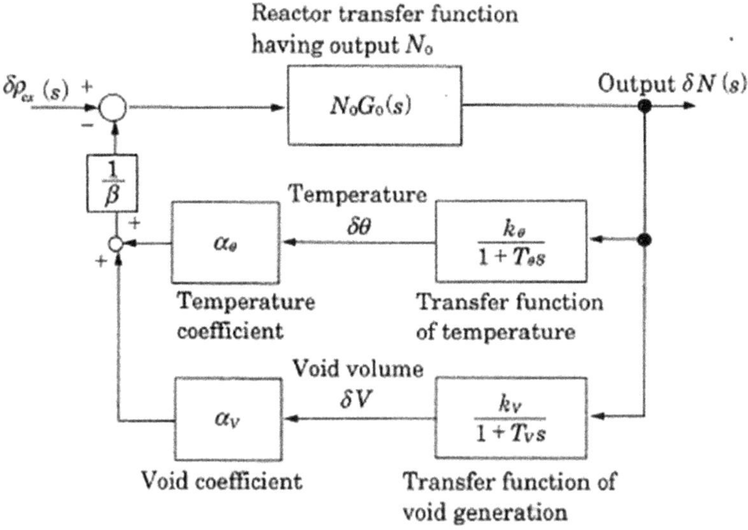

Fig. 6.3 Block diagram of high-power reactor kinetics having the feedback reactivity

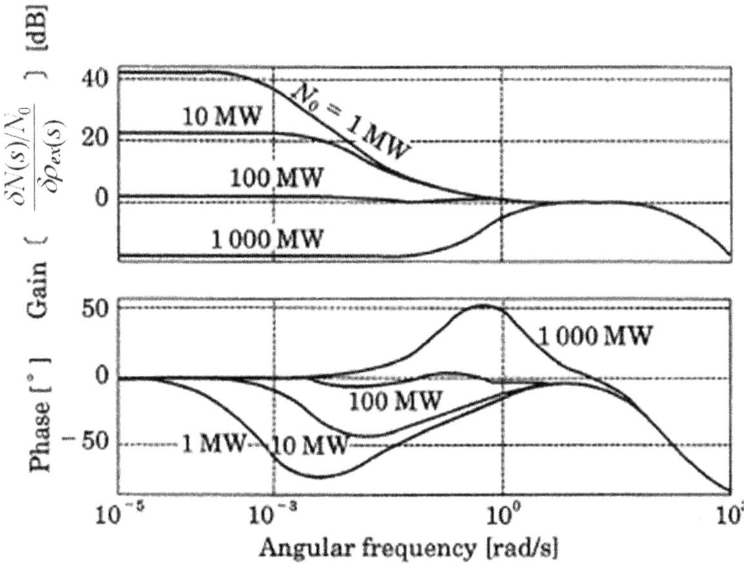

Fig. 6.4 Frequency responses of high-power reactor for various outputs (the unit of $\delta\rho_{ex}$ is $)

Figure 6.4 shows the frequency response when output N_0 of transfer function $G_{N0}(s)$ is changed from 1 to 1,000 MW, by using the nuclear kinetics parameters of Table 6.1 and thermal kinetics parameters of Table 6.2.

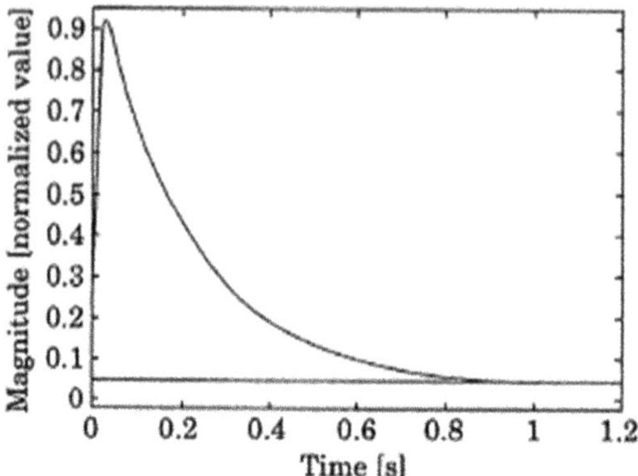

Fig. 6.5 Step response of reactor in power operation ($N_0 = 3{,}000$ MWt)

6.1.3 Design Example of Constant Output Control System of a High-Output Reactor

If transfer function (6.21) is used, the step response of reactor $\delta N/N_0$ output for step reactivity disturbance $\delta\rho_\mathrm{d}$ can be obtained as shown in Fig. 6.5.

Although the output rapidly increases immediately after an input of disturbance, the accompanying increase of fuel temperature and the increase of coolant void fraction cause a negative reactivity feedback. The output gradually decreases as time advances. However, the reactor output is not returned to its original power by this physical feedback effect of the reactor alone. The output may have approximately 5 % of stationary error. Also, the output stabilization takes approximately 1 s.

The following gives a design example of the control system to maintain the constant output against the reactivity disturbance in a BWR reactor operating with the 3,000 MWt output. Also, this system can control the reactor output quickly as possible by following the change of set values. The following explains the design procedure of the control system.

[1] Design specifications

An ideal output control system can adjust the reactor output by following up any change of target value, cancel a disturbance immediately, and maintain the reactor output at the set value. However, it is actually impossible to design such ideal control system and we need to satisfy to have the response close to our ideal system.

Here, we assume to have the design specifications of our control system as follows. (1) Cancel a stepwise disturbance or a low-frequency range reactivity disturbance immediately and maintain the constant output. (2) Limit the overshoot within 25 % against a stepwise change of output set value, and adjust the reactor output by eliminating almost all stationary (steady-state) error.

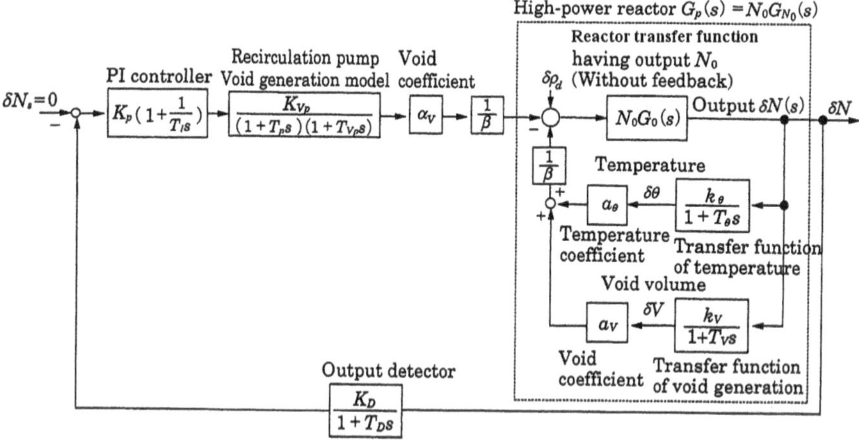

Fig. 6.6 Block diagram of the control system for the rated power operated reactor

[2] Selecting an operation quantity for the control target

The control target is a BWR reactor having the reactivity feedback of fuel temperature and voids. Its transfer function can be obtained as $N_0 G_{N0}(s)$ of (6.21). Assume that core's void fraction V is the function of coolant flux m and output N of the core. If microscopic fluctuation of flow and output (δm, δN) occurs in the stationary operation state, the amount of change of void fraction δV can be approximated by the following equation.

$$\delta V \cong \frac{\partial V}{\partial m}\delta m + \frac{\partial V}{\partial N}\delta N \tag{6.22}$$

The second term of this equation is the variation of void fraction by temperature transfer of fuel rods. The reactivity caused by this variation can be handled as the internal feedback reactivity of the control target, and it is handled as the primary delayed system in (6.17). The first term is the variation of void fraction that is caused by the change of flow rate, and we can control the feedback reactivity by changing the flow rate of recirculation pump.

[3] Block diagram of the control system

Figure 6.6 shows a block diagram of the control system whose output is fixed by the PID controller. In Table 6.3, the constants of the recirculation pump, void generation, and the output detector models are given.

[4] Selecting the PI controller

In Fig. 6.6, when stepwise reactivity disturbance $\delta\rho_{\mathrm{d}}$ is entered, stationary error ε_{d} of control amount δN can be obtained by the second equation of formula (5.6) and by the formula of the last value of Laplace transform as follows.

Table 6.3 Data on pump, void generation model, and detector

Parameters	Symbols	Values
Pump time constant	T_p	3.0 [s]
Pump gain	K_p	1.0 [kg/(s·m^2)/MW]
Void time constant	T_{vp}	3.0 [s]
Void gain	K_{vp}	8.29 [% voidratio/kg/(s·m^2)]
Detector time constant	T_D	0.1 [s]
Detector gain	K_D	1[-]

$$\varepsilon_d = \lim_{\varepsilon \to 0}\left[s \cdot \frac{G_p(s)}{1+G(s)} \cdot \frac{\delta\rho_d}{s}\right] = \frac{G_p(0)}{1+G(0)}\delta\rho_d \tag{6.23}$$

Also, if set value δNs is stepwise changed only for a, stationary error ε_v of the control variable can be obtained from the first term of (5.8) as follows.

$$\varepsilon_v = \lim_{s \to 0}\left[s \cdot \left\{1 - \frac{G_F(s)}{1+G_F(s)H(s)}\right\}\frac{a}{s}\right] = \left\{1 - \frac{G_F(0)}{1+G_F(0)H(0)}\right\}a \tag{6.24}$$

where the open loop transfer function of the control system is $G(s) = G_F(s)H(s)$. $G_F(s)$ is a forward transfer function that contains the PI controller and high power reactor $G_P(s)$. And $H(s)$ is a backward transfer function.

As the reactor having feedback elements is a static system, $G_P(s)$ has no pole in its origin. Therefore, if the PI controller is adopted, the open-loop transfer function $G(s)$ has a pole in its origin and two stationary errors (ε_d and ε_v) can be cleared to zero simultaneously.

[5] Adjusting the controller parameters and stability

It is known that the maximum value (M_p) of the closed-loop frequency gain characteristics is 1.3 for the corresponding overshoot (25 %) when the set value is changed. The frequency responses of the open-loop transfer function for two sets of PI control parameters (a pair of $K_p = 10$ [(kg/s m^2)/MW] and $T_I = 0.5$ [s], and a pair of $K_p = 2.0$ [kg/s m^2)/MW] and $T_I = 0.5$ [s]) is drawn on the Nichols chart as shown in Fig. 6.7.

The former pair gives $M_p = 6$ dB $= 1.99$. The latter pair, however, gives $M_p = 1$ dB $= 1.12$, and the overshoot satisfies the specifications. The Bode diagram of $G(s)$ shown in Fig. 6.8 indicates approximately 20 dB of gain margin and approximately 30° of phase margin. Also, Fig. 6.9 shows the Nyquist locus of $G(s)$. Both figures show that the PI controller with the latter pair of parameters is capable of making the high-power reactor sufficiently stable.

[6] Simulating the control system kinetics

Finally, we confirm the behavior of the control system by numerical simulation. Figure 6.10 shows the transient response when the set value is changed stepwise only for 1.0 MW.

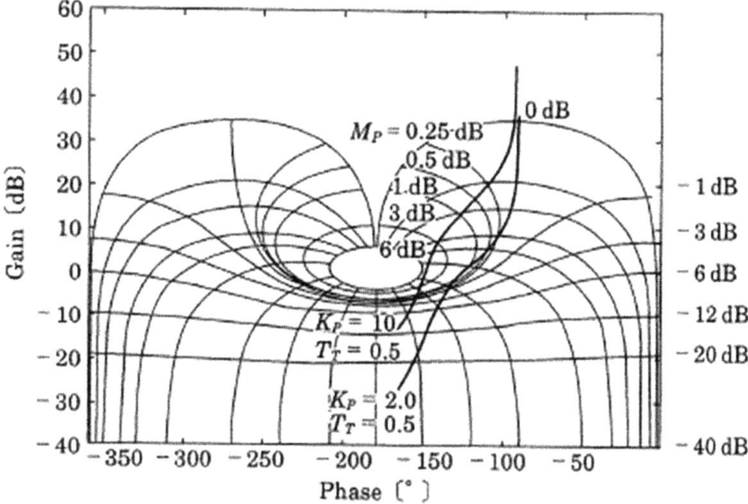

Fig. 6.7 Frequency responses of the open-loop transfer function G(s) on Nichols chart

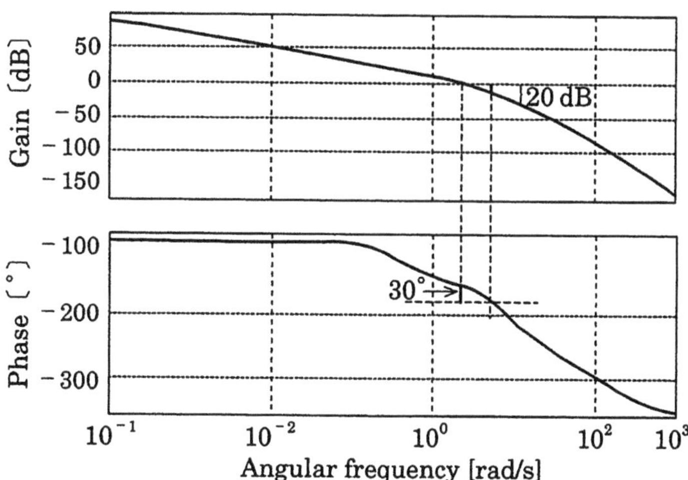

Fig. 6.8 Bode diagram of the open-loop function $G(s)$

Here, the overshoot of reactor output is suppressed to approximately 10 % and the target value is achieved after approximately 20 s. Also, this figure shows the transient response of the total reactivity if the unit reactivity disturbance is added stepwise. The affection by this disturbance is compensated for quickly, and the output is recovered after approximately 12 s.

Figure 6.11 compares the frequency response from the disturbance to the output if the control system is provided and if not provided.

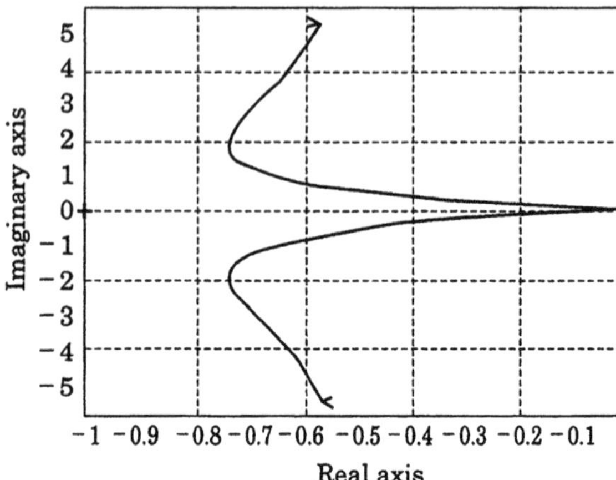

Fig. 6.9 Nyquist locus of $G(s)$

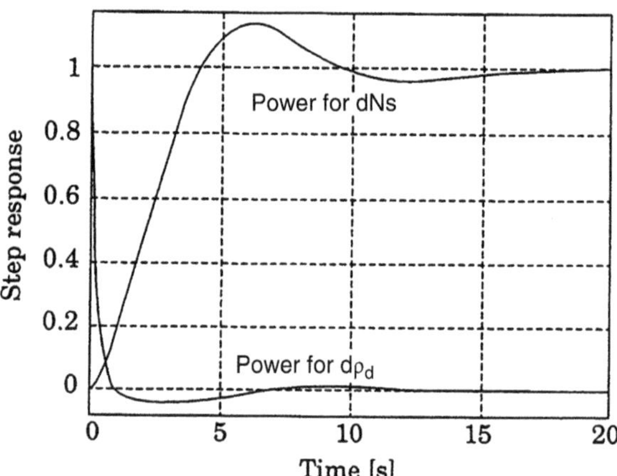

Fig. 6.10 Transient responses to stepwise unit disturbances (δN_s, $\delta \rho_d$)

Although the feedback control system can suppress the affection by the low-frequency area disturbance, this system is ineffective for the high-frequency area disturbance. If this high-frequency area disturbance exists and if its amount cannot be ignored, we need to redesign the control system in order to reduce the gain of this frequency area. To redesign such a control system, the compensation method such as phase delay or phase advance compensation is used.

6.2 Nuclear Thermal-Hydraulic Stability of the Boiling Water Reactor

The BWR core and its related systems have the inherent output suppression characteristics, and they are designed to have the enough attenuation characteristics not to exceed the allowable design limit of the fuel or to allow control of the output vibration.

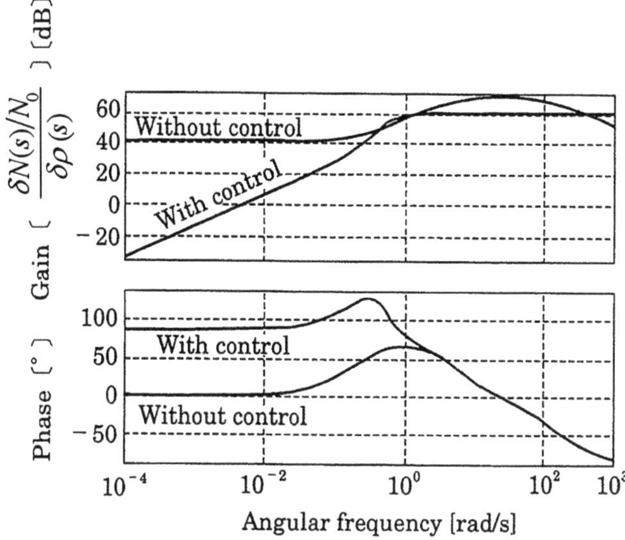

Fig. 6.11 Frequency responses from disturbance $\delta\rho_d$ to output δN

When designing a boiling water reactor (BWR), the following five stability items are assessed.

<1> Channel hydraulic stability (hereafter called the "channel stability")
<2> Core stability
<3> Area stability
<4> Xenon stability
<5> Plant stability

This paragraph especially explains the stability assessment methods of items <1>, <2>, and <3> that are caused by the nuclear thermal-hydraulic feedback of core voids.

Differing from the BWR, the pressurized water reactor (PWR) has almost the single-phase liquid of coolant in the core during normal operation. Therefore, the thermal-hydraulic stability is relatively high and this problem may become negligible.

6.2.1 Stability of BWR Plant

The stability of a BWR plant can be considered in the similar way as for the stability of the normal feedback control system. That is, the system is basically configured by the general negative feedback loop as shown in Fig. 6.12.

If such feedback loop is thought for the nuclear thermal-hydraulic stability, a change of differential pressure is fed back and the channel stability has the behavior

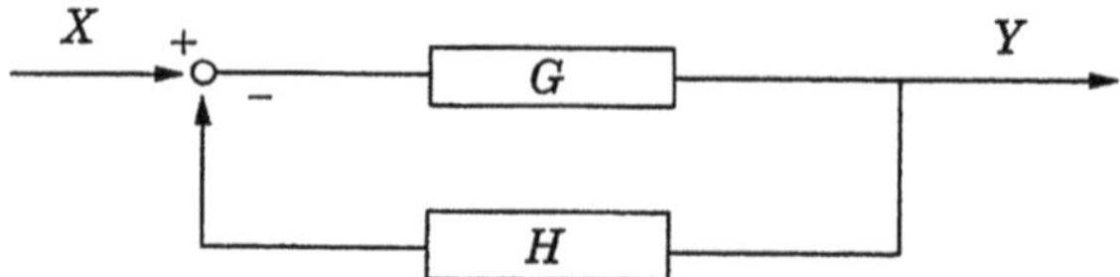

Fig. 6.12 Feedback control system

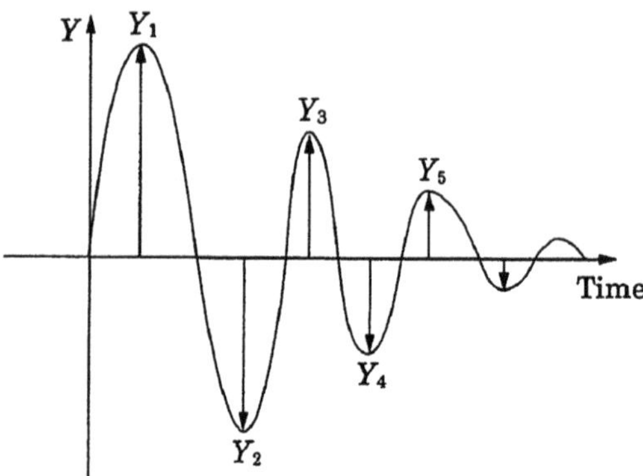

Fig. 6.13 Definition of damping rate

to maintain the channel differential pressure to a fixed level. The core stability can be thought to be the feedback so that the change of reactivity due to a change of voids in the core behaves in a way that maintains the output at a fixed level. If a change occurs in such a negative feedback system, the system tries to recover to the original state. The plant parameters such as output do not change excessively, and these characteristics are desirable. However, the general negative feedback system may have reduced stability in some combinations of gain and phase delay (transmission delay). Its stability design is one of the key items in the control system design.

As the stability index of BWR plant, the amplitude reduction factor is usually used. As shown in Fig. 6.13, the amplitude reduction factor is expressed by the ratio of the amplitude of system response to the adjacent amplitude (Y_3/Y_1) when a stepwise input disturbance is applied.

If the amplitude reduction factor is less than 1.0, the amplitude of vibration components is reduced and it is called stable. If the amplitude reduction factor is 1.0 or larger, it is called unstable. Generally, if the transfer function of the system is determined, the amplitude reduction factor can be calculated.

The stability of the BWR plant relates to the stability of complex system where various parameters such as nuclear characteristics, thermal-hydraulic characteristics, heat transfer characteristics of fuel rods, and recirculation system characteristics outside of the core mutually affect on each other. Here, we think the stability is affected only by thermal-hydraulic factors (the channel stability) and by a combination of nuclear factors and thermal-hydraulic factors (the core stability and area stability) separately, based on the configuration factors of the feedback mechanism that we have noticed.

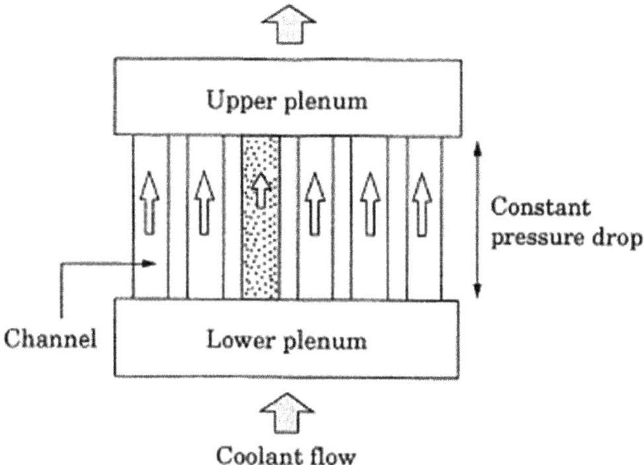

Fig. 6.14 Channel stability in parallel flow path system

6.2.2 *Channel Stability*

In the core of BWR, the cooling water is fed from the bottom of the core and it flows upward through the Zirconium alloy frames (hereafter called the "channels") that surround the fuel rods. The water is boiled by heat from fuel rods. Figure 6.14 is an illustration.

Although the temperature of cooling water is lower than the saturation temperature at the core inlet, the water is heated by fuel rods in the channel. This water contains a large amount of steam (hereafter called the "voids") that is sent as the two-phase flow into the upper plenum. The reactor core has many fuel assemblies that are surrounded by channels, and these assemblies form a parallel flow path system. Also, these fuel assemblies are surrounded by the upper and lower plenums of the core. As we can see that the pressure of fuel assembly is identical in each plenum, the pressure difference in each channel that is surrounded by the upper and lower plenums (hereafter called the "pressure loss") can be thought identical in all channels.

In such parallel passage, many other channels exist in the core and the pressure loss between the upper and lower plenums is kept constant even if the water flow changes in a specific channel and the flow rate changes in the channel. In such case, the mutual effect between the channel flow and the pressure loss affects on the feedback system. The flow rate of the cooling water may fluctuate in channels under certain conditions, and the core may become unstable.

The thermal-hydraulic stability in channels is called the channel hydraulic stability (or channel stability).

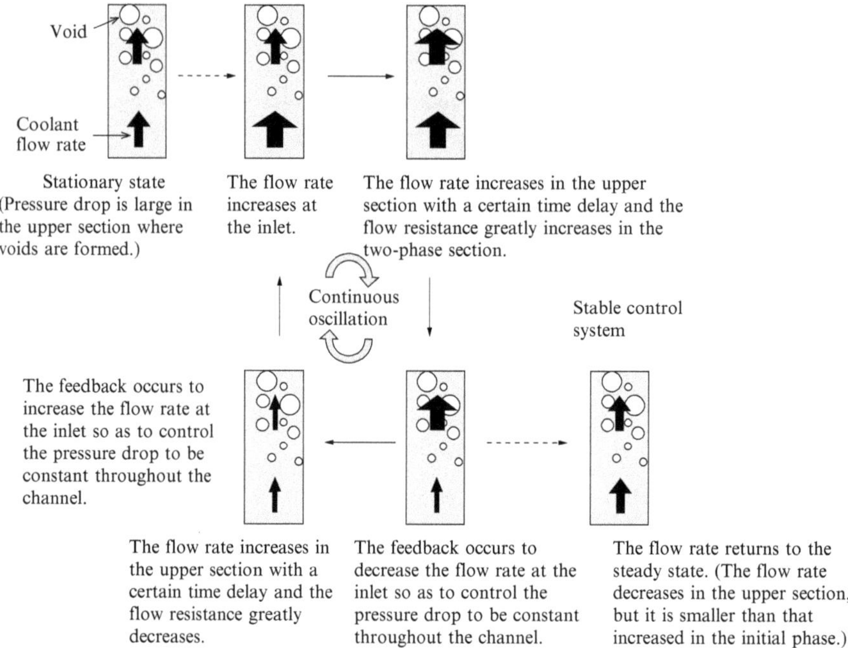

Fig. 6.15 Generating mechanism of channel flow rate fluctuation

[1] Generating mechanism of flow rate fluctuation in channels

If the power output increases in some of many channels, there are a lot of voids in those channels. Both the effect when they move up and the pressure loss caused by the two-phase flow of cooling water and voids affect on (feedback) the flow rate at channel inlet. The so-called density-wave oscillation phenomenon may occur.

Figure 6.15 shows the generating mechanism of channel flow rate fluctuation.

At first, the flow rate of cooling water is constant in channels, and it is in the steady state. At this time, voids are generated in the upper section of channels and the pressure loss in the upper section is larger than that of the single-phase section. Then, if the flow rate increases at the channel inlet due to some reasons, the flow rate increases in the upper section with a certain time delay and the flow resistance greatly increases in the two-phase section.

In the parallel passage having the plenum in the top and bottom sections, the pressure loss between the top and bottom of channels is constant. If the pressure loss changes in some channels, an effect occurs to return the loss to the original level. In the example of Fig. 6.15, the flow rate at channel inlet decreases to return the pressure loss to the original level. After a while, the flow rate at the top section also decreases and the pressure loss drops. However, if the pressure loss drops significantly at this time, feedback causes the flow rate to increase significantly and the continuous fluctuation may occur. In contrast, if the dropped flow rate is not lower than the original flow rate, the fluctuation gradually decreases and the water flow rate returns to the steady state.

When the flow rate increases in the channel inlet as described above, the behavior of water flow can be characterized by the delayed change of flow rate at outlet, the time change of pressure loss, and the time change of flow rate at inlet. Their response is described below.

1) Time delay of outlet flow variance

Because voids are formed in channels, an increase of the inlet flow does not cause an immediate increase of outlet flow. Most of the increase has a time delay (approximately 2 s during normal water circulation). This can be thought that the variance of inlet flow is converted into the variance of density around the boundaries between single-phase cooling water and two-phase voids, and that the density variance (or density waves) is transferred to the outlet by the flow of voids.

2) Variance of pressure loss in time

The pressure loss of the single-phase section can be the pressure loss by friction with fuel rods, local pressure loss by spacers and others, and these loss changes nearly proportional to the square of flow rate. Meanwhile, because voids are formed in the two-phase section, the friction pressure loss further changes depending on the quality. Therefore, when the inlet flow rate increases, the quality (the steam mass/coolant total mass) increases in the area closer to the outlet and the pressure loss also increases. Meanwhile, the increased flow is sent to the outside from the outlet, and a time when the pressure loss reaches the highest level exists.

3) Variance of inlet flow in time

If the pressure loss is fixed in each channel of the core, the flow rate changes in the channels so that it compensates the pressure loss variance. In other words, when the pressure loss increases above the fixed value, the flow rate drops. And when the pressure loss decreases below this value, the flow rate increases and the channel pressure loss is maintained.

[2] Factors affecting on the channel stability

The stability, which is maintained by the feedback of flow rate and pressure loss as described above, is affected by factors such as channel output, channel flow rate, axial output distribution, and others.

1) Channel output

When the channel output is high, more voids are formed in each channel. As the two-phase area and quality increase, the pressure loss increases in the two-phase area. Therefore, the variance of pressure loss of the channel increases when the inlet flow rate changes. This can be thought that the feedback gain of the channel stability has increased, and it causes to drop the stability.

2) Channel flow rate

If the channel output is the same but the flow rate is low, the quality increases in the two-phase section and, therefore, the pressure loss decreases nearly proportional to the flow rate. While in the single-phase area, the pressure loss decreases nearly proportional to the square of flow rate and, therefore, the pressure loss of the two-phase section becomes relatively large. That is, the ratio of two-phase section to the single-phase section tends to increase when the flow rate is low. It can be considered that the feedback gain of channel stability has increased and it causes the stability to drop.

3) Axial output distribution

If the lower part of relative output in the vertical channel direction (that is, the axial output distribution) is deformed to increase, the boiling inception point is also moved downward and the two-phase area increases. In this case, because the two-phase pressure loss increases, we can think that the feedback gain of channel stability is increased in the similar way as described above. This causes to drop the stability.

6.2.3 Core Stability

Although the channel stability is only based on the thermal-hydraulic characteristics, the core stability is primarily the entire core stability based on the nuclear kinetics. If its instability phenomenon occurs, the neutron flux of the entire core starts to oscillate in the same phase. Such phenomenon occurred in reactor No. 2 of the La Salle plant in the USA as well as in other plants.

[1] Core stabilization mechanism

A core stabilization mechanism is shown in Fig. 6.16.

If a positive reactivity applies to the core in the steady state due to some reasons (such as the increased pressure of nuclear core), the core output increases. Although heat of fuel rods is transferred to the coolant, the heat flux increases with a delay due to the heat capacity of fuel rods and others (the time constant is approximately 4–6 s). Because heat transferred to the coolant causes an increase in voids, a negative reactivity occurs and the core output drops. Then, the heat flux decreases with a delay. Because the increased voids are fed to the outside of the core, voids in the core start to decrease. The reactivity reaches a minimum (negative reactivity—ΔK_1) immediately before this time, and then it starts to increase again. However, the reactivity is still negative and, therefore, the output continues to drop. When the void fraction drops below the initial value, the reactivity becomes positive and the output starts to increase. Thus, the heat flux increases with a delay, and the void fraction starts to increase again. Here, the reactivity has the maximum value (ΔK_2). After that, the fluctuation repeats and if the maximum value (ΔK_2) is larger than the absolute value (ΔK_1) of the first negative reactivity, the amplitude gradually increases and the system becomes unstable.

Figure 6.17 illustrates the above mechanism in the feedback loop form.

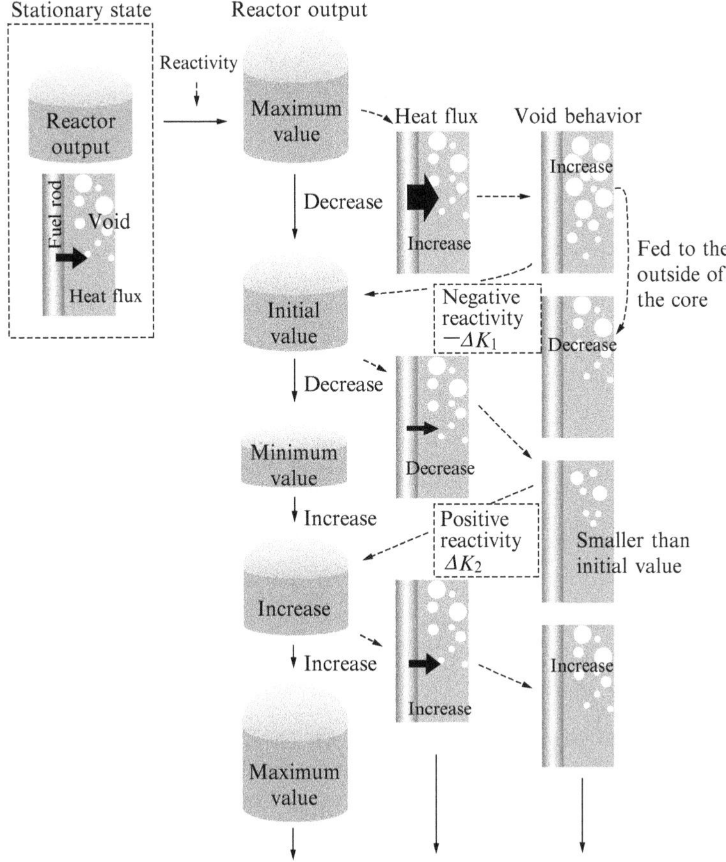

Fig. 6.16 Mechanism of generating oscillation for core stability

The core stability mechanism of Fig. 6.16 is shown as the main feedback loop in Fig. 6.17, and this is a transition loop from the nuclear kinetics, to the fuel thermal transfer characteristics, to the thermal-hydraulic characteristics, and to the nuclear kinematics again. Figure 6.17 also has a transition sub-loop from the pressure loss, to the flow rate, and to the pressure loss again. This is equivalent to the thermal-hydraulic feedback described in the paragraph of channel stability, and the core stability can be said to be the fluctuation mode with nuclear heat coupling.

[2] Factors affecting on the core stability

The factors affecting on the core stability are the void reactivity coefficient, nuclear reactor output, core flow rate, and longitudinal and axial core output distribution.

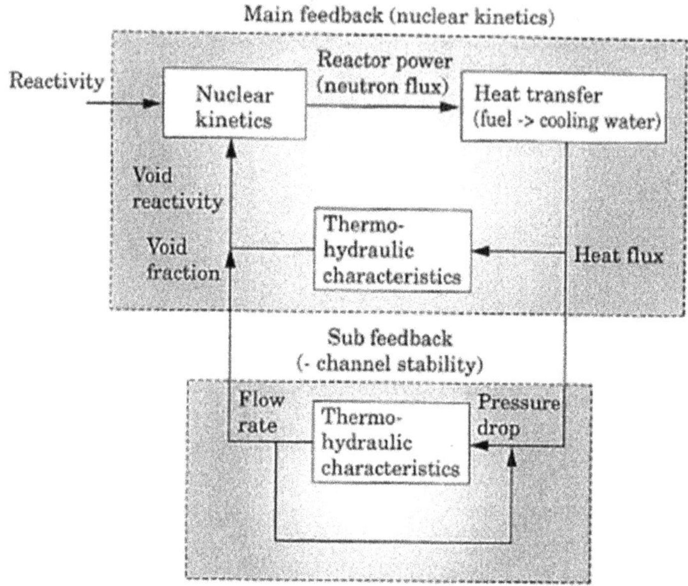

Fig. 6.17 Feedback loop for core stability

1) Void reactivity coefficient

If the void reactivity coefficient is large, the feedback to the nuclear kinetics of void fraction variance increases as shown in the feedback of Fig. 6.17 and the core becomes unstable.

2) Reactor output

When the reactor output is increased, the stability is lost. This is caused by the nuclear factors and by the thermal-hydraulic factors. As the nuclear factors, if the reactor output is high, the void fraction increases in the core. Thus, the absolute value of void reactivity coefficient increases. If the void reactivity coefficient is large, the feedback to the nuclear kinetics of void fraction variance increases as shown in the feedback of Fig. 6.17 and the core becomes unstable.

As the thermal-hydraulic factors, if the reactor output is high, the quality of the two-phase section increases and therefore the two-phase pressure loss increases. If the two-phase pressure loss is large, the channel stability drops in the sub-feedback loop of Fig. 6.17. It affects on the main feedback loop, and the core stability drops.

3) Core flow rate

If the flow rate in the core is low, the core becomes unstable. This is caused by the following two thermal-hydraulic factors. The first one is that the ratio of two-phase section to the single-phase section tends to increase when the flow rate is low as described in the paragraph of channel stability. Thus, we can think that the sub-feedback gain of channel stability, shown in Fig. 6.17, has increased. This factor causes the core stability to drop.

The second one is that the coolant takes a longer time to pass through the core from its bottom to the top. If the coolant passage time is long, we think that the capacity of heat transferred from fuel rods increases and the gain of heat transfer characteristics has increased in the main feedback loop of Fig. 6.17.

4) Core's longitudinal output distribution (the distribution of fuel assembly output)

If the core has a high rate of higher output fuel assembly, the number of fuel assemblies having the poor channel stability increases and the core stability drops even if the entire output of the reactor is the same.

5) Axial output distribution (in vertical direction of core)

In the axial output distribution of the core, the core stability is affected by different factors. The first one is that the two-phase pressure loss area increases if the bottom section is deformed as described in the paragraph of channel stability. And the thermal-hydraulic stability is lost (in sub-loop). If the distribution is flat, the feedback by the void reactivity increases (in main loop) around the center of high neutron flux when compared with the core having the deformation in its bottom section. Generally, a core with a flat output distribution tends to lose stability.

6.2.4 Area Stability

The unstable area phenomenon is the oscillation due to the phase shift of neutron flux between core areas (for example, between the right half and the left half of the core). An example is given in Fig. 6.18.

During a core stability test at a foreign nuclear plant, the unstable phenomenon was observed due to the phase shift of neutron flux oscillation between core areas. An example is the stability test of the Caorso furnace in Italy, and the neutron flux oscillation with an opposite phase that occurred in half of the core.

This type of instability is called the area instability. From the viewpoint of space-dependent kinetics, we can say that the core stability is the stability in the basic mode but the area stability is the stability in the high-order mode. The instability of this high-order mode attenuates and disappears immediately in the normal core if we consider the nuclear characteristics only. However, if a channel with poor thermal-hydraulic stability exists around the core, we can think that the mode that usually attenuates is thermal-hydraulically excited and a continuous oscillation of high-order mode may occur. As described above, the area stability is affected by both the nuclear characteristics and thermal-hydraulic characteristics.

Although our analysis uses the model that is basically the same as the core stability analysis model, we evaluate the area stability by considering the space distribution in the high-order mode, nuclear kinetics, and thermal-hydraulic characteristics. This evaluation differs from the case of core stability evaluation.

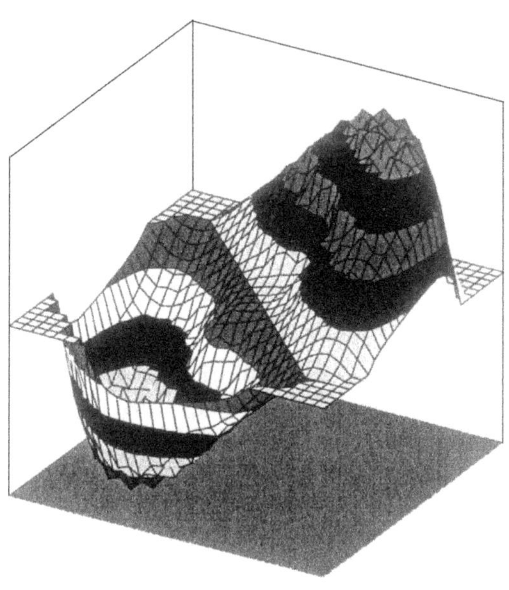

Fig. 6.18 Oscillation between core areas

[1] Area stabilization mechanism

An area stability fluctuation mechanism is shown in Fig. 6.19.

If we assume that the high-output fuel assembly (the hot channel) exists around the core, the channel flow (output) fluctuates when this channel becomes unstable. As the result, adjacent fuel assemblies coupled by nuclear energy also starts oscillating in the same phase. The flow rate of a hot channel of another area starts to oscillate in the reverse phase in order to compensate the fluctuation of flow rate in the oscillation area. Similarly, the peripheral channels coupled by the nuclear energy also start oscillating in the reverse phase. We think the continuation of these oscillations causes a reverse-phase oscillation in different areas of the core. This nuclear coupling corresponds to the high-order mode (in this case, the primary mode in the circumferential direction), and they are characteristics of oscillation pattern during area oscillation.

As described above, the area stability is generated by combination of thermal-hydraulic oscillation and nuclear oscillation effects. Regarding the thermal-hydraulic characteristics, we can think that these oscillations are similar to the oscillation of channel stability. For nuclear kinetics, they are based on the space distribution of the high-order mode. They also correspond to the facts that the core stability is similar to the channel stability for thermal-hydraulic characteristics and that the oscillation (the integrated core oscillation) is based on the space distribution of the basic mode for nuclear kinetics. These are shown in Fig. 6.20.

The high-order mode distribution can exist not only in a pair of core half areas but also in other objects. The familiar examples are a guitar string (one dimension) and a drum panel (two dimensions), and Fig. 6.21 shows a high-order mode image of the drum panel.

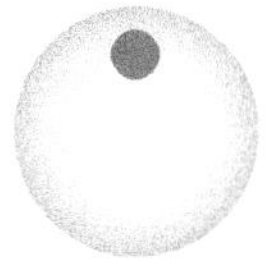

The flow rate passing through the hot channel becomes thermal-hydraulically unstable (channel stability), and oscillation starts.

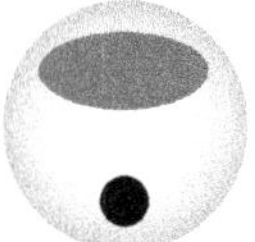

The flow rate of a hot channel of symmetric area starts to oscillate in the reverse phase in order to compensate the fluctuation (increase/decrease) of flow rate in the oscillation area.

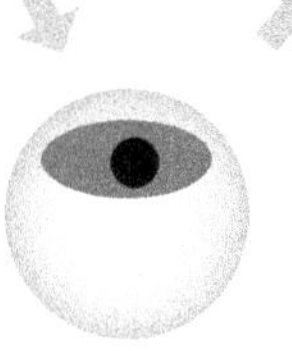

The peripheral area of the hot channel starts to oscillate in the same phase by nuclear coupling.

The peripheral area of the hot channel in the symmetric position also starts to oscillate in the reverse phase by nuclear coupling.

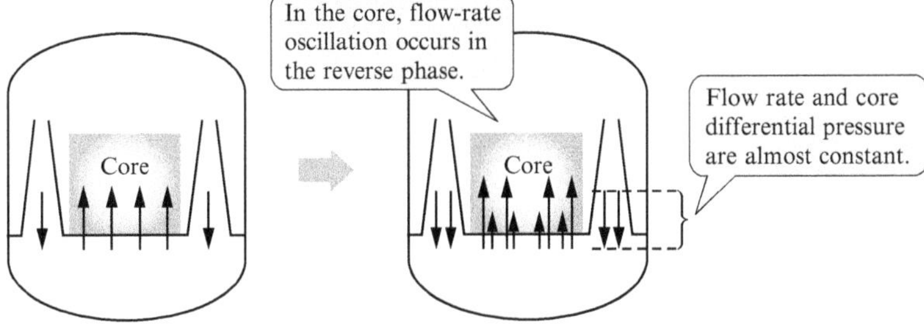

Fig. 6.19 Mechanism of generating area instability

The radial primary mode is the basic mode, and the entire drum panel vibrates in the same phase. It corresponds to the core stability of nuclear reactor. In contrast, there are various types of high-order oscillation mode, the circumferential (azimuth angle) primary mode causes two halves of core to oscillate in reverse phase. It is the most frequently observed area stability in the reactor. In addition, the high-order mode of circumferential secondary oscillation and radial secondary oscillation can exist.

[2] About the subcriticality

If the nuclear model is simulated by one-point approximation, subcriticality ρ_n (where "n" is the nth mode) is defined as an index that shows an easy occurrence of high-order mode oscillation. This is based on the fact that a shape in high-order mode corresponds to the shape of area oscillation. The subcriticality (ρ_n) of the high-order mode can be expressed by the difference between the inherent value (λ_n) of high-order mode and the inherent value (λ_0) of basic mode as follows. It indicates

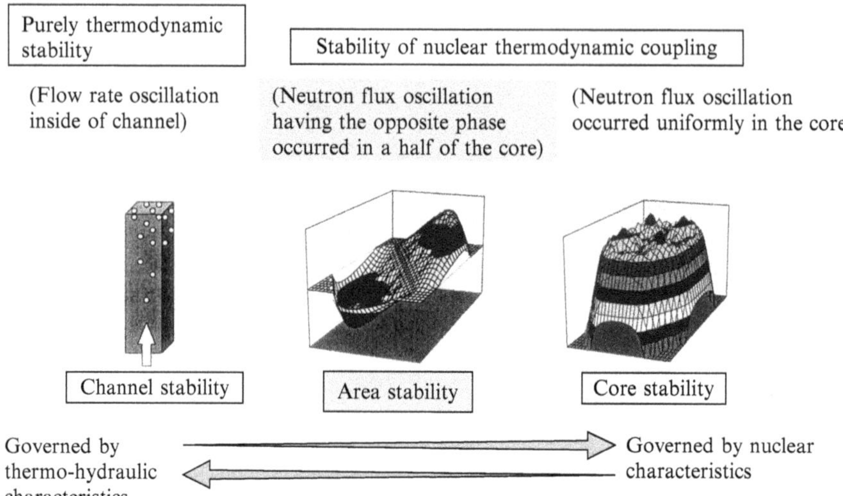

Fig. 6.20 Positioning of area stability

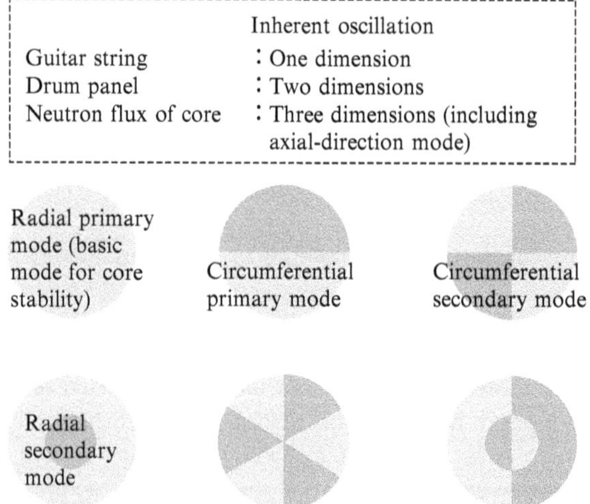

Fig. 6.21 High-order oscillation mode

a degree of proximity of the high-order mode to the basic mode. If value ρ_n is reduced, the affection by the high-order mode increases.

$$\rho_n = \frac{1}{\lambda_n} - \frac{1}{\lambda_0} \tag{6.25}$$

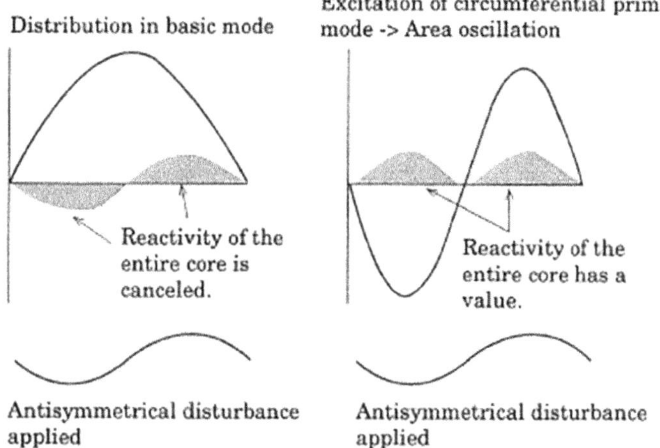

Fig. 6.22 Application of antisymmetrical disturbance and excitation of circumferential primarymode

The transfer function of the nuclear one-point approximation, including the high-order mode, can be expressed as follows.

$$G_n(s) = \frac{N_n(s)}{\frac{\delta\rho_n}{\beta}} = \frac{1}{s\left(\frac{\Lambda}{\beta} + \sum_{i=1}^{6} \frac{\beta_i/\beta}{s+\lambda_i}\right) + \rho_n/\beta} \tag{6.26}$$

This is the same as transfer function (6.11) of the integrated core oscillation, except for value of ρ_n. If the value of ρ_n is reduced, the gain of transfer function increases.

[3] Modeling of area oscillation

Figure 6.22 compares the basic mode distribution with the circumferential primary mode distribution when antisymmetrical disturbance is applied. In the basic mode, it is clear that the reactivity of the entire core is canceled even when the antisymmetrical disturbance is applied.

If the antisymmetrical disturbance is applied to the circumference primary mode distribution, it is clear that the reactivity of the entire core has a value. Figure 6.23 shows a model if such reactivity disturbance is applied.

If the antisymmetrical disturbance is applied to channels 1 and 2 simultaneously, the void fraction changes in these channels. If reactivity weight W_1^+ and W_2^- for the high-order mode are multiplied and added to each change of void fraction, the void reactivity is fed back to the nuclear heat characteristics of the high-order mode.

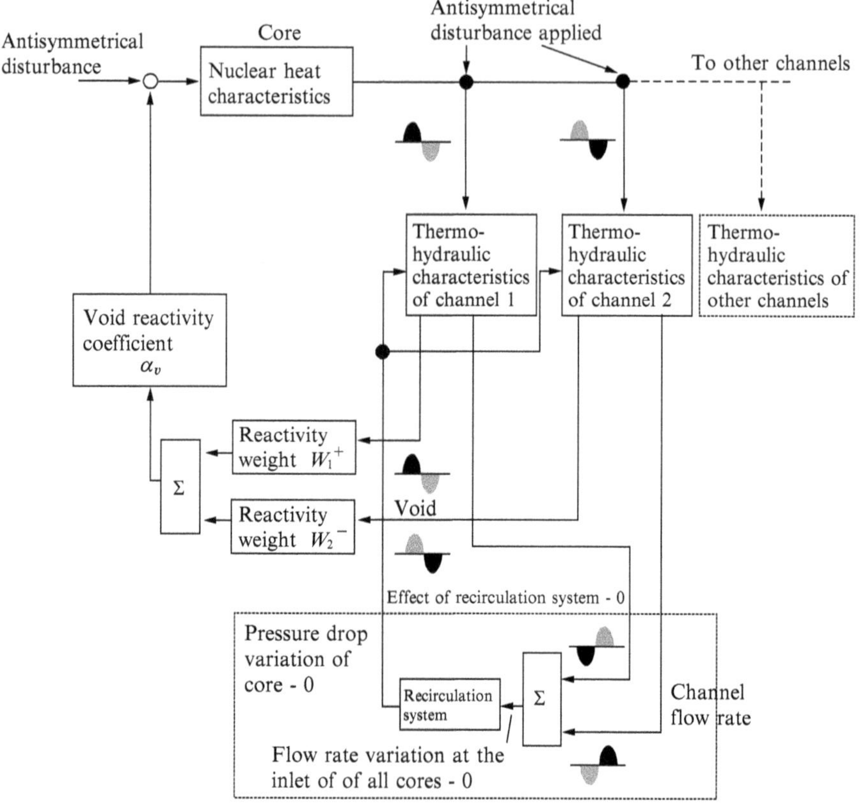

Fig. 6.23 Entire model of area stability

6.2.5 *Stability Measures*

If the BWR is operated with the excessively high output but with the low core flow, the core stability may not be kept. To avoid such a status, the selective control rod insertion system (SRI system) and the stability limit curve are used.

[1] Selective control rod insertion system

As shown in Fig. 6.24, if two recirculation pumps are used and if one or both of these pumps trip (at point indicated by "×") when the reactor is operating, the core flow rapidly drops and the negative feedback causes the output to drop. However, the system enters the relatively high-output and low-flow area. In this area the core may become unstable. If the system enters the selective control rod operating area that is shown in Fig. 6.24, the selective control rods are inserted and the output is reduced so that the core stability is maintained. This is an example of BWRs and the interlock details vary depending on the plant type and others. However, the SRI insertion is the same for all reactors in order to avoid the high-output and low-flow area.

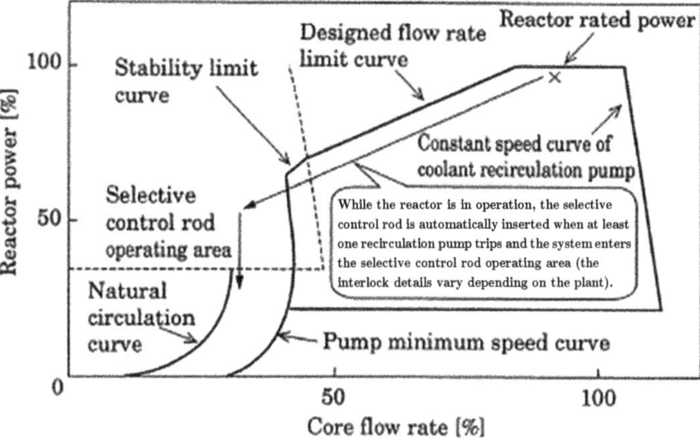

Fig. 6.24 Selective control rod operating area and stability limit curve

[2] Stability limit curve

Figure 6.24 shows the stability limit curve. It is used to keep the safety margin if the core closes to the high-output and low-flow operation area when the reactor is started or stopped. The control rods are restricted in order to be removed, and the recirculation flow control system is restricted in order to reduce the core flow.

6.3 Xenon Stability

There are two types of generation of poisonous xenon-135: the one directly generated by nuclear fission, and the other generated by the decay of fission product iodine-135 and tellurium-135. This poison has a large thermal neutron absorption cross section, and it may cause the core reactivity to change greatly. Therefore, poisonous xenon-135 plays an important role in reactor kinetics and space oscillation. The samarium-149 is another poison, but its time constant is very long and it is negligible for the normal core stability. So, the xenon-135 is only discussed below.

6.3.1 Stability Based on One-Point Core Approximation Model

Although both neutron flux and xenon-135 concentration distribute in the core space, we ignore this space distribution and conduct xenon-135 kinetics equations. By considering the process of xenon-135 generation and extinction, we can express the kinetics equation as follows:

$$\frac{\mathrm{d}X(t)}{\mathrm{d}t} = \lambda_{\mathrm{I}} I(t) + \gamma_2 \sum\nolimits_{\mathrm{f}} \phi(t) - \lambda_{\mathrm{X}} X(t) - \sigma_{\mathrm{X}} X(t)\phi(t) \tag{6.27}$$

where the first term is the xenon-135 generation rate by decay of fission product iodine-135. The second term is the direct generation rate by nuclear fission. The third term is the decay of xenon-135 itself, and the fourth term is the extinction by neutron absorption and it is a nonlinear term.

The kinetics equation of $I(t)$ contained in (6.27) can be expressed by the following equation in the process of iodine-135 generation and extinction.

$$\frac{dI(t)}{dt} = \gamma_1 \sum\nolimits_f \phi(t) - \lambda_I I(t) \tag{6.28}$$

Actually, the iodine-135 is directly generated by nuclear fission and by the decay of fission product tellurium-135. However, as the half-time of tellurium135 decay ($T_{1/2} = 1$ m) is short, the first term is a generation term by assuming that the fission product of tellurium is generated directly as iodine-135. The second term is the extinction term by decay of iodine-135 itself.

First, we determine the equilibrium solution of (6.27) and (6.28) for neutron flux ϕ_0. If the left part of both equations is assigned 0, the equilibrium solution of iodine-135 and xenon-135 can be obtained as follows:

$$I_0 = \frac{\gamma_1 \Sigma_f}{\lambda_I} \phi_0 \tag{6.29a}$$

$$X_0 = \frac{(\gamma_1 + \gamma_2)\Sigma_f \phi_0}{\lambda_X + \sigma_X \phi_0} \tag{6.29b}$$

The equilibrium concentration of iodine-135 is proportional to the neutron flux. In contrast, the concentration of xenon-135 is proportional to ϕ_0 if the neutron flux is in the low range ($\sigma_X \phi_0 \ll \lambda_X$), but in the high range ($\sigma_x \phi_0 \gg \lambda_x$), it can be:

$$X_0 \approx \frac{(\gamma_1 + \gamma_2)\Sigma_f}{\sigma_X} \tag{6.30}$$

and it does not depend on ϕ_0.

Next, we determine the transfer function from $\phi(t)$ for $X(t)$ and $I(t)$. If (6.27) is linearized for variances $\delta X(t)$ and $\delta I(t)$ from the equilibrium solution, we can obtain the following equations:

$$\frac{d\delta X(t)}{dt} = \lambda_I \delta I(t) + \gamma_2 \Sigma_f \delta\phi(t) - \lambda_X \delta X(t) - \sigma_X \phi_0 \delta X(t) - \sigma_X X_0 \delta\phi(t) \tag{6.31}$$

$$\frac{d\delta I(t)}{dt} = \gamma_1 \Sigma_f \delta\phi(t) - \lambda_I \delta I(t) \tag{6.32}$$

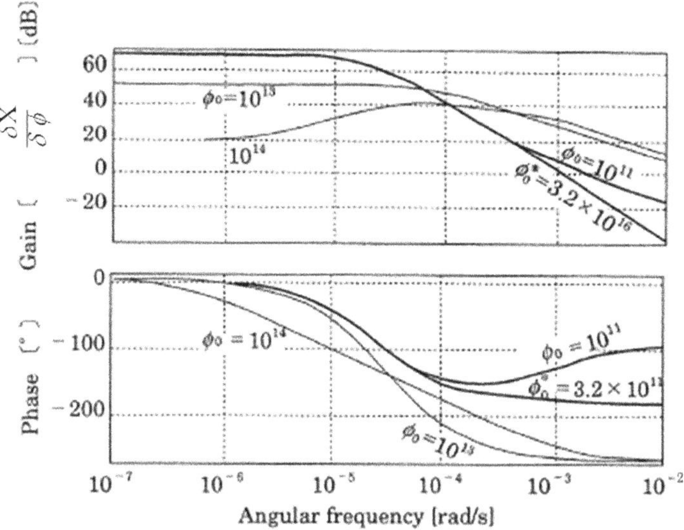

Fig. 6.25 Bode diagram of xenon poisoning

If they are converted by Laplace transform, the transfer function from the neutron to xenon concentration can be obtained by the following equation.

$$G_x(s) \equiv \frac{\delta X(s)}{\delta \phi(s)} = \frac{(\gamma_2 \Sigma_f - \sigma_X X_0)s + \{\lambda_I(\gamma_1 + \gamma_2)\Sigma_f - \lambda_I \sigma_X X_0\}}{\{s + (\lambda_X + \sigma_X \phi_0)\}(s + \lambda_I)} \tag{6.33}$$

It is interesting in (6.33) that the time constant of numerator $T = (\gamma_2 \Sigma_f - \sigma_X X_0)/\lambda_I\{(\gamma_1 + \gamma_2)\Sigma_f - \sigma_X X_0\}$ is indefinite to positive.

Although the denomination is positive from equation (6.29b), the numerator $(\gamma_2 \Sigma_f - \rho_X X_0)$ can be positive or negative depending on ϕ_0. If the numerator is 0, neutron flux ϕ_0^* is given by:

$$\phi_0^* = \frac{\lambda_X \gamma_2}{\sigma_X \gamma_1} \tag{6.34}$$

Here, if $\lambda_I = 2.9 \times 10^{-5}$ s, $\lambda_X = 2.1 \times 10^{-5}$ s, $\gamma_1 = 0.056$, $\gamma_2 = 0.003$, and $\sigma_X = 3.5 \times 10^6$ barn are given, $\phi_0^* = 3.21 \times 10^{11}$ n/cm^2 s. Figure 6.25 shows a Bode diagram of (6.33) for various neutron flux values.

We can see that the diagram shape changes greatly on the boundary of ϕ_0^*, that is, between positive and negative values of time constant T.

The xenon-135 absorbs the neutron and changes the reactivity. The high-output reactor, including the reactivity feedback by xenon poisoning, can be shown by the block diagram of Fig. 6.26. K_X is the reactivity coefficient of xenon-135.

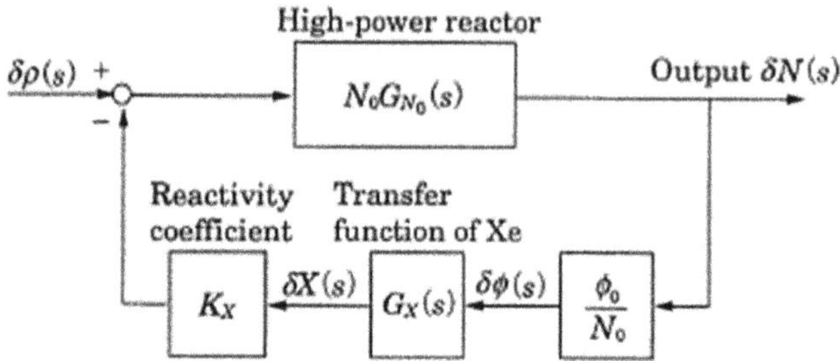

Fig. 6.26 Block diagram of high-power reactor including xenon poisoning

For example, if (6.21) is the transfer function of a high-output reactor, it can be used for the high-output BWR core model. If we determine the closed loop transfer function and check the reactor stability using this figure, it is known that the reactor may become unstable in some cases. However, this unstable phenomenon occurs very slowly because the time constant is very long for $I(t)$ and $X(t)$, the operator can easily make stable the reactor by adjusting the control rods. Usually, this phenomenon does not cause a problem in automatic reactor control.

However, xenon poisoning is a serious problem if high output operation of the reactor is stopped. Now consider that the reactor operating with neutron level ϕ_0 is stopped at $t = 0$. This is mathematically the solution of (6.27) and (6.28), using initial values of I_0 and X_0 of (6.29). This can be dissolved analytically, and the time behavior of xenon-135 concentration can be given by the following equation.

$$X(t) = \frac{-\lambda_{\mathrm{I}}}{\lambda_{\mathrm{I}} - \lambda_{\mathrm{X}}} I_0 \, \mathrm{e}^{-\lambda_{\mathrm{I}} t} + \left(X_0 + \frac{\lambda_{\mathrm{I}}}{\lambda_{\mathrm{I}} - \lambda_{\mathrm{X}}} I_0 \right) \mathrm{e}^{-\lambda_{\mathrm{X}} t}$$

$X(t)$ for various ϕ_0 values can be shown as in Fig. 6.27.

If value ϕ_0 exceeds above 10^{13}, a large transient peak appears in $X(t)$ approximately 10 h after reactor shutdown. The negative reactivity by this peak is very large and, actually, it may be impossible to restart the reactor. With this motivation, an appropriate reactor shutdown process has been researched to minimize this peak.

6.3.2 Xenon Stability by Considering the Space Distribution

[1] Xenon stability of BWR

The xenon-135 oscillation is greatly affected by nuclear binding in the core and by the output factor. Because voids are formed in the core of BWR, it basically has the negative and large output factor. Possibility of oscillation generation by xenon-135 is very low. One of the major advantages of the BWR is the excellent stability for xenon-135.

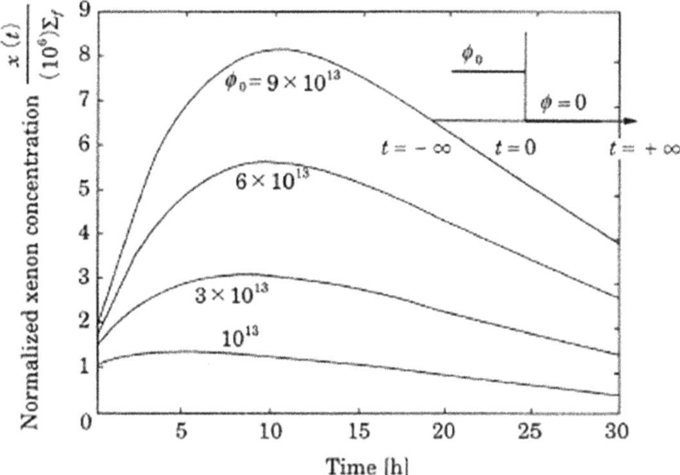

Fig. 6.27 Variation of xenon concentration after reactor shutdown

Figure 6.28 outlines the xenon stability in radial, azimuth, and axial directions, using the ratio of height and diameter of the core.

Because most of the BWRs have a core diameter slightly larger than the core height, the xenon stability in this area is relatively unstable in the axial and azimuth directions than the radial direction. Also, Fig. 6.29 outlines the axial direction xenon stability on a plane of neutron flux and output reactivity factor. It is clear that the xenon stability in axial direction, that is relatively unstable, is quite stable during rated power operation.

[2] Xenon stability of PWR

In the PWR, as the coolant in the core is almost in the liquid single phase, the pressure loss of coolant passage changes very few after the output and reactor pressure change. This is the very stable thermo-hydraulic system. Even if voids are formed transiently in a part of the core, this open channel structure does not have a channel box and, therefore, a feedback to the fuel assembly inlet flow does not occur. The instability caused by the thermo-hydraulic system does not occur. The phenomena that affect on the stability of output distribution of the PWR core is the xenon oscillation due to the generation and extinction of xenon.

The PWR core, especially, its size (the equivalent diameter) is designed to have the strong convergence against the neutron flux distribution oscillation by xenon in radial direction. The xenon oscillation in radial direction does not become an issue in commercial reactors. In contrast, regarding the neutron flux distribution oscillation in axial direction, it is known that the core height, output distribution, Doppler coefficient, and moderator temperature coefficient generally have the following affects.

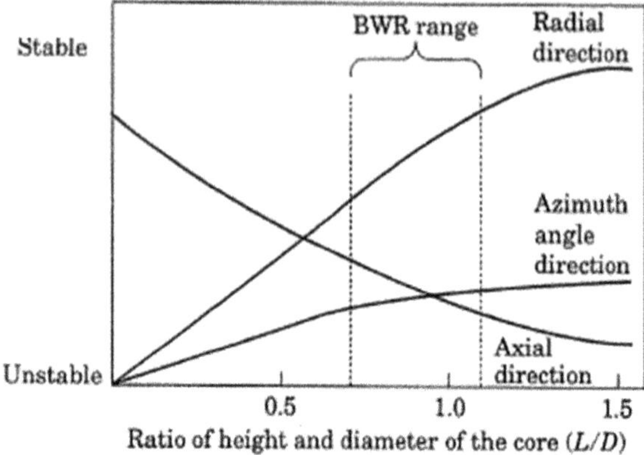

Fig. 6.28 Core shape and xenon stability

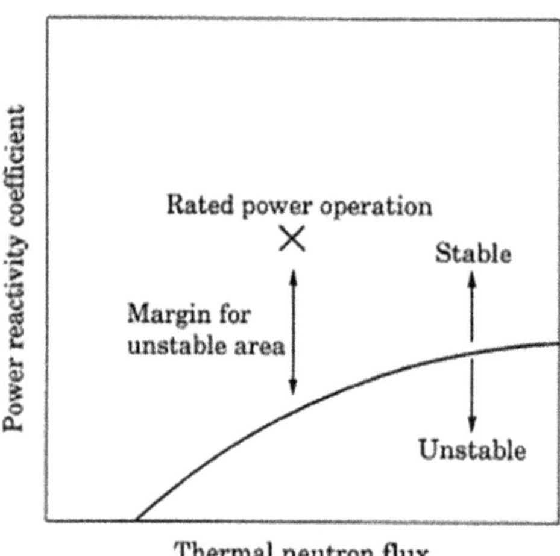

Fig. 6.29 Xenon stability in axial direction

<1> The convergence of xenon oscillation becomes drops if the core height is increased.

<2> The convergence of xenon oscillation is affected by the shape of axial direction output distribution, and if the distribution is flat or if the center of distribution shape is concaved, its convergence becomes worse when compared with the sine distribution.

<3> The Doppler coefficient is the primary attenuation mechanism, and if the absolute value of Doppler coefficient is large, the xenon oscillation is converged quickly.

<4> The moderator temperature coefficient affects on the behavior of xenon oscillation relatively small.

In the realistic PWR core, if the xenon oscillation occurs, the control rod cluster bank of the control group is manually operated based on the indication of output distribution indicator and, therefore, such oscillation can easily be suppressed.

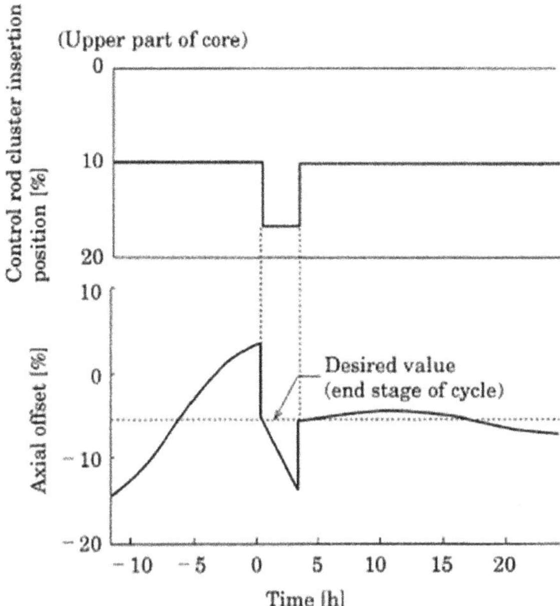

Fig. 6.30 Example of operation for suppressing axial xenon oscillation

In order to explain this operation, we assume the full-power operation at the end of cycle with the typical core characteristics. Figure 6.30 shows our analysis result if the xenon oscillation in axial direction is suppressed by operation of control rods.

Here, the axial offset (AO) is an indicator of axial direction output distribution to be shown by the ex-core neutron flux detectors that are divided into the top and bottom parts of the core.

In this analysis, we first insert and remove control rods to intentionally generate the xenon oscillation. Next, when the axial offset (AO) is transferred to the positive side by xenon oscillation, we insert control rods until the AO reaches the target value. Then, when the AO is changed to the negative side for the same variance from the target value, we pull out control rods to their original insertion positions and we can suppress almost all xenon oscillation.

As described above, if the axial xenon oscillation is excited due to the quick output change by control rods, its oscillation cycle is an order of hour and it occurs very slowly. Therefore, we can easily suppress the oscillation by monitoring the indication of the output distribution indicator and by manually operating the control rod cluster bank appropriately.

Chapter 6 Exercises

1. Prove equilibrium (6.3) and (6.4).
2. Prove that the exact equilibrium of (6.1) does not exist if the neutron source exists and the core is in critical state.

3. Derive (6.11).
4. Derive approximate (6.13a)–(6.13d) for the critical reactor.
5. Draw a block diagram of high-output reactor having the negative feedback output and determine its transfer function. The output coefficient is denoted by α [$/MW].
6. Select appropriate terminologies from the bottom list and complete the following sentences for a BWR.

   (1) The channel stability of BWR plant is based on (<1>). When the channel inlet flow changes, the feedback is made to maintain the (<2>) between the top and bottom plenums to be fixed. Because the channel inlet flow is tried to be returned to the original rate, the oscillation occurs.
   (2) The core stability is mainly based on (<3>). When the output is changed, the reactivity by the change of (<4>) in the core is fed back. Because the output is tried to return to the original level, the oscillation occurs.
   (3) The area stability is a combination of effects of (<5>) and (<6>). If a thermal-hydraulically unstable channel exists around the core, the (<7>) which is normally attenuated quickly can thermal-hydraulically excite the oscillation. This oscillation continues and the area becomes unstable.
   (4) In the BWR, if the (<8>) status occurs, the stability tends to drop. To solve this, the system is provided to drop the output by inserting (<9>) after trip of recirculation pump.

   nuclear characteristics, high-output and low core flow, voids, subcriticality, selective control rods, pressure loss, thermo-hydraulic characteristics, high-order mode

7. Derive (6.33).

Bibliography

1. Suda N (1969) Nuclear reactor kinetics and control. Dobunshoin, Tokyo
2. Ash M (1965) Nuclear reactor kinetics. McGraw-Hill, New York
3. Yoshikawa T (2004) Classic control theory. Shokodo, Tokyo
4. Enomoto, Tanabe et al (1985) The boiling water reactor: recent knowledge and forward perspective for stability. J. At. Energy Soc. Jpn 27(10):890–903
5. Anegawa, Ebata et al (1996) Recent topics relating to the BWR thermal-hydraulic stability. J. At. Energy Soc. Jpn 38(5):348–356
6. Takigawa Y et al (1987) Caorso limit cycle oscillation analysis with three-dimensional transient code TOSDYN-2. Nucl Technol 79:210–227
7. Nariai H et al (2001) Atomic Energy Society of Japan: "Present situation and issues for BWR thermal-hydraulic stability assessment"

Chapter 7
Actual Operation Control of Boiling Water Reactor

Koichi Kondo, Yasuo Ota, Hiroshi Ono, Masahiko Kuroki, Yuji Koshi, and Masayoshi Tahira

7.1 Overview of the Plant

The boiling water reactor (BWR) is divided into two main systems: the nuclear steam supply system (NSSS) that generates steam, and the turbine system (or turbine island) that uses the stream to rotate turbo-generator and generate electric power. The NSSS consists of various subsystems and equipments. They include basically the reactor pressure vessel that houses fuel, control rods, and other nuclear reactor equipments, a reactor auxiliary system that handles circulation of coolant, generated steam, and feed water returned from turbines, the engineered safety features that are required for securing safety, and a reactor auxiliary system that is required for operating the plant. The fuel handling and storage equipment, the instrumentation and control system, and the electrical system, as well as the radioactive waste treatment system that is unique to nuclear power station are also included in the NSSS. The turbine system consists mainly of turbines, generators, the condenser that condenses steam, the feed-water system that resupplies the reactor with the condensate. Figure 7.1 shows the overview configuration of BWR systems.

7.1.1 Reactor Pressure Vessel and Reactor Equipments

Steam generated in the core, from which moisture is eliminated with the steam water separator and steam dryer at the top of the pressure vessel, flows out of the reactor pressure vessel direct into turbines through the main steam pipe. Inside the reactor vessel are the fuel, shroud, control rods, steam water separator, and steam dryer, as well as a jet pump that is used to forcibly circulate coolant (see Fig. 7.2).

The jet pump applies pressure to coolant that flows out of the reactor pressure vessel using a recirculation pump, blows it out of the discharge nozzle as a jet flow, and suctions coolant around the suction nozzle to increase the circulation rate. Coolant that flows out of the jet pump is fed into the bottom of the core, flows through fuel support fittings via holes in control rod guide pipe, and is then fed into the fuel.

Y. Oka and K. Suzuki (eds.), *Nuclear Reactor Kinetics and Plant Control*, An Advanced Course in Nuclear Engineering,
DOI 10.1007/978-4-431-54195-0_7,

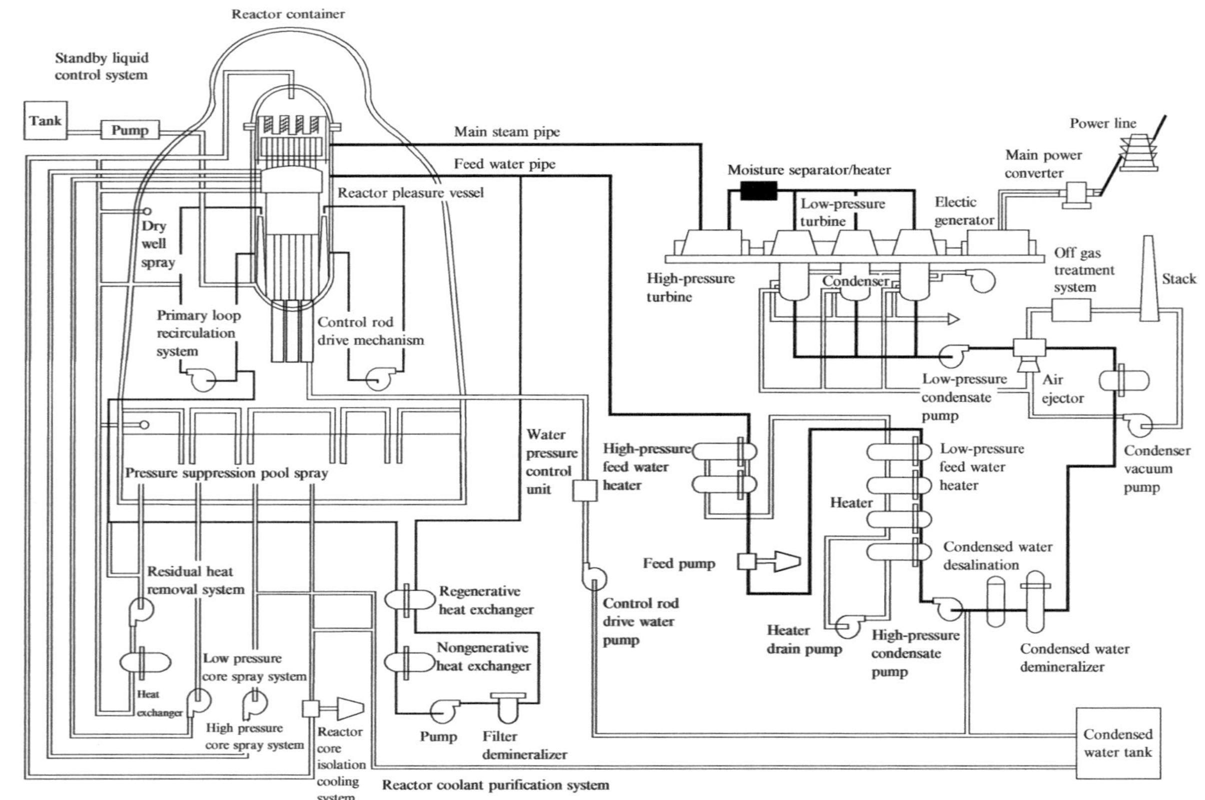

Fig. 7.1 Outline configuration of a BWR plant

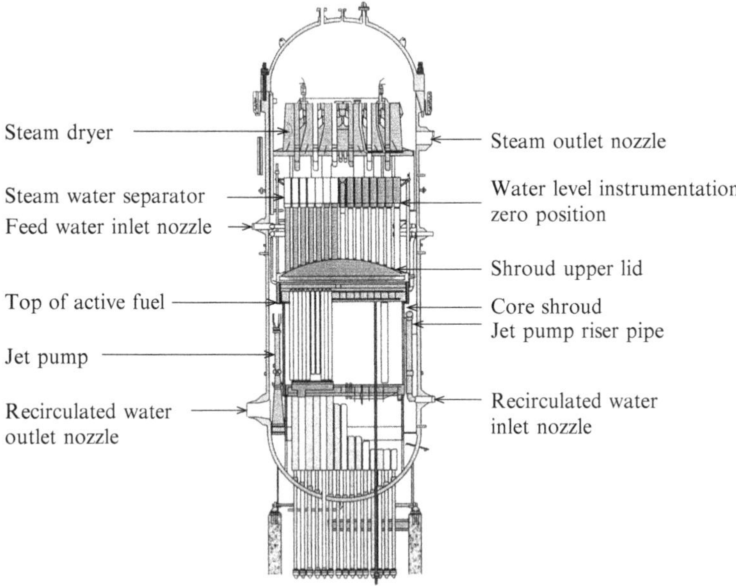

Fig. 7.2 Cross section of reactor pressure vessel

Instead of the jet pump, the ABWR is equipped with multiple internal pumps inside the pressure vessel. The internal pump eliminates the need for recirculation piping outside the pressure vessel.

For control of reactor power and emergency shutdown (scram) of nuclear reaction in abnormal states, a number of control rods are installed, which are inserted or withdrawn as necessary. Boron carbide or hafnium (for long life) is used as neutron absorption material for control rods. Control rods are placed in control rod guide pipe installed in the bottom of the reactor pressure vessel, and driven by control rod drive (CRD) placed in the CRD housing under the reactor pressure vessel.

7.1.2 The Reactor Primary System

A system that circulates coolant (high-quality pure water) in the fuel region, a source of heat, to absorb heat and cool down the fuel is called the reactor coolant recirculation system. A main piping system through which turbine driving steam flows is called the main steam system. The reactor coolant recirculation system and main steam system, and feed-water piping inside the reactor container, which will be described later, are collectively called the reactor primary system.

[1] Main steam system

Steam generated inside the reactor pressure vessel flows through the main steam pipe and is fed into turbines. In the main steam system are main steam safety relief valves used to protect the reactor pressure vessel from over-pressure under abnormal plant conditions, and main steam isolation valves (MSIV) used to prevent coolant from flowing out of the reactor container when a steam pipe rupture accident occurs, which is one of the design basis accidents outside the reactor container. On the outlet side of a number of main steam safety relief valves is an exhaust pipe, leading whole spurted steam into water in the suppression pool, where the steam is condensed. The MSIVs are closed rapidly 3–5 s after an accident signal is received.

[2] Primary loop recirculation system

The primary loop recirculation system circulates coolant to eliminate heat generated by fuel assemblies. The system withdraws the part of coolant from annulus section (between reactor pressure vessel and core shroud) inside the reactor pressure vessel into the primary loop reactor recirculation system, applies pressure to the coolant with a recirculation pump, then returns it as a flow driving a jet pump located in the annulus section. The jet pump pulls in the coolant from a suction inlet and doubles the driving flow rate to circulate the coolant.

The BWR is a thermal neutron reactor and its nuclear reaction rate changes according to the neutron moderating effect. For the moderator, light water is used that is also used as coolant. Therefore, the power can be changed by varying the flow rate of the coolant without changing the position of a control rod. To optimize use of it, the reactor recirculation pump can change its speed to control the circulation rate. An instantaneous guillotine break (loss of coolant) piping accident in the primary loop recirculation system is the toughest event to manage among design basis accidents. With ABWRs, because they adopt internal pumps and have no piping in the primary loop recirculation system, an accident involving loss of coolant is not the toughest event to manage as is the case with conventional BWR.

7.1.3 Engineered Safety Features

[1] Emergency core cooling system

When a design basis (loss of coolant) accident occurs involving a possible rupture of piping containing reactor coolant, multiple systems activate to assure the integrity of the fuel by using water stored in the suppression pool as coolant and injecting it into the reactor pressure vessel. The systems are collectively called the emergency core cooling system (ECCS). The system consists of the high pressure core spray (HPCS), low pressure core spray (LPCS), low pressure core injection (LPCI), and automatic depressurization systems. Heat can be removed also by injecting water through the residual heat removal (RHR) system into the pressure vessel.

[2] Reactor container

When an accident involving loss of coolant occurs that is caused by piping breaking in the reactor primary system, the ECCS injects cool water to prevent fuel rods from being damaged. And, upon the loss of coolant accident, radioactive material generated is confined and constrained so that only a sufficiently low amount is released outside the plant. For this purpose, the reactor container seals the whole reactor primary system. Every boiling water plant has a reactor container that is designed to withstand pressure and high temperatures resulting from an accident and to maintain airtightness at a sufficient level. Even if reactor water leaks from the reactor primary system during normal operation, it is confined within the container.

The reactor container for boiling water plants consists of the drywell that houses the reactor pressure vessel and other equipments, the suppression chamber that stores a large amount of water, and the vent pipe that connects them. If an accident involving loss of coolant occurs caused by piping rupturing in the reactor primary system, steam emitted inside the drywell is led through the vent pipe into the suppression pool where it is condensed and cooled. This suppresses rise in pressure inside the container. The standby gas treatment system and flammability control system (FCS) are also provided as a precaution against accidents. The reactor building that houses the reactor container is a secondary housing structure built to suppress the release of radioactive material into the environment.

<1> Standby gas treatment system
The standby gas treatment system prevents radioactive material, which could be leaked from the container in the case of an accident, from flowing directly out of the reactor building.

<2> FCS
After an accident involving loss of coolant, flammable gases may be generated inside the reactor container. The gases can react with each other (radiation degradation and zirconium–water reaction produces hydrogen and oxygen) to produce a large amount of heat, causing pressure and temperature inside the container to increase. To prevent this, the FCS limits the density of flammable gases to a level lower than the ignition limit. The system has a function of leading gases inside the container to a hot vessel in which hydrogen and oxygen are combusted.

7.1.4 Reactor Auxiliary Systems

Reactor auxiliary systems are responsible for supporting important types of equipment that generate steam or power, supporting those operations, and securing reactor safety when the plant enters an abnormal state. The reactor coolant purification (CUW), RHR, reactor core isolation cooling (RCIC), standby liquid control (SLC), fuel pool cooling and filtering (EPC) are reactor auxiliary systems.

[1] Reactor coolant purification system

Iron rust flows into coolant inside the reactor (or reactor water) from the feed-water system, causing irresolvable material and ions in the coolant to become radioactive. This system is intended to eliminate those impurities using ion exchange resin and to maintain quality of the reactor water. The reactor coolant purification system takes coolant out of piping for the primary loop recirculation system, cools it down, purifies it in the filtrating and desalinating device with ion exchange resin, increases its temperature, and then combines it with the feed-water system flowing back to the reactor. It also has a function to adjust water level according to the water density variation caused by increasing/decreasing temperature of coolant upon startup or shutdown of the reactor.

[2] Residual heat removal system

Even after fission reactions have been stopped by insertion of control rods, generation of heat (decay heat) continues inside the fuel due to the decay of fission products. The RHR system is intended to eliminate the heat. During reactor shutdown, the heat is eliminated by cooling part of coolant ramified from piping for the primary loop recirculation system with a heat exchanger in the RHR system, and returning the coolant to the reactor via the primary loop recirculation system.

[3] Reactor core isolation cooling system

In preparation for the reduction of reactor coolant that could be resulted from reactor isolation from any cause, a system is installed that injects cooling water into the reactor to cool it down. This is called the RCIC system. The system drives a pump with a compact turbine driven by steam generated from the reactor to send cooling water to the reactor.

[4] Standby liquid control system

Control rods are usually used to control nuclear reaction in the reactor. It is assumed that all of the control rods (one per four fuel assemblies) could be inactivated. If it should happen to a BWR, boron, a neutron absorber, is injected into reactor in the shape of sodium pentaborate solution in order to shut down the reactor without fail even in a low temperature state. The SLC system is a backup system of control rods.

[5] Fuel pool cooling and filtering system

This system cools down spent fuel stored in the spent fuel storage pool and purifies water in the pool. It eliminates impurities from pool water by passing the water through a filtrating and desalinating device containing ion exchange resin.

7.1.5 Turbine and Generator Equipments

Nuclear power is generated using steam produced by nuclear fission energy in the reactor to rotate the turbine and generator. The principle of power generation system is the same as that of thermal power except that a boiler is substituted by a nuclear reactor. However, regarding steam conditions, the latest turbines for thermal power use superheated steam, while those for nuclear power uses saturated steam due to various restrictions. Because of this low condition of steam, the size of a turbine for nuclear power is considerably large. The turbine for a BWR also differs from that for thermal power generation in that it has a radiation shielding wall to alleviate the influence of radioactivity contained in steam.

7.1.6 Feed-Water System

Cooling water in the reactor pressure vessel exits the reactor in the form of steam, and so continuous refill of cooling water is required. This is referred to as feed water. After fulfilling its purpose, steam is cooled down with the turbine condenser using sea water via a heat-transfer pipe, and turns into condensate. This condensate is purified by passing it through the filtrating and desalinating device (a device eliminating impurities and impure ion from water), subjected to pressure from a condensate pump and feed pump, heated with a feed-water heater, and then returned to the reactor pressure vessel.

7.1.7 Instrumentation and Control Devices

Instrumentation and control devices for a nuclear power plant are responsible for monitoring necessary parameters vs. variation during normal operation, variation of operational conditions, variation of load, and disturbances, and controlling the parameters properly. We establish the central control room to centrally control the monitoring of various parameters required for operating main plant systems and operation of main equipments. The following lists the main instrumentation and control systems.

<1> Reactor control systems
<2> Safety protection system
<3> Reactor neutron monitoring system
<4> Reactor process instrumentation system

The following describes summary of respective systems.

<1> The reactor control systems consist of the reactor power control system that controls reactor power, the reactor pressure control system that controls reactor pressure, and feed-water control system that controls reactor water level.

<2> The safety protection system works to maintain safety by suppressing any abnormal transient state or design basis accident that may affect the safety of the reactor and to prevent preventable accidents. The system consists of the reactor protection system that activates emergency shutdown (scram) system and the engineered safety feature activation circuits that activates the ECCS and other engineered safety features.

<3> The reactor neutron monitoring system measures reactor power using appropriate neutron flux detectors in the whole range from startup neutron source to output ranges, covering the range of approximately nine digits. This is intended to maximize the sensitivity of detectors vs. movement of control rods and accurately measure neutron fluxes in an intermediate range. All the neutron flux detectors are placed inside the reactor.

<4> The reactor process instrumentation system measures and instructs temperature, pressure, flow rate, water level, and other parameters. For correct and safe operation of the reactor, the process instrumentation is installed on every important system of the reactor. Most of the indicating and recording instruments are placed in the central control room. The reactor process instrumentation system consists of reactor pressure instrumentation, recirculation system instrumentation, feed-water system instrumentation, main steam system instrumentation, control rod driving system instrumentation, etc.

7.1.8 Electrical System

To assure safety of reactor facilities, the electrical system consists of multiple external power supplies and emergency power supplies inside the plant, so that required electric power will not be completely lost in any circumstances including normal operation, shutdown, abnormal events, and accidents.

7.2 Operation Control Methods of the BWR

7.2.1 Principle of Power Control

Two types of methods are provided for reactivity control that increases or decreases reactor power and maintains the rated output: one uses control rods and the other changes the recirculation flow rate.

The BWR core structure enables one cruciform control rod to be inserted for every four fuel assemblies. The reactivity control uses control rods to insert or withdraw the control rods into/from the core during operation. More specifically, inserting control rods while the reactor is operating usually adds negative reactivity, which decreases power. On the other hand, withdrawing control rods adds positive reactivity and increases power.

During steady operation, the reactivity is controlled not by operating all control rods, but by limiting the operation to control rods at between several to more than a dozen positions. The control rods are usually inserted halfway at first, and further operation is dependent on the degree of reactivity in the core.

Next, let us explain power control using the recirculation flow rate. Assume that, during steady operation with a constant flow rate and output, we have increased the recirculation flow rate and therefore the core flow rate. Then, thermal-hydraulic and nuclear variations in the core are:

<1> The flow rate of cooling water in the core increases.
<2> Cooling water flowed in from the bottom of the core moves to the upper of the core while temperature rises at a low rate.
<3> The steam starting point in the core moves up.
<4> The void fraction decreases and the fuel temperature drops.
<5> Positive reactivity is added as effects of void reactivity and Doppler reactivity.
<6> Output increases, causing the void fraction to increase and the fuel temperature to rise.
<7> Negative reactivity is added as effects of void reactivity and Doppler reactivity.
<8> Output settles at the point where reactivity is balanced between <5> and <7>. (The output level is higher than that before the recirculation flow rate increases.)

Thus, we can increase output by increasing the recirculation flow rate. If we decrease the flow rate, output can be decreased in inverse order (negative reactivity is added first, and output decreases until reactivity is balanced). This is an example of power control using the recirculation flow rate.

For the BWR, we can control reactivity to a certain extent by adjusting this recirculation flow without moving control rods. The advantages of this control method are that it is easier than moving control rods, variation of output distribution affected by the control is comparatively small, and impact on the fuel is low.

Actual steady operations combine the reactivity control using control rods and that using the recirculation flow rate. Sections 7.3 and 7.4 of this chapter illustrate the example.

7.2.2 *Plant Control*

In the BWR, fission energy generated in the core directly boils cooling water and converts the water into steam. Then the steam rotates the turbine to generate power. To allow for stable energy conversion, three control systems are provided, controlling the reactor pressure, the reactor water level and the recirculation flow rate, respectively. Figure 7.3 shows the configuration of three control systems.

Stabilizing pressure denotes the balance control between steam volume generated in the core and that consumed by turbines. This also helps suppress reactivity variation of the core caused by fluctuation of the pressure. In the reactor

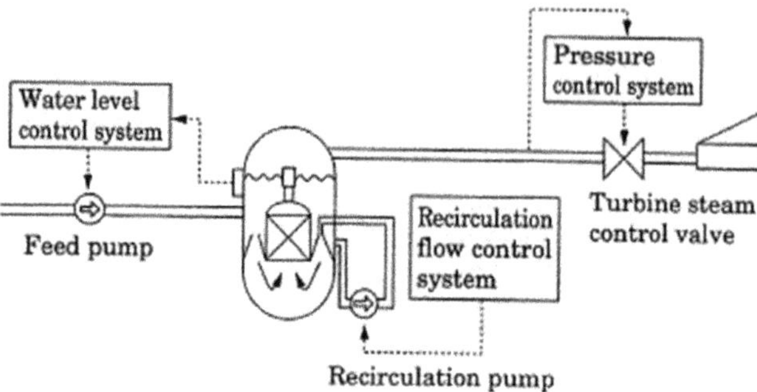

Fig. 7.3 Main control systems of BWR

pressure vessel, a two-phase exists consisting of water and steam, forming water surface. This raises the need for suppressing vertical variation of the surface, and the water level control system (or feed-water control system) is provided. Furthermore, in order to cool down the fuel in the core and maintains steady the ratio of steam generation (hereafter called void fraction), the flow rate of cooling water must be kept at a certain level. This is realized by the recirculation flow control system. The steady void fraction in the core is required because its variation causes fluctuation of the fission rate. To put it the other way around, we can change output by adjusting the recirculation flow rate. The following explains the mechanism of each control system.

[1] Pressure control system

Steam generated in the core is filled in the dome of the reactor pressure vessel. The steam is led through the main steam pipe into turbines to rotate generators. The generators are connected to external power systems and affected by fluctuation of the power load. Irrespective of fluctuation of the power systems, the control method differs depending on whether to stabilize the reactor power or to change the reactor power based on load fluctuation of the power systems. The former is called "reactor-main and turbine-sub," adjusting opening of the governing valve of the turbine so as to maintain constant reactor pressure.

On the other hand, the latter adjusts the opening of the governing valve of the turbine to control the speed of the turbine so that output from the generator matches external power load. BWRs adopt the reactor-main and turbine-sub control. This is because the core of a BWR has a positive feedback characteristic for the pressure change, and the core output change caused by pressure must be suppressed.

Let us explain this control method more specifically.

Figure 7.4 shows that a desired pressure value is specified in direct proportion to the main steam flow rate (or equivalent to reactor output), and opening of the

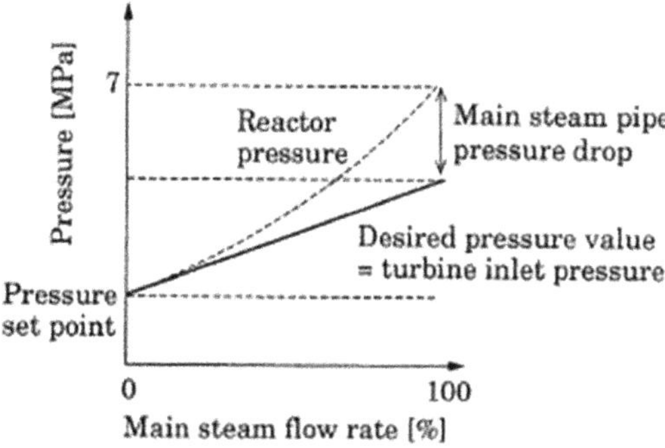

Fig. 7.4 Pressure control technique of BWR

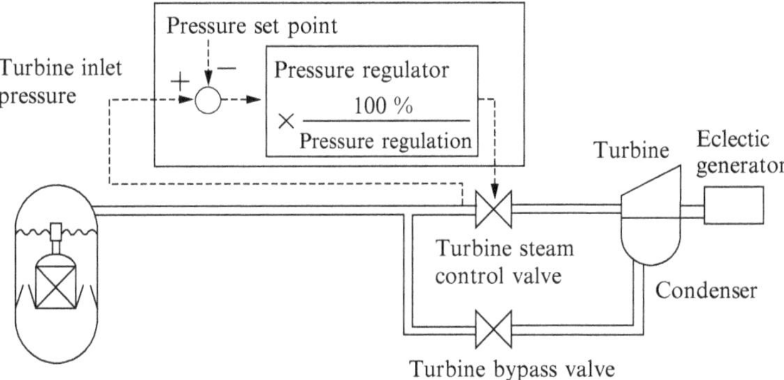

Fig. 7.5 Pressure control system of BWR

governing valve installed in the inlet of the turbine is adjusted so that the turbine inlet pressure can match the desired value.

This control method is called proportional control. The difference between the turbine inlet pressure at zero output (called pressure set value) and the turbine inlet pressure at the rated output is called pressure regulation ratio. As shown in Fig. 7.5, inside the pressure controller, opening of the turbine governing valve is adjusted by the value obtained from dividing the difference between a pressure set value and a turbine inlet pressure by a pressure regulation ratio.

The reactor pressure increases corresponding to pressure loss of the main steam piping that connects between the turbine inlet and the reactor pressure vessel. As a result, BWR pressure is hold at approximately 7 MPa during rated output. The latest ABWR adopts the dome pressure control method that controls the turbine governing valve directly using the measured value of reactor pressure.

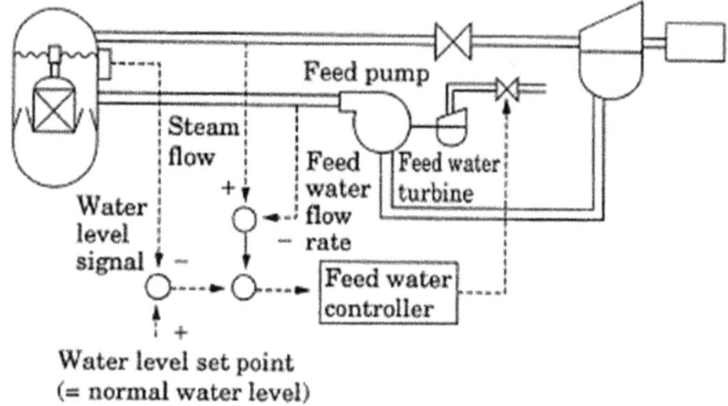

Fig. 7.6 Feed-water control systems of BWR

If the power system load drops to a considerably low level, the method is switched to governing valve control based on turbine speed in order to prevent over-speed of the turbine. As the governing valve begins closing at that point, steam volume will become excessive in the reactor. The excessive steam can be led directly to the condenser via a turbine bypass valve (TBV).

[2] Feed-water control system

The pressure vessel of a BWR is filled with cooling water and steam generated in the core, and their interface forms water surface. The reactor water level must always be higher than the core in order to sufficiently cool down the fuel. On the other hand, the water level must sufficiently be lower than main steam piping so that a volume of moisture will not penetrate the turbine. In actual, a normal water level is specified with sufficient margins of the upper and lower limits. The water level is controlled at a location near the steam water separator located at a point higher than the core. If the water level should drop drastically due to any abnormal state, the reactor scrams. On the other hand, if the water level rises significantly, feed pumps are tripped. Because the volume of steam generated in the reactor is balanced with that consumed in turbines during usual operation, the water level does not fluctuate significantly. However, the water level fluctuates as output, pressure, or core flow rate changes. The feed-water control system (or water level control system) is provided for suppressing this kind of water level fluctuation by controlling the flow rate of feed pumps.

Figure 7.6 shows a pattern diagram of the feed-water control system.

The system controls the feed-water flow rate by adjusting the opening of the turbine governing valve installed on a turbine driving reactor feed pump based on measuring the reactor water level. The turbine driving reactor feed pump uses steam

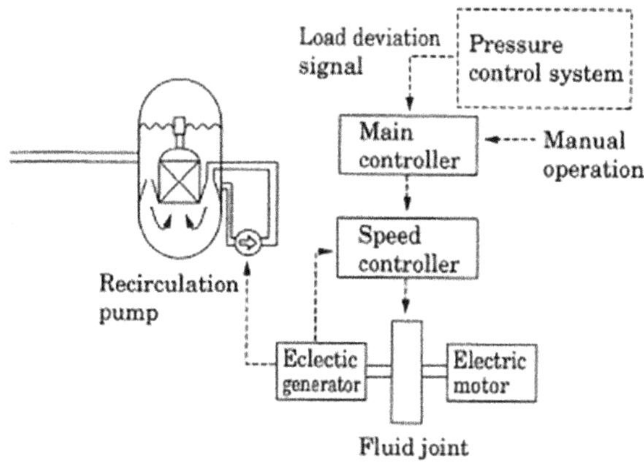

Fig. 7.7 Recirculation flow control systems of BWR

generated in the reactor as a driving source. The system measures three input signals: the steam flow rate signal, the feed-water flow rate signal, as well as the reactor water level signal. This is a kind of predictive control. The amount of water flowing out of the pressure vessel increases as steam flow rate increases. Then the system increases the feed pump flow rate in prospect of water level drop. When the feed pump flow rate increases, on the contrary, the system reduces the feed-water flow rate to suppress the rise of the water level. In summary, the system controls the water level with balance of the rates coolant flowing into or out of the pressure vessel. This water level control technique is called three-factor control. Unlike pressure control system that adopts proportional control, the feed-water control system adopts the PI control method.

[3] Recirculation flow control system

The recirculation flow control system changes the recirculation flow rate by adjusting the speed of the recirculation pump, and changes the density of moderator (water) of neutrons to control output. This recirculation control can change the reactor output at high speed while maintaining output distribution in the core at an approximately specified level. Figure 7.7 illustrates the recirculation flow control systems.

The flow rate to the core inlet (core flow rate) is supplied by a jet pump located on the periphery side (downcommer) of the core. At the inlet, the jet pump discharges cooling water to the bottom of the core after driving water supplied from the recirculation loop sucks water in downcommer. The recirculation loop is a pipe loop located outside of the pressure vessel, driving cooling water with the recirculation pump. The recirculation flow control system controls the speed of this recirculation pump. The M–G set is used to control the speed. The M–G set is a combination of a driving motor and a variable frequency generator, jointed with a fluid coupling. It transmits torque of the constant speed driving motor to the generator by adjusting transmission volume using the fluid coupling to control speed of the recirculation pump. In this system, a speed controller is provided to maintain the constant speed of the variable frequency generator. If the generator speed deviates from a preset value, the speed controller adjusts the flow rate inside the fluid coupling to change transmitted torque, so that the speed falls within the desired range. While the preset value of the speed controller can be changed manually, it can be interlocked with load deviation signals from the turbine control system (pressure control system) to automatically adjust speed of the recirculation pump, so that the reactor output is adjusted corresponding to a load setting value. This automatic adjustment is called the automated recirculation control operation (Master Auto), which is effective for the constant power operation, or for load-fis following operation.

The recent BWR uses an inverter instead of the M–G set for the speed control. In addition, the ABWR is equipped with ten internal pumps for recirculation that have lower rotation inertia and higher response characteristics than conventional recirculation pumps.

7.3 Operation Control During Plant Startup/Shutdown Operation

7.3.1 Basic Points About Plant Startup/Shutdown

The most critical operation required for plant startup/shutdown is to change the reactor power, including the reactor's criticality and nuclear heating. A reactor's criticality, nuclear heating, and other outputs at a low power level are changed only by operating control rods. Outputs at a higher power level are changed by operating control rods and changing the core flow rate. While changing the reactor power, we must pay attention to maintaining the integrity of the fuel.

Changing the reactor power, which fluctuates a flow of steam generated in the reactor, is a disturbance to pressure and the water level in the reactor. Controlling the reactor pressure and the reactor water level is critical to help ensure stable operation of the reactor. The reactor pressure is controlled at a constant level by correctly handling the flow of the reactor-generated steam in turbines and condensers. On the other hand, the reactor water level is controlled at a constant level by correctly balancing the volume of feed water flowing into the reactor and the volume of steam generated in and flowing out of the reactor.

[1] Controlling the reactor power

The reactor power is controlled using two methods: one is by operating control rods and the other by adjusting the core flow rate using the recirculation flow control system to change the speed of a recirculation pump. The most important point to note during the power change is to assure the integrity of the fuel.

Output of thermal limit logs is used to monitoring integrity of the fuel. We obtain the thermal limit from two computations: one is the computation of reactor heat output, regularly operated with a process computer for the whole core, using neutron flux, control rod position, the core flow rate, the reactor pressure, the feed flow rate, and other input signals on the reactor. The other is the computation of core performance that is obtained from thermal-hydraulic computations, operated for each fuel bundle in the core. The computations can be operated optionally before or after changing power while power is rising. For local monitoring during control rod operation, a rod block monitor is used into which only neutron flux signals around operated control rods are input. If a signal exceeds a predefined threshold, withdrawal of control rods is stopped. In the process of adjusting the core flow rate in automatic mode, if a thermal limit obtained from the core performance computation becomes difficult to obtain, the mode is changed from automatic to manual, and the adjustment operation is stopped. For the BWR, the operating range of the reactor is defined on an operating characteristics diagram for the core flow rate and reactor heat output. If the control rods or the core flow rate operation is going to deviate from the operating range, a command is output to stop the control rods operation or to stop core flow rate operation.

For actual power rise, we define procedures for control rods operation and the core flow rate adjustment at designing stage using offline core thermal-hydraulic computation codes, so that integrity of the fuel is checked and the powering up procedure is predefined. The operating procedure of a control rod is loaded on the rod worth minimizer (RWM), and the operation guide is output to ensure the correct actual operating procedure. Should an erroneous procedure be actually taken, the operation is stopped.

[2] Maintaining the reactor pressure

Fluctuation of the reactor pressure is a disturbance for the reactor water level and the reactor output. The reactor pressure is controlled by handling flow of steam generated in the reactor in turbines and condensers. The electro hydraulic control (EHC) system is responsible for the reactor pressure control. Whenever a condenser is available irrespective of plant status, EHC uses the turbine inlet pressure as a process signal (EHC does not control the turbine valve based primarily on turbine load) and stabilizes the flow rate of steam generated in and flowing out of the reactor by controlling opening of the TBV or regulator valve to flow steam into the condenser. The turbine is in a trip state before it starts, and the turbine regulator valve is forcibly closed. Therefore, the reactor pressure is controlled with the TBV. Even after the turbine start, opening of the turbine regulator valve is limited to small until parallel operation of generator begins, limited by a limiter (load limiter or load setting device). After parallel operation of the generator starts, the opening of the turbine regulator valve becomes wider by the gradual increasing of output of the limiter (load setting device). As the opening of the turbine regulator valve becomes wider, the TBV closes automatically. The TBV closes completely on the spot when the limiter (load setting device) value exceeds the steam volume generated in the reactor, and the pressure control is switched to the turbine regulator. After the pressure control is switched from the TBV to the regulator valve, the limiter (load setting device) value is set to a level slightly higher than the actual reactor-generated steam volume so that pressure can be controlled with the turbine regulator valve as long as the pressure fluctuates within a normal range.

[3] Maintaining the reactor water level

Unusually higher reactor water level may increase carryover of moisture to turbines, while unusually lower level may expose fuel from coolant. The reactor water level is controlled by using the condenser as a water source, and applying pressure to water with the condenser system and feed-water system to adjust the feed-water rate so that the water supply volume balances with the volume of steam generated in and flowing out of the reactor. Before the reactor reaches a critical state, the reactor water level rises due to coolant and purge water flowing from auxiliaries reconnected to the reactor (reactor recirculation pumps, circulation pumps for reactor coolant purification system, and control rods driving system). At this stage, therefore, the reactor water level is adjusted by draining coolant using the blowdown regulator valve in the reactor coolant purification system. Once the reactor reaches a critical state and steam is generated, the reactor water level drops, requiring feed water. Water is fed to the reactor by correctly switching feed pumps according to the feed-water rate. Usually, a motor-driven feed pump (M/DRFP) is used for feed water at a low power level. When we use M/DRFP, the feed-water rate is controlled by adjusting the opening of the feed regulator valve arranged on the outlet side of the pump. For feed water at a high power level, a turbine-driven feed pump (T/DRFP) is used. Feed capacity of this pump is 50 %. In the process of powering up, therefore, two units shall be operated. When we use T/DRFP, the feed-water rate is controlled by adjusting the speed of a turbine used to drive the pump. Also for feed-water control, we can carry out the three-factor control, an advanced control. This method of control adds to the reactor water level control signal the deviation (mismatch flow rate) between the main steam flow corresponding to the reactor-generated steam flow and the feed-water flow. We switch the control to this method when the power is at a level of 25 % or higher where the accuracy is ensured with a main steam flow rate signal.

7.3.2 Plant Startup Operation of the BWR

BWR startup operation follows the startup diagram shown in Fig. 7.8.

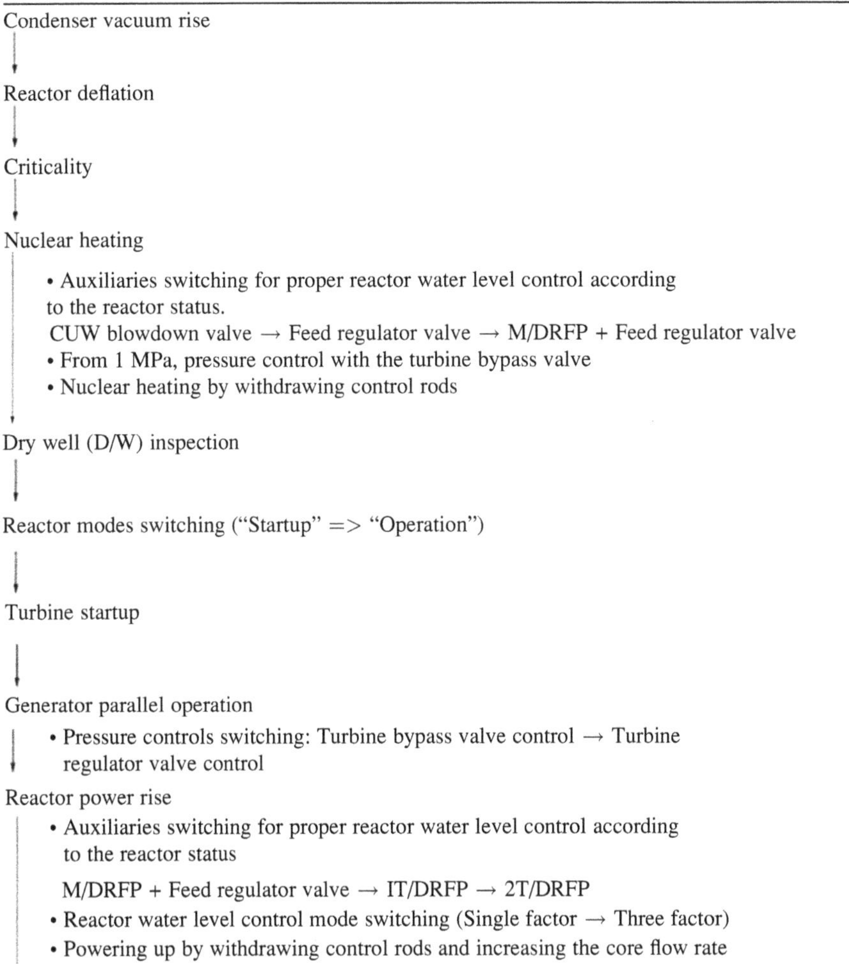

Plant startup operation is outlined as follows:

<1> Prior to the reactor startup, we carry out a condenser vacuum up operation for preparation of sending steam to the condenser and for deflation of the reactor water using condenser atmosphere.

<2> Once the condenser vacuum up is completed, we operate the MSIV to lower the dissolved oxygen (deflation operation) rate of reactor water for the purpose of preventing stress corrosion cracking (SCC) of reactor structure material.

<3> We start up the reactor first by switching the reactor mode switch from "Stop" to "Start," and releasing the interlock that blocks withdrawal of control rods, and then withdrawing control rods. Through analysis, the procedure (withdrawal sequence) of withdrawal of a control rod is predefined, which conforms fully to the thermal limit of fuel, and is

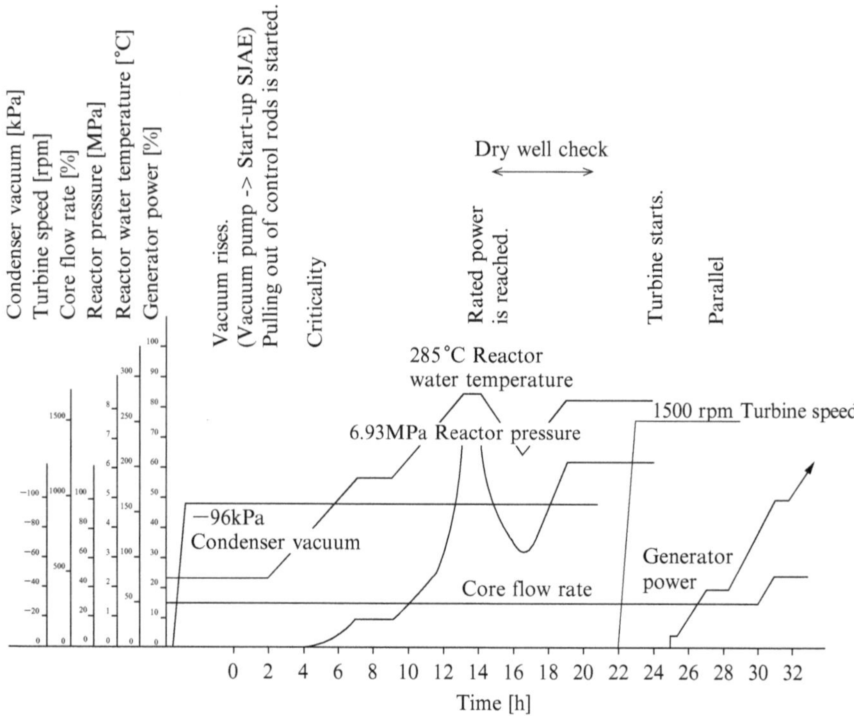

Fig. 7.8 Startup curve of BWR

registered with the Rod Worth Minimizer (RWM) to be indicated to operator. If an operation deviates from the procedure, a signal is output to stop withdrawing a control rod. A state before criticality is monitored using output from a neutron instrumentation that monitors the startup range (counter rate range of startup range neutron monitor (SRNM) or source range monitor (SRM)). Control rods are withdrawn carefully by keeping an appropriately long period.

<4> Once the reactor enters the critical state, nuclear heating begins in the reactor and the reactor water temperature rises. Affected by a temperature coefficient resulting from the rising reactor water temperature, negative feedback is applied. In order to balance the neutron flux level with a desired temperature rise rate, therefore, control rods are gradually withdrawn according to rising temperature with reference to the RWM guide. Thus, the reactor pressure is adjusted to the rated pressure. The neutron flux is monitored using output from a neutron instrumentation that monitors intermediate rage [the MSV range of SRNM or intermediate range monitor (IRM)]. Control rods are withdrawn carefully by maintaining an appropriately long period.

<5> Until the reactor reaches the critical state and generates steam, a certain volume of purge water, etc. is flowed from the control rods driving system into the reactor. To maintain mass balance and control the reactor water

level, we must discharge the water from the reactor. In this state, the reactor water level is controlled by adjusting the regulator valve on the blowdown line of the reactor coolant purification system. If nuclear heating begins, and the reactor generates steam, the reactor water level is regulated by adjusting the feed rate of water flowing into the reactor. During nuclear heating, the reactor pressure changes from atmospheric to approximately 7 MPa, and so we must operate pumps of feed-water and condensate systems by changing the combination to supply water to the reactor. When the reactor pressure is low, the water level is controlled with the feed regulator valve located at the outlet of M/RFP by using pressure of a condensate pump. M/DRFP is started before the reactor pressure exceeds the pressure of the condensate pump, so that the water level can be controlled with the feed regulator valve. Each actuator has a controller at the lower part to directly adjust a manipulated variable. The controller feeds back the reactor water level located at the upper part, combines the output of a controlling manipulated variable from the controller and the output from a control device, and connects with an upper control device, and then disconnects the controller on the actuator to be stopped from the upper control device.

<6> During nuclear heating, the reactor pressure is controlled by adjusting the opening of the TBV according to the deviation between a preset reactor pressure value and an actual reactor pressure value. To increase the reactor pressure to the rated value through nuclear heating, we adjust the preset pressure value to be slightly higher than the reactor pressure so that the reactor pressure can increase at the maximum change rate of nuclear heating through control rods manipulation. Once the reactor pressure is set to the rated value, the set value is retained. If the reactor pressure exceeds the preset value during nuclear heating, the reactor pressure is regulated to the preset value by opening the TBV based on a manipulated variable corresponding to the deviation.

<7> Control rods are withdrawn additionally after the reactor pressure reaches the rated output, and if the neutron flux level further rises, the neutron instrumentation is changed from a neutron instrumentation that monitors intermediate range (the MSV range of SRNM or IRM) is changed to average power range monitor (APRM) to monitor the neutron flux. Because interlocks of the reactor protection system are also switched, we change the reactor mode switch from "Start" mode to "Operation" mode. The reactor power rises up to approximately 10 % in preparation of turbine start and generator parallel operation. Steam generated by the heat is disposed of with the TBV.

If the reactor pressure reaches the rated output during startup after a usual periodic inspection, we regard that the reactor is in a sub-critical state and inspect devices in dry well (for leak, vibration, unusual noise, interference, etc.).

<8> While the turbine is starting, its speed is controlled by adjusting the opening of the turbine regulation valve, so that the speed rises at a constant rate. Varying opening of the turbine regulator valve can change the flow rate of

steam from the reactor and cause the reactor pressure to fluctuate. To regulate the fluctuation of reactor pressure, the TBV is configured to open to compensate for the variation on the turbine regulator valve side.

<9> After the turbine has completed the start process, a magnetic field is applied to the generator to generate voltage and activate parallel operation. The parallel operation is done fully automatically. For the parallel operation, the voltages, frequencies, and phases must be matched between the generator and the system. The operation to match voltages between the generator and the system is referred to as voltage matching. The generator voltage is controlled by monitoring the deviation between them with an automatic synchronous system. The operation to synchronize frequencies between the generator and the system is referred to as speed matching, operated by adjusting the opening of the turbine regulator valve with EHC. To match phases, frequency of the generator is set slightly higher than that of the system for speed matching, and the deviation of phases is monitored with a synchroscope. Once the phases are matched, a generator breaker is automatically turned on. After parallel operation is established, the generator power is raised by increasing the load setting and by widening opening of the turbine regulator valve. The TBV closes according to the degree of opening of the turbine regulator valve, so that the reactor pressure is maintained constant. When the TBV completely closes, the reactor pressure is controlled with the turbine regulator valve. Usually, generator power is not controlled with the turbine regulator valve, which is used only to control the reactor pressure.

<10> The feed-water flow rate increases as the reactor power rises. To cope with the increasing feed-water flow rate considering the limited feed pump capacity, a unit of M/DRFP is switched to a unit of T/DRFP. If power further rises, it is shifted to two units of T/DRFP. Thus feed pumps are switched or additionally started. Each actuator has a controller at the lower part to directly adjust a manipulated variable. The controller feeds back the reactor water level located at the upper part, combines the output of a controlling manipulated variable from the controller and the output from a control device, and connects with an upper control device, and then disconnects the controller on the actuator to be stopped from the upper control device.

<11> The reactor power is raised by withdrawing control rods, or by increasing the speed of a recirculation pump to increase the core flow rate. The reactor power can also be increased automatically. This is carried out by first increasing a load setting at the preset change rate, and second computing a manipulated variable based on the signal of deviation between the load setting and the signal of total steam flow rate obtained from EHC pressure deviation, and then by adjusting the speed of the recirculation pump to the computed values of manipulation.

7.3.3 *Plant Shutdown Operation of the BWR*

BWR shutdown operation follows the shutdown diagram shown in Fig. 7.9.

Rated output

↓

Reactor power down

- Auxiliaries switching for proper reactor water level control according to the reactor status.
  2T/DRFP → 1T/DRFP → M/DRFP + Feed regulator valve
- Reactor water level control mode switching (Three-factor → Single factor)
- Powering down by inserting control rods and decreasing the core flow rate

↓

Generator parallel-off

- Pressure control switching: Turbine regulator valve control → Turbine bypass valve control

↓

Turbine stop (by turbine trip operation)

↓

Reactor mode switching ("Operation" => "Start")

↓

Sub-criticality (Inserting all control rods)

↓

Lowering pressure

- Auxiliaries switching for proper reactor water level control according to the reactor status
  M/DRFP + Feed regulator valve → Feed regulator valve → CUW blowdown valve
- Reactor depressurization: 1 MPa or higher => Lowering pressure setting operation
  Lower than 1 MPa => The turbine bypass valve opening jack, and turning on cooling mode while RHR stops

↓

Condenser vacuum break

Plant shutdown operation is outlined as follows:

<1> The reactor power is lowered by inserting control rods or by decreasing the speed of a recirculation pump to decrease the core flow rate. The reactor power can also be lowered automatically. This is carried out by first decreasing a load setting at the preset change rate, and second computing a manipulated variable based on the signal of deviation between the load setting and the signal of total steam flow rate obtained from EHC pressure deviation, and then adjusting speed of the recirculation pump to the computed values of manipulation.

<2> The feed-water flow rate decreases as the reactor power drops.
To cope with the decreasing feed-water flow rate considering the limited feed pump capacity, the combination of feed pumps is correctly switched. Two units of T/DRFP are switched to a unit of T/DRFP. If further drops, a unit of

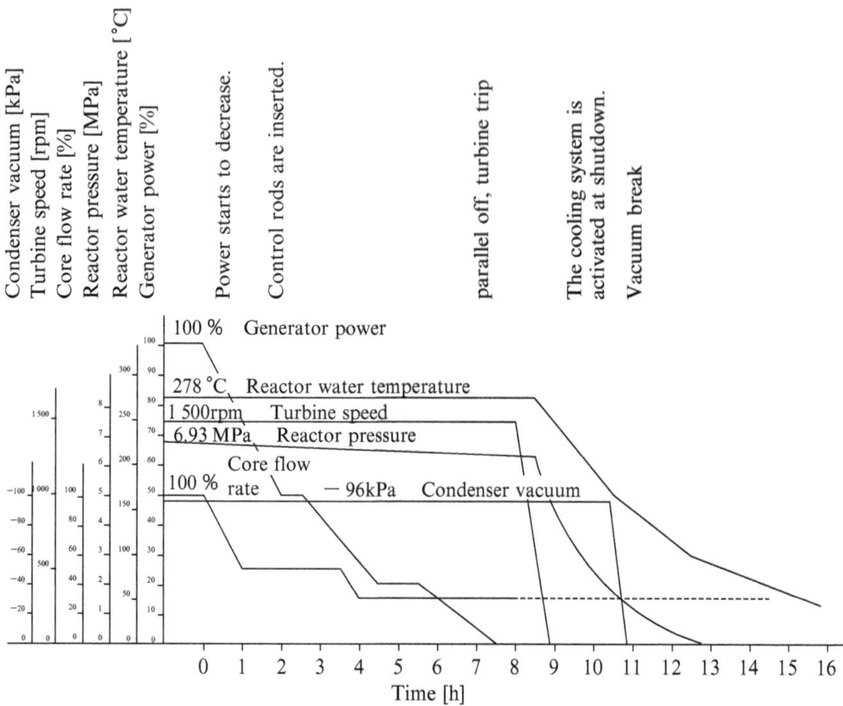

Fig. 7.9 Shutdown curve of BWR

T/DRFP is switched to a unit of M/DRFP. Each actuator has a controller at the lower part to directly adjust a manipulated variable. The controller feeds back the reactor water level located at the upper part, combines the output of a controlling manipulated variable from the controller and the output from a control device, and connects with an upper control device, and then disconnects the controller on the actuator to be stopped from the upper control device.

<3> When plant power drops to the initial load (approximately 5 % output), the generator is paralleled off. Simultaneously with the parallel-off, load setting is lowered and the opening speed of the turbine regulator valve is reduced to approximately the rated speed level. The TBV opens according to the degree of closing of the turbine regulator valve so that the reactor pressure is maintained constant.

<4> When the reactor pressure is stabilized with the TBV, the turbine is stopped. The turbine is stopped by operating the turbine trip. The turbine trip closes all valves including the turbine regulator valve and the main steam shutoff valve, stopping steam supply to the turbine trip. Steam previously supplied to the turbine is sent to the TBV to hold the reactor pressure constant.

<5> Control rods are inserted additionally after the turbine is tripped, and if the neutron flux level drops, the neutron instrumentation is changed from APRM

to a neutron instrumentation that monitors intermediate rage (the MSV range of SRNM or IRM) to monitor the neutron flux.
Because interlocks of the reactor protection system are also switched, we change the reactor mode switch from "Operation" mode to "Start" mode. Control rods are further inserted until the sub-critical state is reached, and then all control rods are inserted.

<6> Even after all control rods are inserted, the reactor pressure does not drop because of decay heat in the reactor. Depressurization is carried out by gradually lowering the pressure setting and disposing of steam with the TBV while monitoring the change rate of reactor water temperature. The pressure setting can lower pressure to 1 MPa. For further depressurization, the TBV opening jack operation is carried out, as well as cooling mode operation while RHR system stops. This lowers the reactor temperature to 100 °C or lower.

<7> When the reactor water temperature drops to 100 °C or lower, MSIV is fully closed and vacuum break of the condenser is executed.

7.4 Operation Control During Steady Operation

Let us first explain excess reactivity during steady operation. During steady operation, fissile material in fuel is decreasing as the fuel is combusted, and reactivity of the core drops. To maintain the reactor in a critical state and operate at a constant power during a steady operation period, we provide the core with excess reactivity in advance. To compensate the excess reactivity, we insert control rods or decrease the recirculation flow rate to keep the constant power. Because unusually higher excess reactivity cannot be controlled with control rods or recirculation flow, we must suppress the excess reactivity below a certain level. For the BWR, gadolinia (Gd_2O_3), a burnable poison is mixed in the fuel to suppress excess reactivity. Figure 7.10 shows a pattern of effect of gadolinia on an infinite multiplication factor.

Without gadolinia contained in the fuel, the infinite multiplication factor of fuel drops monotonously with progress of combustion process. With gadolinia, on the contrary, the infinite multiplication factor is suppressed to low level in the early stage of combustion owing to neutron absorption effect. Because the neutron absorption effect decreases as combustion progresses, the infinite multiplication factor rises once and then drops as a solid line in Fig. 7.10 indicates.

Figure 7.10 shows variation of the infinite multiplication factor when new fuel is loaded into the core and combusted. In fact, however, various burnup fuel exist in the core, and combination of these fuels results in the variation of excess reactivity during operation.

Figure 7.11 provides an example of variation of excess reactivity during power operation.

Figure 7.11 shows that excess reactivity during power operation increases once due to the effect of burning gadolinia and, as combustion progresses further, the

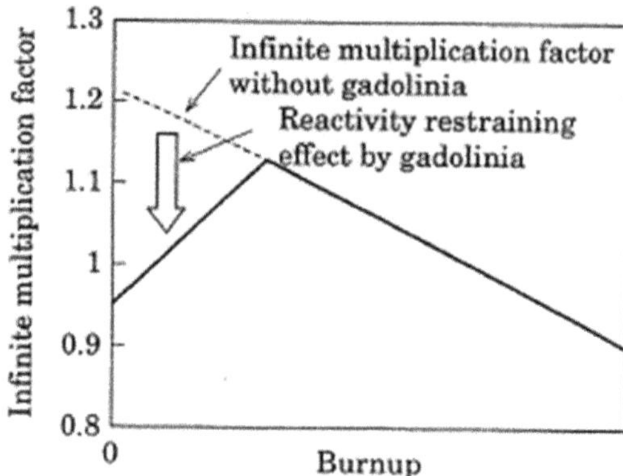

Fig. 7.10 Effect on infinite multiplication factor by gadolinia

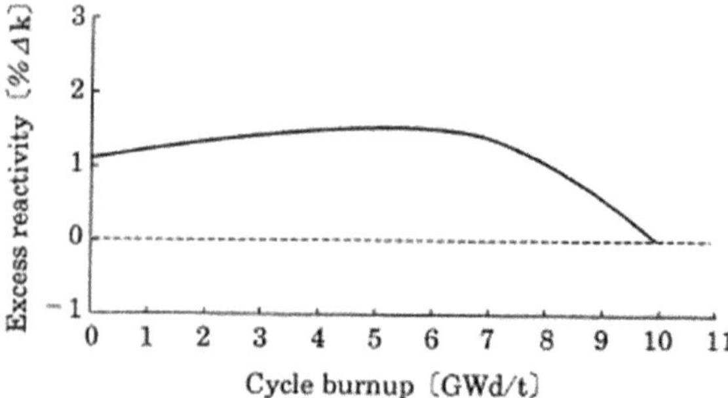

Fig. 7.11 Example of excess reactivity change

excess reactivity in the core decreases. To stabilize the power, an appropriate operation is needed corresponding to this variation of excess reactivity.

Figure 7.12 provides an example of variation of the control rods pattern and the core flow rate.

[1] Method to compensate reactivity while excess reactivity is increasing

In this stage, power is gradually increasing if control rods are inserted in a predetermined pattern and the flow rate are maintained at a constant level. So, power is stabilized by decreasing the core flow rate depending on the excess reactivity in the core. The core flow rate is controlled in the range of the upper and lower limits. If it reaches to the lower limit (point A in Fig. 7.12), it cannot be decreased further. Then, control rods are inserted to a certain extent to raise the flow rate by the upper limit (point B in Fig. 7.12). After then the core flow rate is decreased again.

[2] Method to compensate reactivity while excess reactivity is decreasing

Next let us consider the stage where excess reactivity is decreasing as combustion progresses. Power is stabilized by increasing the core flow rate depending on the

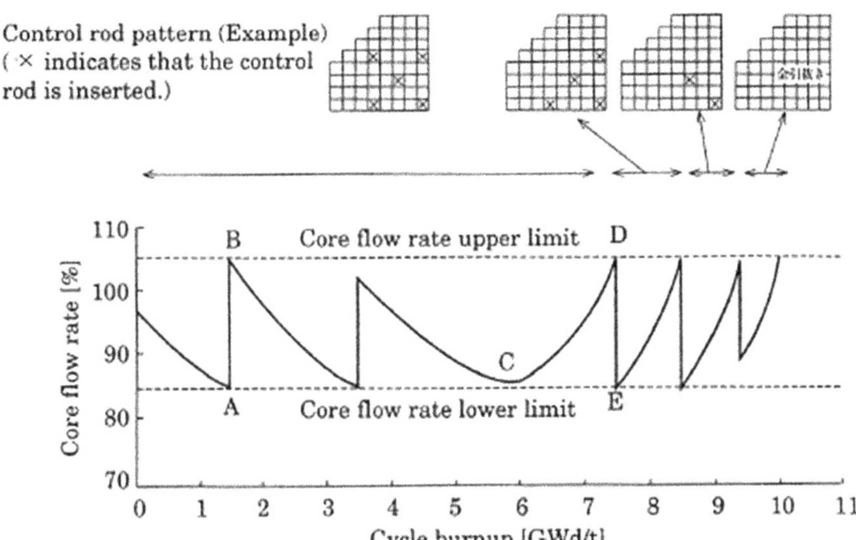

Fig. 7.12 Example of variation of control rod patterns and the core flow rate

decreasing excess reactivity in the core. When the excess reactivity reaches to the maximum point in Fig. 7.11, the core flow rate changes from downward to upward (point C in Fig. 7.12). If the core flow rate reaches the upper limit (point D in Fig. 7.12), it cannot increase further. Then, control rods are withdrawn to a certain extent to drop the flow rate to the lower limit (point E in Fig. 7.12). After then the core flow rate is increased again. This operation is repeated several times until the end stage of a cycle is reached. Thus control rods are gradually withdrawn. If the core reactivity cannot maintain criticality even after all control rods have been withdrawn and the core flow rate has been raised by the upper limit, we can hardly operate the reactor at the rated output and we shall stop the operation.

[3] Spectral shift operation

Described above is an operation technique to compensate reactivity by maintaining a control rods pattern in a certain period and gradually increasing or decreasing the core flow rate. There is another operation technique that manipulates control rods frequently in the early to middle stages of a cycle, and increases the core flow rate in the end stage. This operation is highly cost effective because the void fraction of the core is high in the early to middle stages, accumulating Pu, while the void fraction turns to low in the end stage, enabling the accumulated Pu to be used effectively. This operation technique is referred to as spectral shift operation. For the advanced BWR (ABWR), electric method CRD mechanism has been adopted instead of conventional hydraulic method, enabling reactivity to be adjusted more minutely. This drive mechanism facilitates spectral shift operation. Figure 7.13 provides an example of variation of the core flow rate during spectral shift operation.

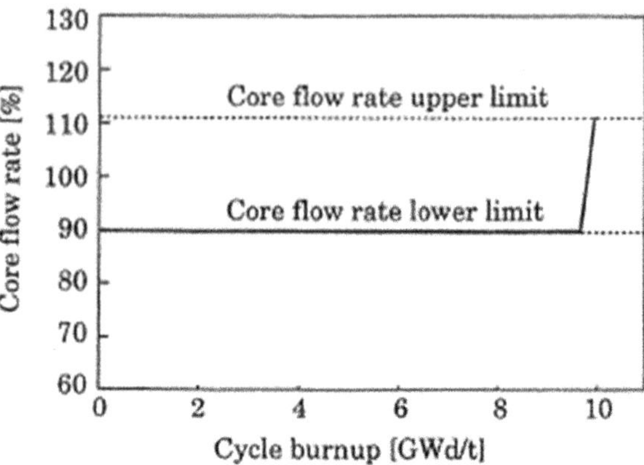

Fig. 7.13 Example of variation of the core flow rate during spectral shift operation

7.5 Control When Load Fluctuation or Abnormal Conditions Occur

7.5.1 Control When Load Fluctuation Occurs

[1] Load-following operation

Currently, nuclear plants in Japan operate at rated reactor power that is maintained at a constant level, which is called based load operation. However, as the proportion of nuclear power generation increased in comparison with the total amount of energy generated in Japan (including thermal electric power generation and hydraulic power generation), the need to adjust the amount of power generation to meet future power demand, which fluctuates from day to day, the so-called load-following operation is projected to become high for nuclear power plants as well.

To cope with this situation, the BWR is designed to provide load-following operation functionality.

The current facility has demonstrated that it is fully capable of performing the so-called daily load-following operation in which, for example, the reactor operates at rated power during daytime when the demand for electric power is high and operates at low power during nighttime when the demand is low.

Figure 7.14 shows an example of the daily load-following operation.

In the example, the rated power operation time is set to 14 h, the low power operation time is set to 8 h, and the power up and down time is set to 1 h each (14-1-8-1 system). The power variation range is set from 5 % to 20 % (maximum).

Figure 7.14 shows the behaviors of the generator power (solid line) and the core flow rate (broken line) during the daily load-following operation. As shown in the figure, the generator power (reactor power) is changed and controlled promptly and stably by the core flow rate. Although not shown in the figure, the operation of the control rod, which requires much time to change the power, is unnecessary.

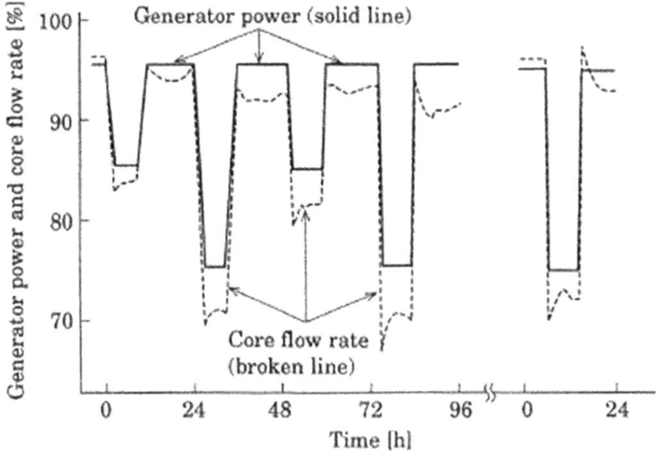

Fig. 7.14 Example of daily load-following operation (14-1-8-1 system)

The power change speed of the BWR is maximum 30 %/min (maximum 60 %/min for the latest model ABWR), which shows high-speed load-following performance.

Meanwhile, the reactor pressure and reactor water level are controlled stably by the reactor pressure control system and the feed-water flow rate control system as described in Sect. 7.2 in this chapter.

[2] In-house standalone operation of full capacity turbine bypass plant

In Japan, there is a BWR plant with the TBV capacity that is capable of accepting the entire steam flow rate flowing through the main turbine while the reactor is operating at its rated power. In case the electrical system malfunctions and the electric load is interrupted, this full capacity turbine bypass plant is designed so that the reactor does not scram (auto shutdown) but continues partial power operation at a level of approximately 30 %. This allows the house-service electric power to be supplied by the main generator, and the operation is switched to so-called in-house standalone operation. The electric power supplied by the main generator is also used to start neighboring BWR plants that have been shutdown automatically.

Figure 7.15 shows the major interlock sequence the plant when the generator load is interrupted and Fig. 7.16 show the behaviors of the reactor major parameters when the generator load is interrupted.

As shown in Fig. 7.15, when the generator load is interrupted, the main turbine steam control valve (TCV) closes rapidly to prevent over-speeding of the turbine. The rapid closing of the TCV causes rise of the reactor pressure and the neutron flux. To prevent this, the TBV with full capacity opens rapidly. Then, the reactor recirculation pump trip (RPT) and the selected control rod insertion (SRI) occur nearly simultaneously, and the reactor power is lowered rapidly to switch to partial power operation at a level of approximately 30 % power.

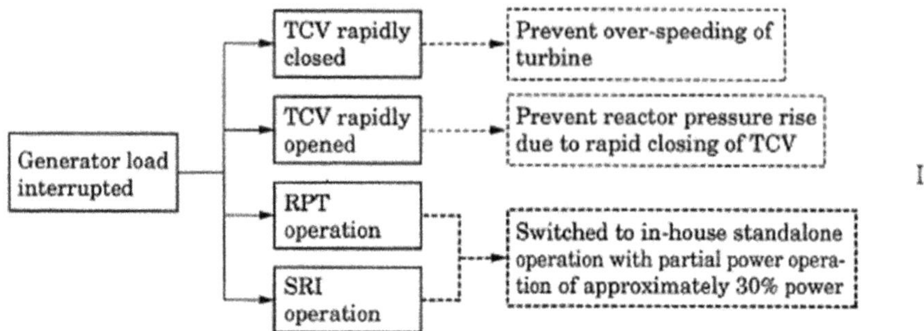

Fig. 7.15 Major interlock sequence when the generator load is interrupted

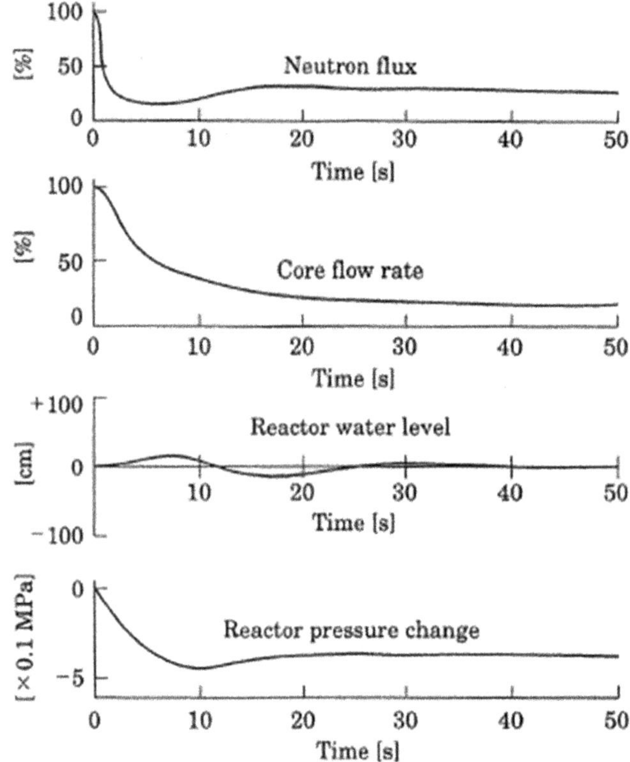

Fig. 7.16 Behavior of the reactor major parameters when the generator load is interrupted

Figure 7.16 shows the behaviors of the reactor major parameters after the generator load is interrupted. A small rise of the reactor pressure is observed even at the point immediately after the generator load is interrupted. The neutron flux does not exceed the initial value at any time, and reactor power lowers down to approximately 30 % of power in a stable manner. The core flow rate is lowered by the RPT to stabilize to natural circulation conditions. No remarkable variation of the reactor water level is observed due to the feed-water flow rate control system, and a near normal level of operation water is maintained.

As described above, none of the neutron flux, the reactor pressure or the reactor water level reaches the set value that causes a reactor scrum, and the parameters are stabilized, and the reactor is switched to in-house standalone operation.

7.5.2 Control When Abnormal Condition Occurs

[1] Concept of Control in Abnormal Condition
If an abnormal condition develops exceeding the specified control range of the control systems as described in Sect. 7.3 of this chapter, the plant is designed to shut down automatically for safety purposes and reactor safety is maintained without depending on the manual operation by the operator (so-called 10-min rule).

1) Reactor power
Heat generation of the reactor is stopped immediately to prevent the fuel rods from breaking.

<1> Reactor scram
Emergency shutdown of reactor by automatic insertion of all control rods.

<2> SLC system
When an anticipated transient without scram occurs, boric acid solution is injected into the core to shut down the reactor.

<3> RPT
The RPT is activated along with operation of the reactor scram function, to rapidly decrease the core flow rate and therefore to lower the reactor power.

2) Reactor water level
When the abnormal condition that the feeding of water by the normal feedwater and condensate system is unavailable, this function ensures the coolant to be fed to the core.

<1> RCIC system
When the feedwater and condensate system function is lost, to prevent the reactor water level from lowering excessively, the RCIC operates to feed water into the core.

<2> ECCS
Even if the reactor water level drops excessively to the extent that the fuel is exposed due to a pipe rupture accident (loss of reactor coolant) around the reactor, the ECCS operates to inject water into the core to reflood the reactor core to satisfy the design permissible value (peak cladding temperature, etc.)

3) Reactor pressure
   Excessive pressure rise is prevented to protect the reactor pleasure vessel from breaking.

   <1> TBV
   To prevent the reactor pressure from rising excessively due to a main turbine trip, etc., the TBV is opened to control the reactor pressure to constant.

   <2> Safety relief valve (SR valve)
   If the reactor pressure rises excessively due to a closure of the main steam valve, etc., the SR valve is opened to prevent the reactor pressure from rising.

4) Containment atmosphere
   When a loss of reactor coolant accident occurs, radioactive materials are prevented from being released to the environment.

   <1> Containment vessel spray
   When a loss of coolant accident occurs, the containment vessel spray is performed to suppress leakage of the atmosphere inside the containment vessel to the permissible value and to remove radioactive materials from the atmosphere.

   <2> FCS
   This system recombines the hydrogen and oxygen gas that are generated after a loss of coolant accident to suppress the hydrogen and oxygen gas concentration inside the containment vessel to the inflammable limit so that the integrity of the containment vessel is ensured.

[2] Typical analysis example of plant control in abnormal condition

The following explains the control (moderation) functions when an abnormal condition develops as described above, based on the typical analysis result of an abnormal event.

1) Analysis example of reactor power/reactor pressure control in abnormal condition (Main turbine trip event)

Figure 7.17 shows the analysis example of the plant behavior when the main turbine trip occurs for some reasons, and Fig. 7.18 shows the progress flow chart of this event.

As shown in Figs. 7.17 and 7.18, when the main turbine trips, the main turbine steam stop valve (Tb stop valve) closes rapidly so that the flow of steam from the reactor to the main turbine is blocked immediately. At the same time, the reactor pressure is raised and the core void fraction is reduced, and the positive reactivity is inserted into the core.

This causes a temporary sharp rise of the neutron flux; however, the "reactor scram" and the "reactor RPT" as described in [1]-1) of Sect. 7.5.2 operate automatically to reduce the neutron flux sharply. Therefore, damage to the fuel

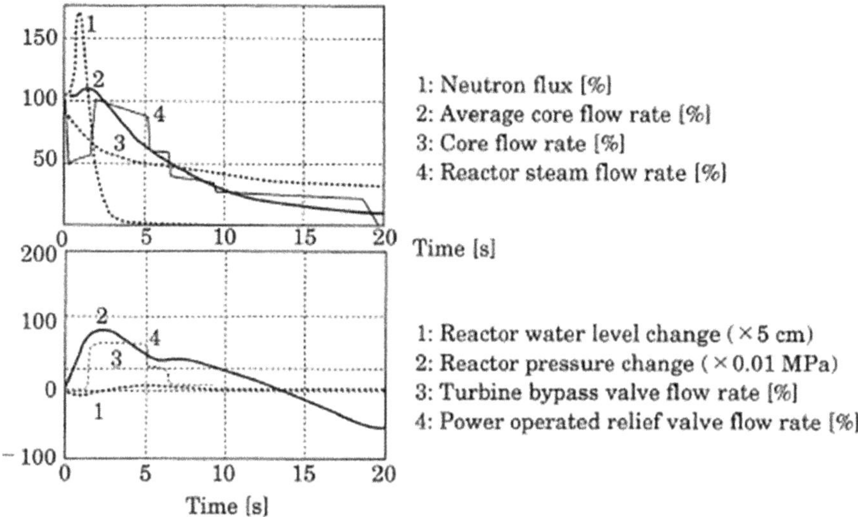

Fig. 7.17 Analysis example of main turbine trip event

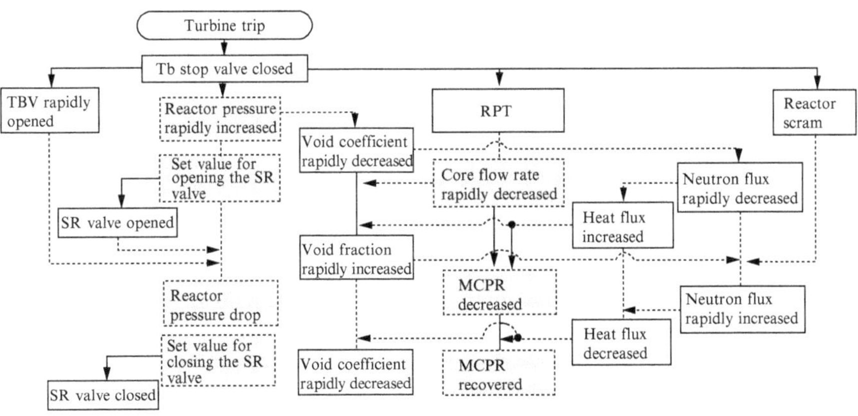

Fig. 7.18 Progress flow chart of main turbine trip event

is prevented. In addition, the RPT operates to decrease the core flow rate as shown in Fig. 7.17.

Although the reactor pressure rises temporarily, it is decreased as the "TBV" and the "safety relief valve" as described in [1]-3) of Sect. 7.5.2 automatically operate. Therefore, the integrity of the reactor pressure vessel and other components is not affected.

2) Analysis example of reactor water level/containment atmosphere control in abnormal condition (Loss of reactor coolant accident: major rupture of reactor recirculation piping)

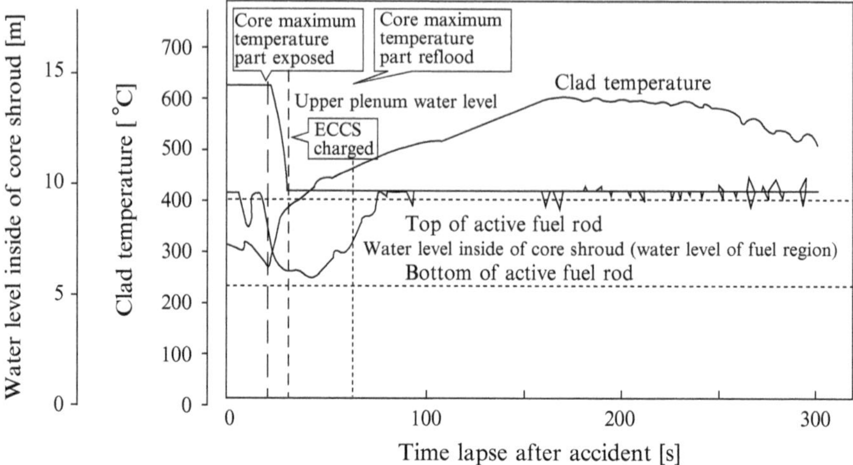

Fig. 7.19 Analysis example of loss of coolant accident (major rupture of reactor recirculation piping) – 1

Figures 7.19 and 7.20 show the analysis example when a major rupture of the reactor recirculation piping occurs (simulated event).

As shown in Fig. 7.19, the coolant rapidly flows out from the reactor and the reactor water level drops, and the fuel region inside the core shroud is exposed temporarily, causing a rise in the fuel cladding temperature. However, as the "ECCS" described in [1]-2) of Sect. 7.5.2 automatically operates to reflood the exposed fuel region to cool down, the fuel cladding temperature is prevented from rising. Therefore, the core integrity is ensured.

On the other hand, the pressure and temperature inside the containment vessel for reactor containment undergo a temporary sharp rise by the major rupture of the reactor recirculation piping as shown in Fig. 7.20. However, then the coolant flowing out through the rupture decreases (the water level drops to the location where the pipe is rupture, and the coolant turns to steam) and the pressure inside the containment vessel lowers. Furthermore, when unsaturated water injected by the ECCS flows out from the rupture, the steam inside the containment vessel is condensed and the pressure drops rapidly. In addition, the containment vessel spray as described in [1]-4) of Sect. 7.5.2 operates to maintain the pressure and temperature inside the containment vessel low to prevent leakage of the containment atmosphere.

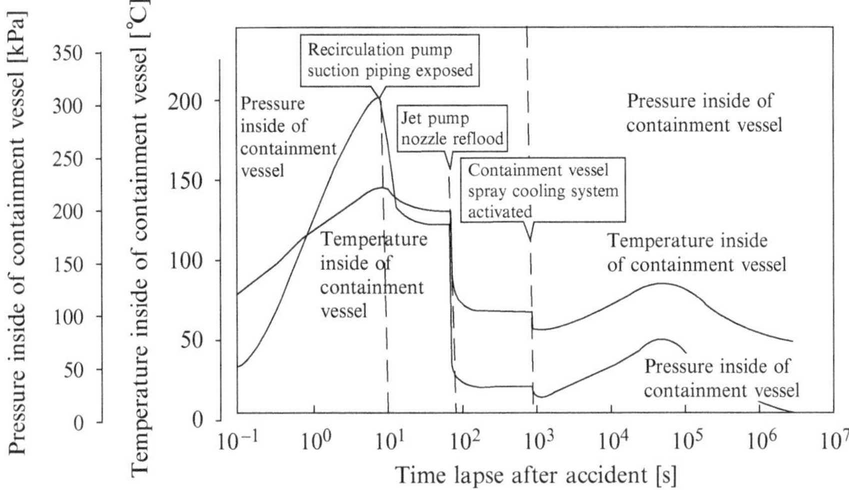

Fig. 7.20 Analysis example of loss of coolant accident (major rupture of reactor recirculation piping) – 2

7.6 Central Control Panel

The central monitoring control panel of the nuclear plant has a critical role to ensure the reliability, and stable and safety operation of the plant. The structure and function of the control panel has been improved.

The first-generation central control panel (1970–) was a bench type control panel, on which all gauges and switches are arranged. The second-generation central control panel (1985–) was developed to reduce the burdens imposed on the operator and to prevent human errors. The control panel was designed based on ergonomics, the computing machine and cathode ray tube (CRT) display terminal devices were adopted, and the man–machine interface was improved. In addition, CRT displays were used extensively, and the start/stop operations of the various components associated with starting and stopping of the reactor were automated. For the third-generation central control panel (1996–), the monitoring performance, reduction of burdens to the operator, and human error prevention were further improved, which is currently used for the ABWR plant operation.

The following explains the latest version of the central control panel (third-generation central control panel).

7.6.1 Operation Monitoring Function

The operation is monitored using the small-size operation console that the operator can monitor and operate while being seated, and the large-size display panel that is located in the back of the console and provides information to all of the operators in the central control room in an integrated fashion. The CRT displays, FDs (flat displays), hard switches, and indicators are arranged in each of the console and the display panel. The third-generation central control panel features the small number of hard switches and extremely large number of CRT displays and FDs. The touch operation device is incorporated into the CRT display, so that it senses a touch input. The FD consists of a 10-in. color liquid crystal display and a controller. The FD has the capability of either recognizing touch-input sense or display only. The FD is used as a backup to the safety system operation and CRT display.

[1] Operation consoles

The safety system components are located in the left wing, the Balance of Plant (BOP system) components in the right and the main control system components in the center of the operation console. Seven CRT displays and 17 FDs are arranged in the console.

Using the CRT displays, operators can monitor the operation of the systems required for starting and stopping the plant except the main control system, safety system, and power supply system. One CRT display is arranged in the left wing, five in the center and one in the right wing. Operators can arbitrary select monitor items on the display.

FDs are used for monitoring and operating various systems in place of switches and indicators equipped with the conventional control panel. The FD provides the display function of the system for indicating the operating conditions, parameter display function for monitoring the process conditions, trend record display function, and touch operation function for operating the equipment.

[2] Large-size display panel

The large-size display panel consists of the critical alarm display section, plant monitoring panel, system batch alarm display section, variable screen display section, FD and hard switch controls as shown in Fig. 7.21.

Critical alarms related to the plant operation are displayed in the critical alarm display section. The display section consists of the following four parts.

1) First hit display
   The first signal that the process calculator detects the change in operation contact is displayed for the cause of the following four major events.
2) Major four events display
   The display part is made up of the following windows that led to a plant trip.

   <1> MSIV
   <2> Scram

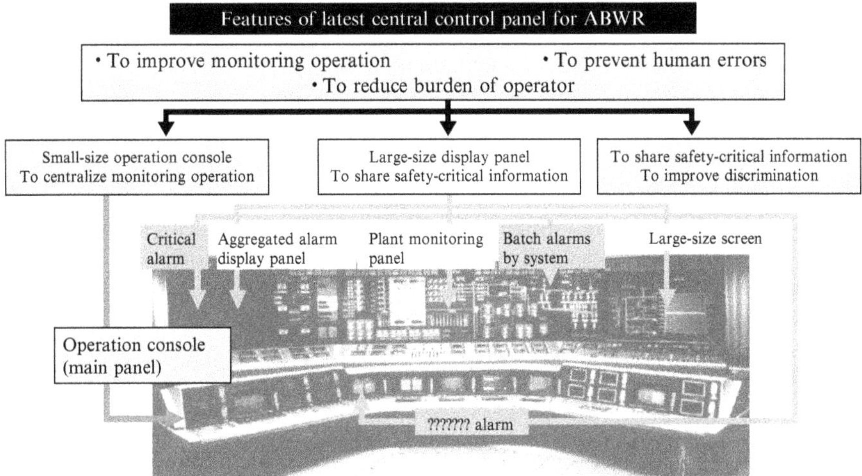

Fig. 7.21 ABWR third-generation central control panel

<3> Turbine trip
<4> Generator trip

3) Plant abnormality/safety system operation alarm
   The display section is made up of windows that show activation of the plant protection logic that accompanies a power control system such as that of the safety system and RPT, or one that activates the safety system.
4) Critical alarm on plant control system
   This display part is made up of windows showing critical alarms that do not activate the safety system of, for example, the high radioactivity and leak detection system but that are critical to ensuring safety and operational control of the plant. In the plant monitoring panel, the major parameters on the plant are displayed by overlaying the plant configuration diagram.

These parameters are the signals input directly from the control calculator without intervening of the process calculator. Therefore, even if the process calculator is shutdown, monitoring can be continued. The following elements are displayed in MIMIC.

1) Symbol display

The major pumps, isolator, generator, D/G states are displayed by symbols in different colors. When the state changes, the symbol color is changed and the symbol flickers for 10 s to alert the operator.

2) Bar chart display

The reactor water level and the neutron flux level are displayed in the bar chart. The bar chart is displayed in green in normal condition. When the parameter exceeds the alarm set value (level 3, level 8), the color is then switched to red.

3) Digital display

The major parameters on the plant operation are digitally displayed on the MIMIC.

4) Others

When a safety-critical signal (all of the control rods are inserted, PCIS isolation, RPT, etc.) is generated, it is displayed with a lamp.

The system batch alarm display section is located above the fixed mimic and the variable screen. It notifies an abnormality and state change of the corresponding system. There are three patterns of the display; major malfunction (red), minor malfunction (yellow), and state display (green). The state display (green) lights up if the system/device state is different from the normal state (bypass, etc.)

The variable screen display section is a 110-in. rear projection-type large-size screen. The data on the CRT display is displayed on this large screen to share information among the operators. There are four patterns of the timing and the screen to be displayed.

1) Screen of the CRT display for which "screen display" is selected.
2) If the screen master of the chief desk is set to "ON," the corresponding screen is displayed every time the break point is selected for checking the progress of automated operation.
3) If the screen alarm display of the chief desk is set to "ON," when an alarm state is generated, the associated alarm screen is displayed.
4) The emergency trend screen (reactor pressure, reactor water level, etc.) is displayed when any of the major four events occurs.

There are 32 units of FDs in the large-size display panel. The FDs function is to back up the operation console FD and CRT displays. The functions of FDs are classified into the safety system FD (10 units), reactor uninterruptible power supply system FD (4 units), BOP system FD (10 units), in-house power supply FD (5 units), and alarm display FD (3 units).

Some of the hard switches that cannot be arranged in the operation console, due to operation frequency or space, are located in the large-size display panel. Examples of those hard switches include the channel bypass switch of the nuclear instrumentation system.

[3] Overall configuration of third-generation central control panel

Figure 7.22 shows the overall configuration of the third-generation central control panel.

The data displayed in the central control room is forwarded from the process computer, safety protection system, power control system, NSSS/in-house power supply system, and turbine generator system via the optical network. These systems are connected to the field site by the RMU (remote multiplexing unit) via the optical network. The RMU and the filed sites are wired for signal-based communication.

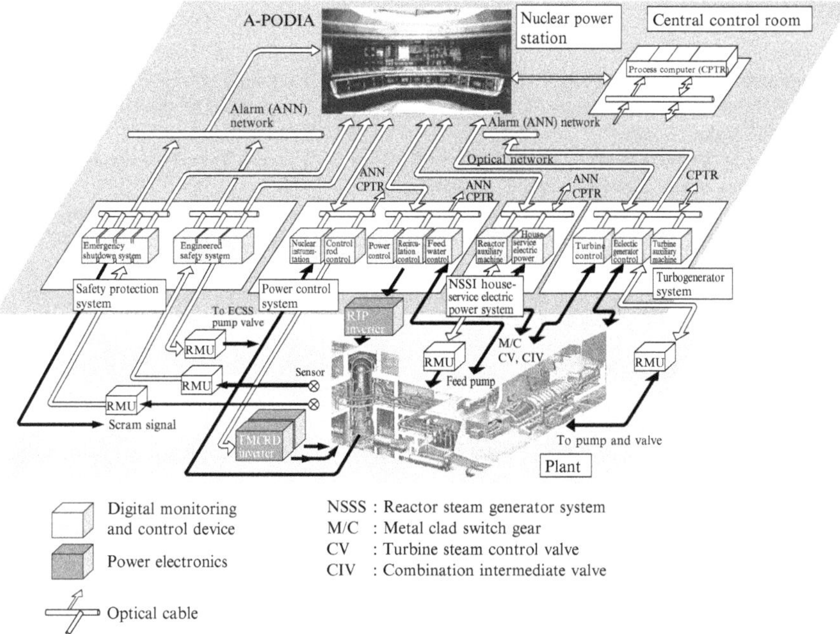

Fig. 7.22 ABWR total digital system

Chapter 7 Exercises

1. The critical operation for staring and stopping the plant is to change the reactor power. Name 3 systems used for monitoring or maintaining integrity of the fuel during the operation.
2. Name 2 means for controlling reactivity for maintaining constant power at rated operation of the BWR.
3. Give 3 major moderation functions for preventing the reactor power from rising when an event occurs that the reactor power rises excessively.
4. Give 2 major moderation functions for preventing the reactor pressure from rising when an event occurs that the reactor pressure rises excessively.
5. The central control panel of the first generation had been improved and the third-generation central control panel is currently being used. Give 2 purposes of the control panel what is improved for the operators.
6. Give 2 display functions equipped with the large-size display panel of the third-generation central control panel for sharing information among operators.

Bibliography

1. Atomic Energy Society of Japan (1984) C43 BWR power plant daily power adjusting operation (1). Results of continuous operation in various mode. In: Proceedings of the 1984 conference, Higashi-Osaka, Japan
2. Atomic Energy Society of Japan (1986) E33 Kinetics of full capacity turbine bypass plant. In: Proceedings of the 1986 conference, Kyoto, Japan
3. Fukushima Daiichi Nuclear Power Station (1992) License application for nuclear power plant (complete copy of plant no. 5), body and attachment (attachment 8 and attachment 10) (Aug 1992)
4. Tomisawa T, Tone Y, Fujii M (1997) Electrical, instrumentation and control system of advanced boiling water reactor (ABWR). Toshiba Rev 52(4):20–23

Chapter 8
Actual Operation and Control of Pressurized Water Reactor

Shuhei Miyake, Toshihide Inoue, and Satoshi Hanada

8.1 Method of Operation Control of PWR

The operation control of a power plant maintains or changes the generator power or the volume of electricity supplied steadily to the grid system as appropriate. Various types of automatic control systems are installed to maintain the processing volume of the plant equipment stably within an appropriate range when the power plant is operated under operation control. The pressurized water reactor (PWR) has automatic control systems built for the reactor and turbine systems separately because the primary and secondary coolants of the reactor and turbine systems are separated from each other by the steam generator.

To change the generator power in a PWR plant, the output from the turbogenerator system is adjusted first, then the reactor system is controlled to follow the change in the turbine load. This is called the "Turbine main, reactor subordinate" control method.

For the reactor system, the change in plant output is equivalent to the change in the steam flow consumed by the turbine, i.e., the turbine load. So the principle of operation control of the reactor system involves adjusting the volume of heat generated by the reactor according to the change of the heat consumed by the turbine.

[1] Overview of the reactor control system

The reactor control system, coupled with the self-regulating characteristics unique to the reactor, maintains stably the heat generation in the reactor and the heat transfer and steam generation in the steam generator, and attenuates transient changes caused by the change in the turbine load without reactor trip, and recovers and maintains the equilibrium condition. The reactor control system is configured by the following individual control systems. Figure 8.1 shows the overall configuration of the system.

<1> Control rod control system
<2> Boron concentration control system (manual)
<3> Pressurizer pressure control system

Y. Oka and K. Suzuki (eds.), *Nuclear Reactor Kinetics and Plant Control*, An Advanced Course in Nuclear Engineering, DOI 10.1007/978-4-431-54195-0_8, © Springer Japan 2013

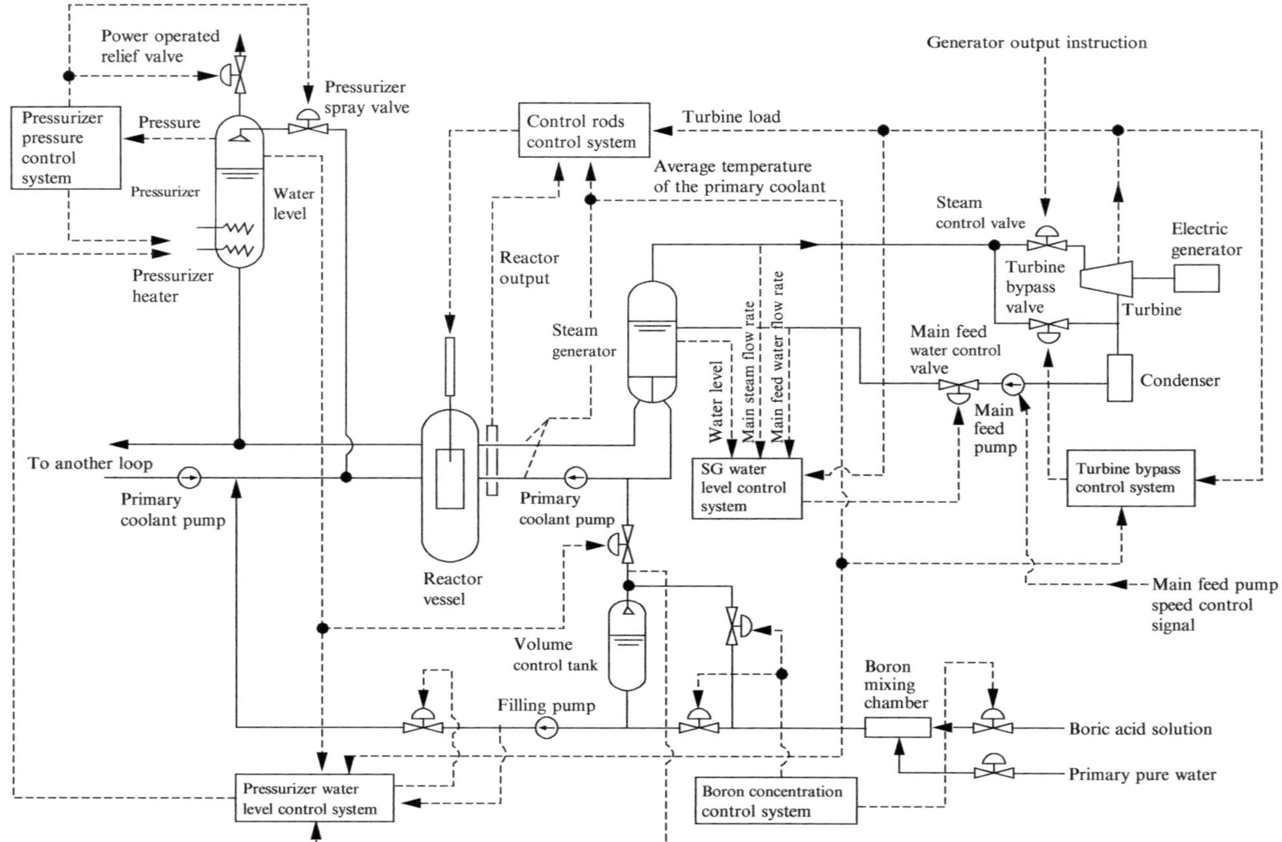

Fig. 8.1 Overall configuration of reactor control system

<4> Pressurizer water level control system
<5> Steam generator water level control system
<6> Turbine bypass control system

The reactivity of the core is controlled by adjusting the position of the control rod cluster and the concentration of boron in the primary coolant. The reactivity is controlled against changes in plant output, including those in the turbine load, by the control rod control system that adjusts the positions of the control rods. The relatively slow changes in the reactivity of the core over a long period, such as combustion of fuels and changes in the concentration of xenon, are compensated for by adjusting the concentration of boron in the primary coolant. The concentration of boron is also adjusted in order to adjust the positions of control rods for the purpose of maintaining the output distribution in the axial direction in the core within a targeted range in normal conditions. The concentration of boron control system is operated manually by the operator for relatively slow and intermittent operations.

The control rod control system makes the reactor output follow the changes in plant output and controls the average temperature of the primary coolant (T_{avg}) to match it to the preset reference temperature. Regarding PWR plants in Japan, the reactors are designed so that they can follow the changes in load as described below within a power range of 15–100 %: these three types of changes in load, called designed load changes, have been established as a standard that meets the requirements of operation of the electricity system.

<1> Stepwise change in load at a 10 % step
<2> Rampwise change in load at 5 %/min
<3> Sharp stepwise decrease in load at a 50 % (or 95 %) step (when the turbine bypass is also being controlled)

The pressurizer pressure control system and the pressurizer water level control system control the pressure and water level in the pressurizer, respectively, in order to maintain the pressurizing function of the reactor system. The steam generator water level control system controls the feed-water flow to maintain the water level on the secondary side of the steam generator at the reference value so that the function to remove heat from the reactor system can be maintained.

The turbine bypass control system secures temporary removal of heat from the reactor system when the turbine load decreases rapidly by a large margin, suppressing excessive increase in temperature of the primary coolant due to imbalance of output between the primary and secondary systems caused by the sharp decrease in the load.

[2] Reactor following system

In the PWR plants, if an instruction of generator output is given, the turbine load of the secondary system is adjusted first, then the reactor output of the primary system follows the adjustment. This is called the "reactor following system."

The PWR applies pressure on the core coolant, which also serves as the moderator, to maintain its single liquid phase, thus minimizing the effect on the reactivity of the core caused by changes in the moderator density occurring when the pressure in the reactor changes. The steam required to drive the turbine needs to be generated by a steam generator because the primary coolant is pressurized in an unsaturated state. However, installing the steam generator has advantages to ease up the changes in the volume of heat removed caused by the changes in the turbine load due to the heat capacity of the water held by the steam generator and the steam generator tube, thus slowing down the changes in the temperature and pressure of the primary coolant. As a result of such a system configuration, the changes in the core reactivity or the temperature and pressure of the primary coolant become slower even if temporary imbalance occurs in the output between the primary and secondary systems, ensuring there is time to control the reactor output after the turbine load changes. Such an easing-up effect on the process changes caused by the steam generator makes it possible to adopt the reactor following system for the PWR plants.

In addition, the balance of output between the primary and secondary systems can be detected in changes in T_{avg} because the coolant has a single liquid phase. For example, a turbine load that is smaller than the reactor output causes the volume of heat to accumulate in excess in the reactor system, thus putting T_{avg} in an uptrend. On the contrary, a turbine load larger than the reactor output causes the reactor system to cool, thus putting T_{avg} in a downtrend. Thus, the difference between the volume of heat generated by the reactor and that consumed by the turbine emerges as the changes in the average temperature of the primary coolant system. This is the reason why T_{avg} is used as the main signal for controlling the reactor output.

[3] Primary coolant average temperature program

To control the reactor output of the PWR, a method is used where the T_{avg} is controlled targeting at the reference temperature (T_{ref}) which is set as a function of the turbine load. This reference temperature is programmed so as to optimize the effect on the primary and secondary system equipment as mentioned below.

The steam generator is required to generate and feed the steam flow meeting the requirement from the turbogenerator at the predetermined level of steam pressure. The steam generated by the steam generator is saturated, so the steam pressure is decided substantially by the saturation temperature on the secondary side of the steam generator. The thermal output from the plant is the difference of temperature between the primary and secondary sides of the steam generator, and it can be roughly expressed as follows:

$$\text{output} = (UA)_{SG} \cdot (T_{avg} - T_s) \tag{8.1}$$

where $(UA)_{SG}$: (steam generator heat transflux coefficient) × (heat-transfer area of heat-transfer tube);
T_{avg}: average temperature of the primary coolant;
T_s: temperature of the steam of the steam generator.

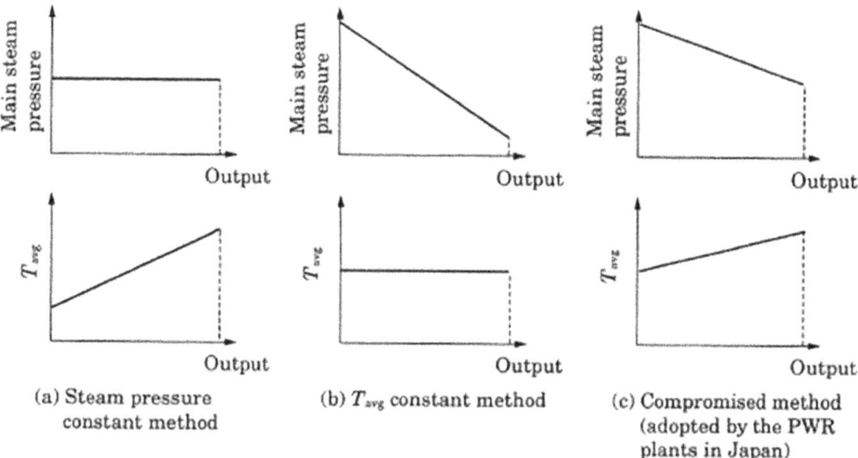

Fig. 8.2 Primary coolant average temperature program method

Also, the reactor output has the relationship with the difference of temperature between the fuel and the primary coolant substantially as follows:

$$\text{output} = (UA)_{\text{f}} \cdot (T_{\text{f}} - T_{\text{avg}}) \tag{8.2}$$

where $(UA)_{\text{f}}$: (equivalent heat transfer coefficient of fuel) × (surface area of fuel);
T_{f}: average temperature of fuel.

Equations (8.1) and (8.2) show the relationship of the plant output with the temperature of the reactor system and steam generator secondary side. The first idea to set the average temperature for the primary coolant is a method which satisfies the constant steam pressure or the constant average temperature of the primary coolant against the change in output.

The former method (Fig. 8.2a) is ideal for the conditions for designing the turbine system in that a constant steam pressure is obtained regardless of output. However, it requires the temperature of the primary coolant and the fuel to increase significantly as the output increases. To compensate for the negative feedback of reactivity caused by such a significant increase in temperature, it is necessary to install a control rod cluster with large enough capacity to control reactivity and a big pressurizer that can contain the expansion and contraction of the coolant caused by the change in the temperature of the primary coolant. This will increase the burden on the primary system equipment. On the other hand, the latter method (Fig. 8.2b) is favorable for the conditions required for designing the primary system in that the equipment of control rods and pressurizer can be minimized. However, it has disadvantages in which the secondary system equipment is excessive against a low level of steam pressure during the rated operation that accounts for almost the entire operation period because the pressure resistance of the equipment must be designed against the pressure when the output is zero. Currently, a compromized method (Fig. 8.2c) harmonizing those two methods has been adopted by the PWR plants in Japan to optimize the primary and secondary system equipment.

[4] Self-regulation and the control of reactor output

The PWR is designed to have negative reactivity coefficients against the rise of the temperature of fuel or the primary coolant. This gives the PWR the self-regulating characteristics where the changes in the reactor output are suppressed because of the negative feedback of reactivity caused by the increase in temperature of fuel, or the moderator carries out in a way to suppress the change in the reactivity against disturbances that cause increased reactor output or temperature of the primary coolant. In that process, the effect of the reactivity feedback caused by the temperature of moderator (moderator temperature coefficient) is affected by the concentration of boron in the coolant. At the beginning of the reactor's life, the concentration of the boron in the coolant is high so as to compensate for the excess reactivity of the fuel. Accordingly, a change in the temperature of the primary coolant in the core causes a significant change in the density of boron due to the change in the density of water. In other words, the negative reactivity effect caused by the decreased density of water (moderator) due to the rise of the coolant temperature is set off by the positive reactivity effect caused by the decreased density of boron (neutron absorber), thus resulting in a small value of the moderator temperature coefficient in total, though it is negative. At the end of the reactor's life, the concentration of boron in the coolant is low, causing the reactivity effect due to the change in the density of water to be dominant. This leads to a large value of the moderator temperature coefficient. As mentioned above, the PWR maintains a negative value for the moderator temperature coefficient during the whole period of power operation although the coefficient may change depending on the concentration of boron during the operation period. This negative coefficient and the effect of the negative reactivity feedback (Doppler coefficient) against raised fuel temperature constitute the self-regulating characteristics of the reactor, serving as the base of the inherent safety of nuclear plants.

Here, we will consider a case where the turbine load is slightly decreased from the equilibrium condition of the plant. T_{avg} is raised by the discrepancy in output between the primary and secondary systems, causing the effect of a negative moderator temperature coefficient, which ultimately reduces the reactor output recovering the balanced outputs from the primary and secondary systems (Fig. 8.3a).

In other words, the balance of the outputs from the primary and secondary systems is maintained autonomously by self-regulation unique to the reactor without any control operation by the reactivity control system. As a result, T_{avg} is stabilized at a temperature higher than that at the initial condition, although it leads to a temperature value that is different to the referential temperature given by the T_{avg} program.

The primary and secondary system equipment of the PWR plant is designed under the design requirement that T_{avg} is controlled to match to the referential temperature that corresponds to power. Accordingly, just expecting the core to self-regulate is a not sufficient expectation with regard to maintaining T_{avg} at the referential temperature; the control operation of the reactivity control system is

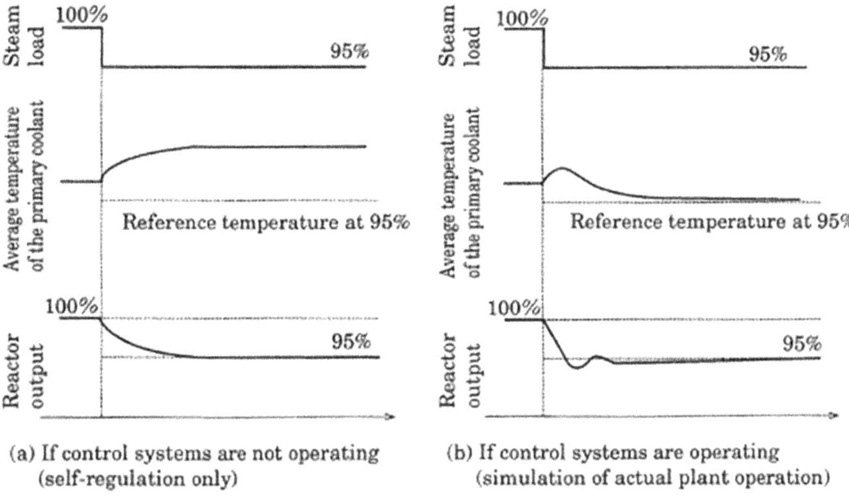

Fig. 8.3 Example of plant response to decrease in load (5 % down)

required to reduce the reactor output to a value lower than the turbine load temporarily, and thus to lower the T_{avg} raised transiently in the beginning period of load decrease to the referential temperature (Fig. 8.3b). The control rod control system is designed to meet such a functional requirement.

[5] Control rod cluster

The control rod cluster has a structure as shown in Fig. 8.4 and uses Ag–In–Cd as the neutron absorber.

It is installed at a rate of approximately one per four fuel assemblies, and approximately 40 % of all the control rod clusters, called the "shutdown group," is held at the fully extracted position during normal operation. The remaining clusters, called the "control group," are used for output control, divided into four banks ranging from banks A through D and driven on a bank basis. As shown in Fig. 8.5, they are extracted or inserted in the order from bank A to D or from bank D to A, respectively.

The reactivity value per one step of control rod position (differential reactivity value) gets smaller around the full extraction or insertion. So, each of the control banks has an overlap with the next in the driving sequence in order to flatten as much as possible the differential reactivity values across the driving sequence of the control group.

In case of the reactor trips, all the control rod clusters are inserted into the core by their own weight with the power source of their drivers cutoff. During the power operation, it is possible to maintain the control rods at any insertion position by, in principle, adjusting the concentration of boron. To secure the shutdown reactivity in case of the reactor trip, however, a "limit of insertion of control rods" is set at a position where the reactor can be put into the hot shutdown condition with an

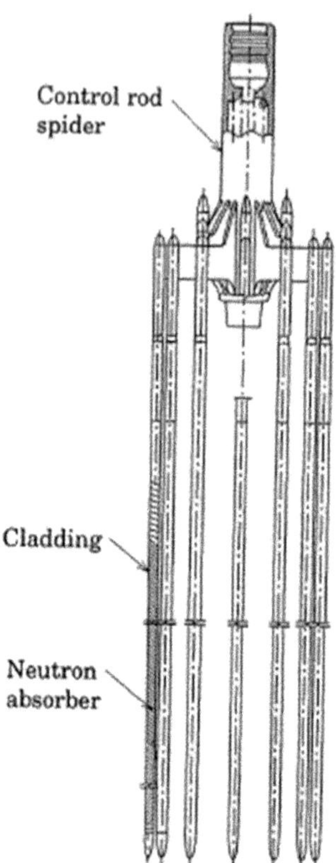

Fig. 8.4 Control rod cluster

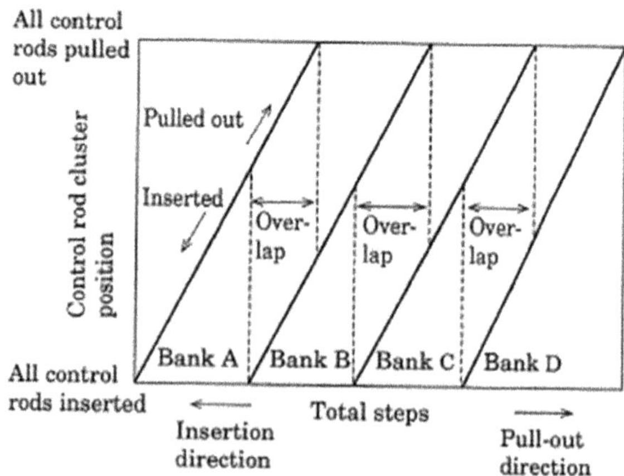

Fig. 8.5 Control rod driving sequence

enough shutdown margin even if a control rod cluster of the maximum value gets stuck and is not able to be inserted. It is necessary to operate the control rods at a position separate from this limit of insertion in the direction of extraction while the

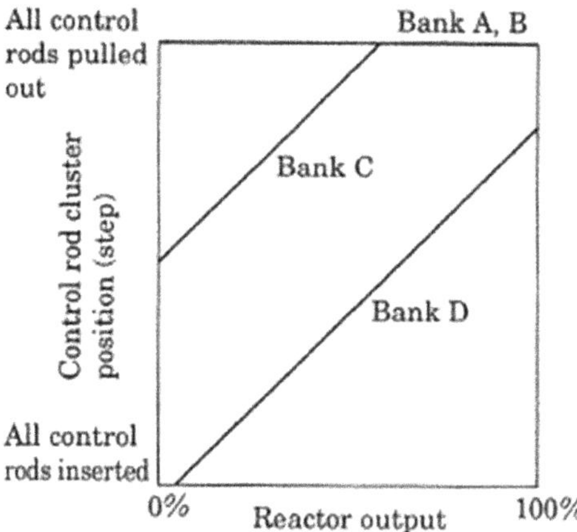

Fig. 8.6 Limit of control rod insertion

reactor is operating. The limit of insertion is given as a function of reactor output as shown in Fig. 8.6 because the reactivity required for the reactor shutdown needs to compensate for the deficiency in output.

8.2 Operation Control When Starting/Stopping the PWR

8.2.1 Plant Startup Operation

The plant startup operation covers the processes from switching its cold shutdown condition to the hot shutdown condition until it reaches the rated output condition after the reactor attains criticality and the generator is integrated.

Before the plant startup operation, the pre-startup check is carried out according to a written procedure. The check is made in the central control room and at the site for the system configuration (recovery of the systems and devices that had been isolated for checking, repair, etc.; completion of water filling and venting), conditions of devices (valves opened/closed, power source configuration, operation conditions, etc.), and various kinds of parameters (major parameters, tank water level, radiation level, etc.). The check also covers the conditions for starting up the plant (primary coolant water quality, etc.) and the functions and operating conditions of the safety system. Following that, further checks are carried out to ensure that the pre-startup check of the measurement devices, power unit, reactor control protectors, the control rod cluster driver, and the control rod cluster position indicator has been completed and that those devices are all in normal operating condition.

Upon completion of the pre-startup check, the plant startup operation is initiated. The pressure-raising operation required to start up the primary coolant pump is started first because the temperature of the primary system of the PWR plant can be raised easily by the Joule heating operated by the primary coolant pump. It should be noted that the temperature or pressure is not raised by nuclear heating starting from the cold shutdown condition in the PWR plant because boron is diluted and then the criticality operation is conducted after the temperature is raised to approximately 292 °C (hot shutdown condition) where the reactivity feedback characteristics comprising the Doppler coefficient, moderator temperature coefficient, etc., are confirmed to have become negative.

[1] Starting up the primary coolant pump

No primary coolant pump is operated so long as the pressure of the primary coolant system is lower than approximately 2.7 MPa so as to secure the seal with the amount of differential pressure required to seal the axis seal soundly.

The pressure of the primary coolant system is raised by regulating the flow filled or extracted by the chemical and volume control system (CVCS). Filling more coolant than the extracting flow raises the pressure because the primary coolant system is filled with coolant at this point. The pressure can also be raised by the pressure control with the steam phase formed in the pressurizer. In this case, however, it is necessary to raise the temperature of the water in the pressurizer up to the saturation temperature at the relevant pressure. This causes an excessive temperature difference to form between the primary coolant system, whose temperature is low, and the water in the pressurizer. This is an unfavorable development from the viewpoint of the evaluation of fatigue of connected piping, and due to this, no plants in Japan have adopted this approach. The gas phase is formed in the pressurizer after the temperature of the primary coolant system reaches approximately 177 °C.

The pressure of the primary coolant system is raised until it reaches approximately 2.7 MPa, then all the primary coolant pumps are started up with that pressure value maintained.

[2] Raising temperature (to approximately 177 °C)

The primary coolant temperature is increased by the pressurizer heater that is put in at the time the primary coolant pumps are started. When the temperature is raised, it is necessary to comply with the pressure/temperature limit curve applicable to the heating of the plant and to control the heating rate with the residual heat removal system so that the predetermined heating rate may not be surpassed from the viewpoint of prevention of inductile destruction of components.

[3] Regulating water quality

The water quality of the primary coolant is regulated to attain the tolerable level by adjusting the oxygen concentration, etc., before the temperature of the primary coolant is raised over approximately 80 °C, because the components of the primary coolant system are susceptible to corrosion in high-temperature and high-oxygen-concentration conditions. Also, the gas phase section of the volume control tank is replaced with hydrogen, then adjustment is carried out so that the primary coolant will have the predetermined value of the dissolved hydrogen. Thus, any oxygen that may be generated by the radiolysis of the primary coolant is returned to water by its reaction with the dissolved hydrogen, suppressing the rise of concentration of oxygen in the primary coolant.

[4] Forming gas phase in the pressurizer

When the temperature of the primary coolant system reaches approximately 177 °C, the pressure control of the primary coolant system is switched from the extracting pressure control to the pressurizer pressure control by forming the gas phase in the pressurizer.

The gas phase is formed in the pressurizer by continuing to raise the temperature of the water in the pressurizer with the pressurizer heater while maintaining the pressure of the primary coolant system at approximately 2.7 MPa. Steam begins to be generated in the pressurizer when the temperature of the water in the pressurizer attains the saturation temperature at the relevant pressure (approximately 231 °C). Then the water in the pressurizer can be maintained at any level lowered by increasing the extracting flow or decreasing the filling flow.

Once the gas phase is formed in the pressurizer, the pressure of the primary coolant system is controlled by controlling the pressurizer pressure with the pressurizer spray and heater, and the regulation of filling flow for raising pressure applied until then is now applicable to other purposes than the control of the pressurizer's water level. Also, the residual heat removal system whose maximum working pressure is low is isolated from the primary coolant system in preparation for the pressure-raising phase thereafter. The steam generator is used for controlling the temperature of the primary coolant system after the residual heat removal system is isolated.

[5] Raising temperature and pressure (to approximately 177 °C and 2.7 MPa)

After the gas phase is formed in the pressurizer, the temperature and pressure are raised continuously by the heat input from the primary coolant pumps and pressurizer heater with the pressure/temperature limit curve applicable to the heating of the plant complied with and the prescribed heating rate not surpassed, until the reactor reaches the hot shutdown condition (approximately 292 °C and 15.41 MPa).

[6] Criticality

Once the reactor reaches the hot shutdown condition, its criticality operation is started. During the operation to raise temperature and pressure, all the control rods of the shutdown group remain extracted outside the core, and those of the control group remain inserted in the core. The critical boron concentration at the target control bank position (higher than insertion limit) when criticality is reached is calculated, then the concentration of boron in the primary coolant is diluted until the critical boron concentration is attained. After that, the control rods in the control bank are extracted manually to attain the criticality.

Almost in tandem, the main steam piping is warmed up to avoid thermal shock to the main steam pipes and the secondary system equipment.

[7] Raising load and rated-power operation

The reactor power is raised by extracting the control rods or diluting the concentration of boron in the primary coolant. Almost in tandem, the turbine speed is raised up to the synchronization speed and the generator is synchronized with the grid system ("generator integration"). The turbogenerators are operated with the reactor output maintained at a rather higher level and any steam more than necessary bypassed to the condenser through the turbine bypass valve beforehand. When the generator output surpasses the load on the in-house auxiliary machine, the power supply is switched from the external source (startup transformer) to the generator (in-house transformer). In the case of a plant having a generator load break switch (GLBS), the in-house load can be met by the in-house transformer even before the generator integration. When the reactor output increases up to 15 % or more, the main feed-water bypass control valve is switched to the main feed-water control valve which has a larger capacity, and the main feed-water pumps are started up additionally corresponding to the feed-water flow required. While the load is increased, the control rods are controlled automatically by the signals of deviation between the average temperature of the primary coolant and the referential temperature determined by the turbine load. Still they need the regulation of position, though, by adjusting the concentration of boron as required so that the axial offset in the core (deviation of output between the upper half of the core and the lower half) may be maintained within a predetermined range. The generator integration and the rated power operation are also described in Sect. 8.3.2 of this chapter. Figure 8.7 shows the standard startup curve.

8.2.2 Plant Shutdown Operation

The plant shutdown operation covers the processes from switching its rated power condition to the hot shutdown condition until its cold shutdown condition is attained. The basic operations are the reverse to those for starting up the plant.

[1] Reducing the load

The turbine load is reduced starting at the rated power condition by closing the main steam governor valve gradually. Following that operation, the control rods are inserted automatically by the control rod control system, lowering the reactor output. When the reactor output reaches 15 %, the automatic control is switched to the manual operation. The feed-water flow into the steam generator decreases in the meantime; therefore, it is needed to stop unnecessary main feed-water pump operation by switching the main feed-water control valve to the main feed-water bypass control valve in the feed-water system. After that, the output is decreased gradually by control rods inserted manually until the reactor is shutdown under the hot shutdown condition. To switch the reactor to the cold shutdown condition for periodic inspection or other purposes, the boron is supplied from the CVCS until it reaches the concentration corresponding to the refueling shutdown condition, or the cold shutdown condition at least.

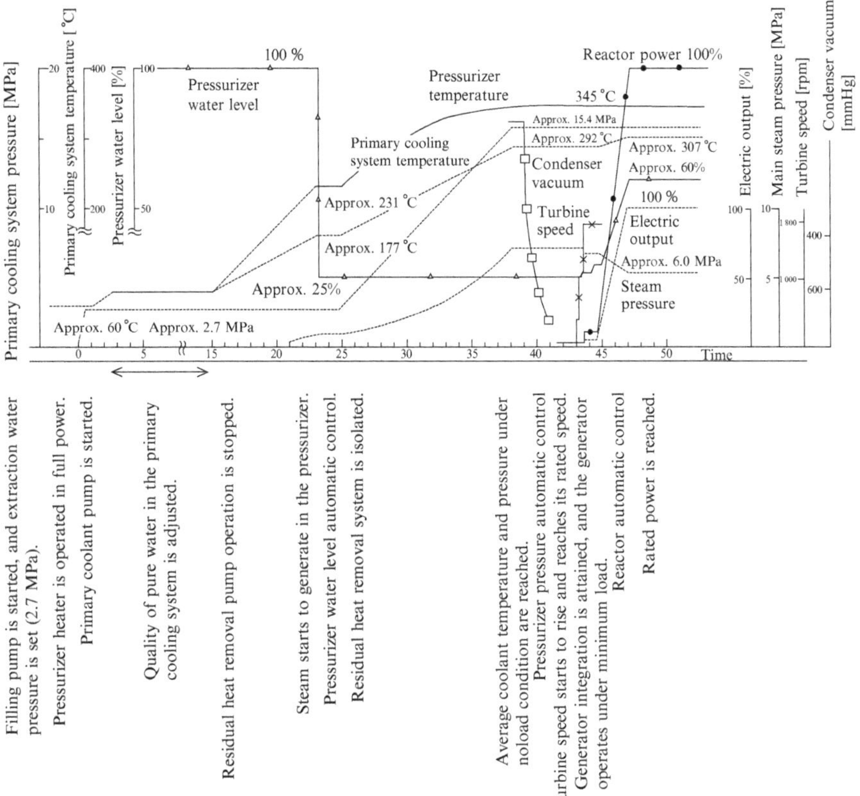

Fig. 8.7 Standard startup curve

[2] Decreasing temperature and pressure (to approximately 177 °C and 2.7 MPa)

The temperature and pressure of the primary coolant system are lowered by the cooling in the steam generator with the turbine bypass valve or the main steam escape valve and by controlling the pressurizer's pressure with the pressurizer spray. At this time, it is necessary to comply with the pressure/temperature limit curve applicable to the cooling of the plant and to control the cooling period rate so that the predetermined rate may not be surpassed from the viewpoint of prevention of inductile destruction of components.

When the temperature and pressure of the primary coolant system reach approximately 177 °C and 2.7 MPa, respectively, the operation of the residual heat removal system is started switching the cooling process with the steam generator to that with the residual heat removal system. This is because the lowered temperature of the primary coolant system decreases the steam generator's pressure, which reduces in turn the efficiency of the cooling with the steam generator.

[3] Extinguishing gas phase in the pressurizer

The gas phase in the pressurizer is extinguished and the primary coolant system is filled with water with the temperature and pressure of the primary coolant system maintained at approximately 177 °C and 2.7 MPa, respectively. The purpose of this procedure is to avoid an excessively large temperature difference between the water in the pressurizer and the primary coolant, as described in the section of startup operation (Sect. 8.2.1 of this chapter).

The gas phase in the pressurizer is extinguished by the operation to increase the filling flow to raise the water level of the pressurizer, whereas the pressure of the primary coolant system after the gas phase is extinguished is controlled by the extracting flow pressure. Also, the pressurizer spray valve is opened fully to accelerate the cooling of the pressurizer by mixing the water in the pressurizer with the primary coolant.

[4] Decreasing temperature (from approximately 177 °C to 60 °C)

After the extinction of the gas phase in the pressurizer, the primary coolant system continues to be cooled by the residual heat removal system until its temperature reaches approximately 70 °C, then the operation of the primary coolant pumps are stopped. It is desirable to operate the primary coolant pumps as much as possible to mix the water in the primary coolant system; because the efficiency of cooling by the residual heat removal system is reduced as the temperature of the primary coolant system is reduced, and the effect of heat input by the pumps becomes nonnegligible. After the primary coolant pumps have stopped, the pressurizer auxiliary spray driven by the filling pump is used to cool the pressurizer because the spray driven by the primary coolant pumps cannot be used.

The cooling operation is accompanied by the degasifying and cleaning operations for the plant shutdown for any work involving the opening of the reactor, the primary cooling system or others, such as refueling, and periodic inspection, to remove radioactive gases and corrosion products, etc., so that workers may be exposed to less radiation dosage.

Figure 8.8 shows the standard shutdown curve.

8.3 Automatic Control System and Normal Operation

8.3.1 Automatic Control System

[1] Control rod control system

The control rod control system regulates the position of a control rod cluster in a control bank automatically to control the reactor output by using three types of input signals representing: average temperature of the primary coolant (T_{avg}), reactor output (Q_n), and the pressure after the first stage of turbine (P 1st). P 1st

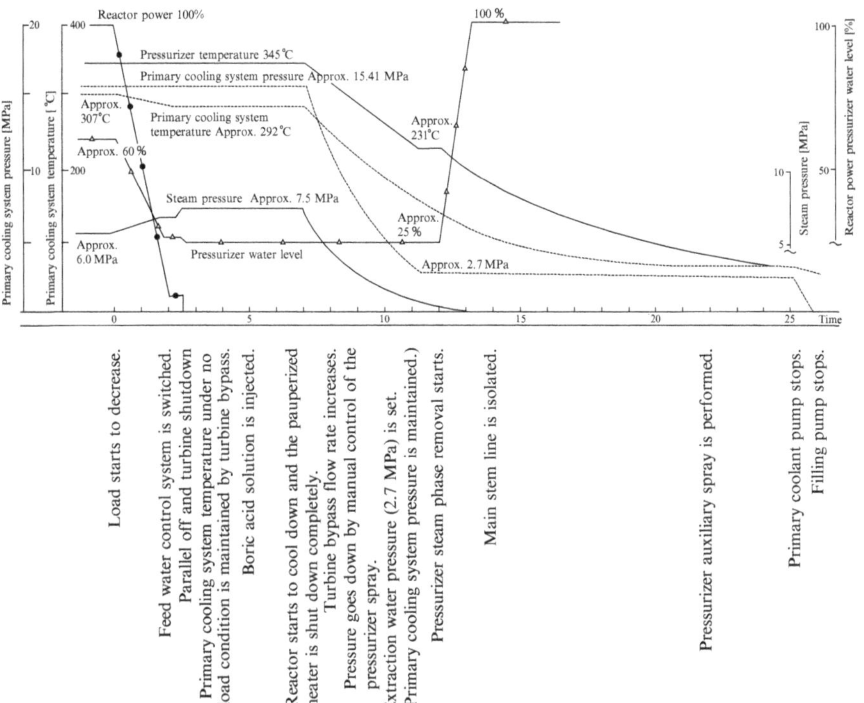

Fig. 8.8 Standard shutdown curve

is a signal proportionate to the turbine load (Q_{tu}). The system uses those signals to configure a circuit for the temperature deviation signal ($T_{avg} - T_{ref}$) and another circuit for the output deviation signal (P 1st − Q_n). The former circuit is called "average temperature channel," and the latter "output discrepancy channel." The average temperature channel, consisting of the signal of difference between T_{avg} and T_{ref}, is the main signal to control T_{avg} to match it with T_{ref}. The output discrepancy channel, consisting of the signal of difference between turbine load and reactor output, serves for improving the response and performance of the control system. Figure 8.9 shows the configuration of the control rod control system.

The average temperature of each primary coolant loop is acquired by calculating the average of the temperature at the reactor outlet piping (high-temperature side) and the reactor inlet piping (low-temperature side) detected by the resistance-temperature detectors (RTDs) installed there. The T_{avg} used for the control system is the second highest signal selected by the signal selection circuit among the average temperature signals from each loop. Using the signal selection circuit makes it possible to expect for continued normal control operation in case of a failure at the high or low side of the T_{avg} signal. This signal is compared with T_{ref} after passing a compensation circuit. The compensation circuit compensates for

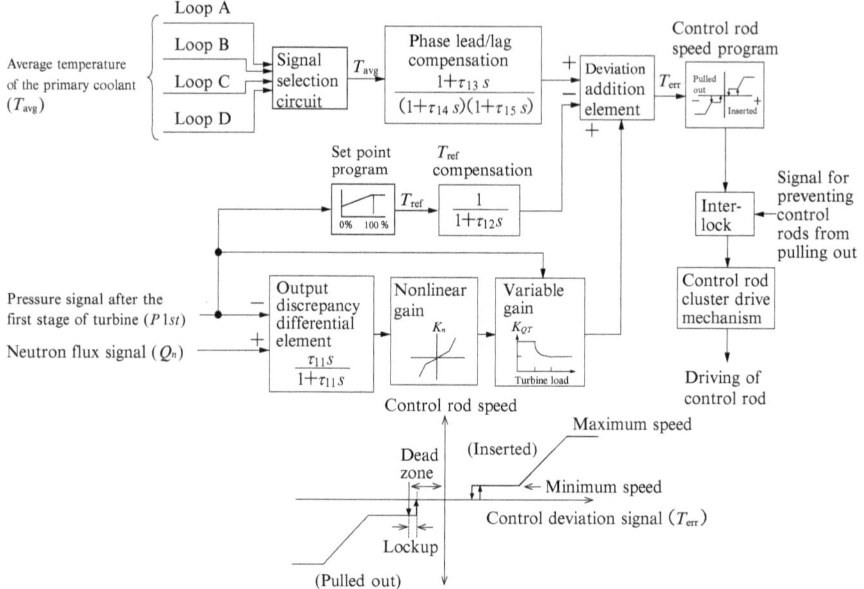

Fig. 8.9 Control rod control system

delay of response caused by the heat capacity of the temperature detector or the coolant system and makes phase-advance compensation to improve control responses.

The T_{ref} is set as a function of the turbine load signal (P 1st). There is no need to operate the control rods against minute fluctuation in the turbine load which can be absorbed by the heat capacity of the plant and the self-regulation of the core; the corresponding T_{ref} is added to the control signal after phase-lag compensation is made for the T_{ref}.

The signal of output deviation between the turbine load signal (P 1st) and the reactor output signal (Q_n: measured by the ex-core neutron flux detector) is added to the control signal through the differential circuit, the nonlinear gain, and the variable gain. The differential circuit outputs the variation rate of the imbalance in output between the primary and secondary systems. If the deviation between P 1st and Q_n is varied, the circuit outputs a signal corresponding to the relevant variation rate. The nonlinear gain (K_n) works against a large output deviation signal in a way to reinforce the effect of the channel. The variable gain circuit (K_{QT}) is set in reverse proportion to the turbine load to compensate for the characteristics of the reactor output gain decreasing under low load against the control rods' reactivity. The output discrepancy channel improves the response and stability when load changes. It detects the change of T_{avg} by the turbine load and operates the control rods before the average temperature deviation increases excessively in effect (feed-forward operation).

The control deviation signal (T_{err}) sent to the control rod speed program is given as a sum of the signals of the average temperature channel and the output discrepancy channel. The sign and magnitude of the T_{err} decide the direction and speed of operation of the control rods. The control rods speed program has a dead zone to avoid continued driving of the control rods under the steady conditions. It also has a lockup to ensure proper timing for stopping the control rods in the settling process.

The control rod control system is designed to make the reactor output follow the turbine load, and it is switched to manual operation if the plant power becomes low at 15 % or less. Any disturbances such as the generators put into parallel or parallel off are absorbed by the turbine bypass valve and main steam escape valve controlled automatically.

In the PWR, the excess reactivity of fuel is controlled by the adjustment of the concentration of boron in the primary coolant, and the control rods are placed at the almost fully extracted position through the cycle. This makes it possible to minimize the distortion of the output distribution in the axial direction. However, the differential reactivity value of the control rods is very small around the fully extracted position, so an interlock preventing excessive extraction is installed to secure the required reactivity effect when the load changes. In addition, another interlock is installed to prevent the reactor output from increasing excessively with automatic or manual extraction of control rods.

[2] Pressurizer pressure control system

If the temperature of the primary coolant changes due to fluctuation in load, etc., the primary coolant is expanded or contracted, and the volume change flows into or flows out from the pressurizer through the surge pipe. This causes the change of the water level in the pressurizer. This causes a change in the volume of the gas section within it, which in turn causes a change in the pressure. The pressurizer pressure control system comprising four control devices (pressurizer spray valve, pressurizer escape valve, pressurizer proportional heater, and pressurizer second reserve heater) suppresses such change in pressure to maintain the control of the set pressure. Figure 8.10 shows the configuration of the control system.

In the steady operational condition, the pressurizer pressure is maintained by partial output of the exothermic heat of the proportional heater to compensate for thermal loss caused by radiation from the pressurizer. If pressure drops due to fluctuation in load, etc., the proportional heater increases its output; if the heater cannot suppress the drop, then the second reserve heater comes into operation. If the pressurizer pressure rises, the proportional heater decreases its output; if the pressure continues rising, the pressurizer spray valve is operated to suppress the pressure rising. The pressurizer spray is a system used to reduce the pressure by spraying the coolant in the piping on the low-temperature side of the primary coolant system into the gas phase section of the pressurizer to condensate steam. The output and opening angle of the pressurizer heater and spray valve are regulated by the PID controller working based on the deviation between the pressurizer pressure and the set pressure. The set pressure value is constant for the whole power range. The pressurizer escape valve releases the steam in the gas

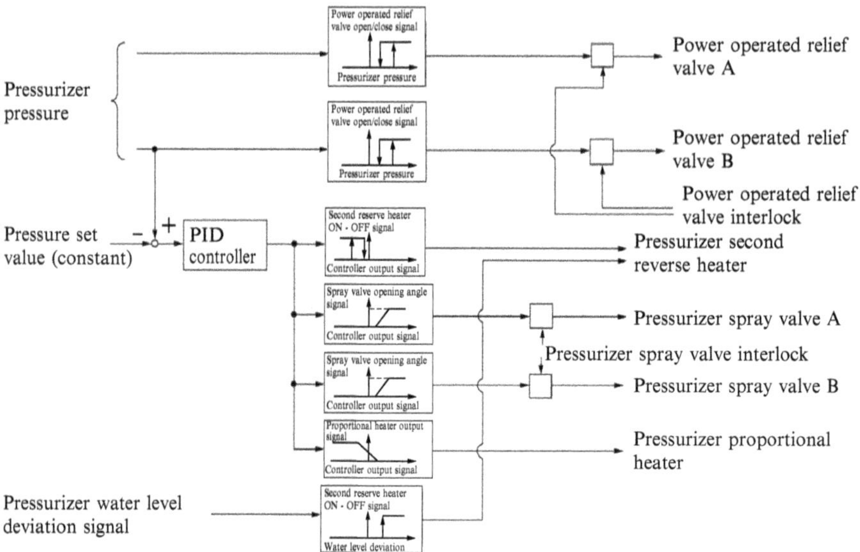

Fig. 8.10 Pressurizer pressure control system

phase section of pressurizer to reduce pressure in case those pressurizer spray valve cannot suppress the pressure rise significantly. The steam released is led to the pressurizer escape tank for condensation. The pressurizer escape valve is on/off-controlled directly by the signal of the actual pressurizer pressure.

Any error in opening the pressurizer escape valve or spray valve may work as a factor for abnormal drop in the reactor pressure. Accordingly, an opening prevention circuit for those valves operated by the pressurizer pressure low signal is installed as an interlock to prevent unnecessary operation of the valves. If the interlock signal is transmitted, supply of driving air to the pressurizer escape valve and spray valve is stopped to close fully all the valves. In addition, the plant has a function to detect the stuck open condition of the pressurizer escape valve and close its main valve automatically. This function works by closing the main valve if the limit switch of the escape valve remains in the open condition for five or more seconds though the pressurizer pressure is lower than the set pressure for operating the valve.

[3] Pressurizer water level control system

The pressurizer water level control system keeps the pressurizer water level at the referential level by regulating the filling flow into the primary coolant system. The volume of water held by the primary coolant system is regulated by adjusting the water flow extracted or filled by the CVCS. The flow is extracted at an approximately constant rate during the power operation unless the extraction line is isolated because of excessive drop of the pressurizer water level. So the volume of water held by the primary coolant system is regulated by adjusting the filling flow continuously. Figure 8.11 shows the configuration of the pressurizer water level control system.

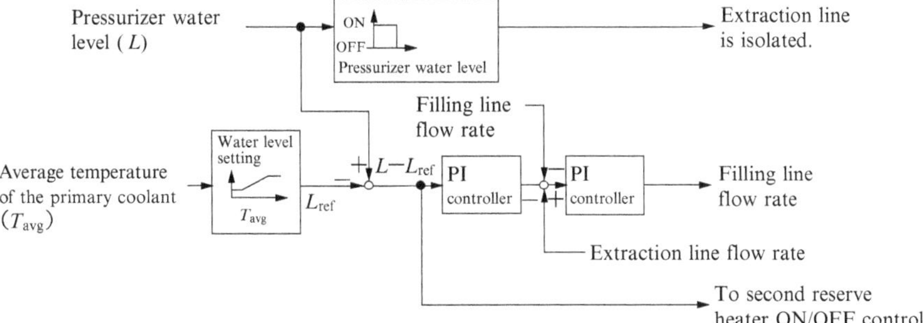

Fig. 8.11 Pressurizer water level control system

A three-element control system is adopted for the pressurizer water level control system, where three input signals are used representing the difference between the pressurizer water level and the referential level, the filling flow, and the extracting flow. The filling flow is adjusted by changing the opening angle of the filling flow control valve to control the pressurizer water level to match it to the referential level.

The referential water level of the pressurizer is set in such a way that it matches the changed water level caused by the change in T_{avg} as much as possible. In other words, the basic concept for the referential water level of the pressurizer is to reduce the operation of the pressurizer's water level control system against the load fluctuation as far as possible to lighten the burden on the CVCS system. For this purpose, a program setting method based on the T_{avg} signal is adopted so that the referential water level is given a change corresponding to the change in the pressurizer's water level caused by expansion/contraction of the primary coolant due to change in T_{avg}. Accordingly, the pressurizer's referential water level changes automatically according to the change in T_{avg} if the load changes. Such change in the referential level is substantially the same as that in the pressurizer's water level, so the control operation of the pressurizer water level control system is relatively minor, causing not so great changes in the filling flow.

The change in T_{avg} when the load changes affects both the water level and pressure of the pressurizer. However, the pressurizer water level control system performs only the minimum control operation even in case of a transient change, so it does not interfere in the control operation of the pressurizer pressure control system.

Continued drop of the pressurizer's water level may cause the heater installed in the liquid phase section of the pressurizer exposed and burnt out. Accordingly, an interlock to suppress the drop of the pressurizer's water level is installed, which is a function to turn off the heater and isolate the extracting line if the pressurizer's water level comes lower than a predetermined value.

[4] Steam generator water level control system

The steam generator water level control system performs control by regulating the main feed-water flow to maintain the water level on the steam generator's secondary side at the referential level so that water held by the secondary side may

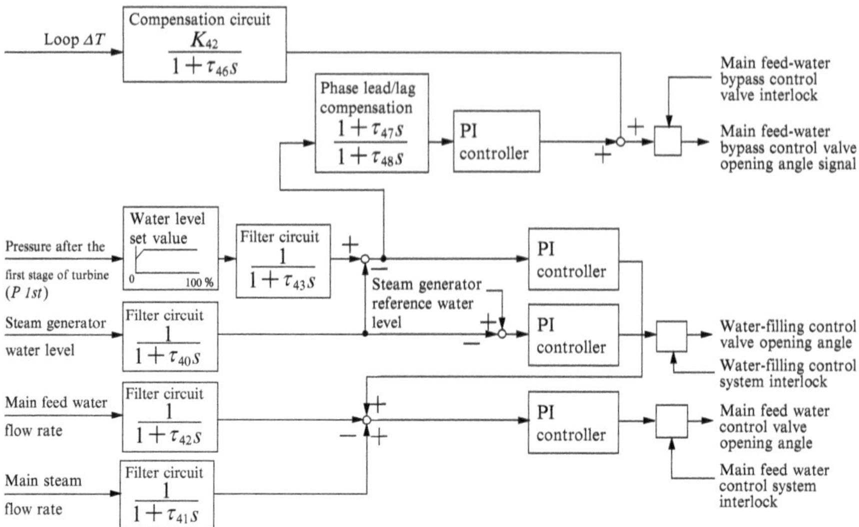

Fig. 8.12 Steam generator water level control system

be secured during the plant operation. The main feed-water system of the steam generator in each loop has three types of flow control valves with different valve capacity corresponding to the plant's power conditions. The main feed-water control valve is used in the high power condition of 15 % or higher, whereas main feed-water bypass control valve in the low power condition of 2–15 %, and the water-filling control valve in the minute flow conditions during the startup, shutdown, or hot shutdown of the plant, etc. The steam generator water level control system has individual control systems for each control valve so that appropriate control operation is attained for each control valve. Also some plants have a system where the main feed-water control valve and the main feed-water bypass control valve are switchable to each other automatically. Still in some other plants equipped with the main feed-water pump driven by turbine, which is convenient to control the pump revolution speed, the discharge pressure of the pump is controlled to assist the main feed-water control system. For the PWR, however, it is necessary to control the steam generator's water level for each loop independently, so the plant has an independent water level control system for each loop's feed-water control valve. Figure 8.12 shows the configuration of the steam generator water level control system.

A three-element control system is also adopted for the main feed-water control system, where three signals are used representing the deviation between the steam generator water level and referential level, the main feed-water flow, and the main steam flow. The water level of the steam generator does not only fluctuate due to the difference between the steam flow and feed-water flow, but is also affected by the change in the steam pressure. A change in steam pressure causes a change in the void volume in the steam generator tube section, which in turn causes a change in

the balance between the two-phase flow pressure loss in the tube section and the static water head in the downcomer section of the secondary side, resulting in a fluctuation of the fluid surface level. Because of this phenomenon, the change in the main steam pressure when the load changes causes large fluctuation of the water level temporarily. However, using the three-element control system makes it possible to control the steam generator's water level matching it to the referential water level ultimately resulting in maintaining the balance between feed-water and steam and securing the water volume held in the steam generator.

A two-element control system, where the water level deviation and the primary coolant temperature difference (ΔT) are used for input signals, is adopted for the main feed-water bypass control system applied to low power plant operation. It works as the one-element control system with the signal of deviation between the steam generator water level and the referential level under a constant power operation condition. Under the changing power condition, it performs the feed-forward control with the ΔT signal that is an index of the reactor output in order to improve the control capability to follow the change of power. While the main feed-water bypass control system is used, the power changes in line with the startup and shutdown of the plant is the main factor causing the fluctuation of the steam generator water level. As the change in output is grasped as the change in ΔT in an early stage, the addition of ΔT signal to the output signal of the PI controller is useful to improve the control capability to follow the power change.

In a range where the water-filling valve is used (lower than approximately 2 % power), very little steam is generated, so it is not inconvenient to use manual control; nonetheless, automatic control is used to alleviate the burden on operators. For the steam generator water-filling control system, the single factor control is adopted where only the deviation between the steam generator's water level and the referential level is used.

A plant equipped with a turbine-driven main feed-water pump has a speed control system for such pump which controls the discharge pressure of the pump. This control system regulates the pump speed to control the differential pressure between the main feed-water header pressure and the main steam header pressure, matching it to the set value of the differential pressure programmed based on the total sum of the main steam flow of each loop. This control gives the differential pressure of the main feed-water control valve corresponding to the power, resulting in not excessively great change in the opening angle of the valve and thus alleviated burden on the feed-water control performed by the main feed-water control system.

An interlock to force the feed-water control valve to close is installed to avoid a plant trip caused by an abnormally high water level in case the steam generator's water level is raised abnormally by a failure in the control system, etc.

[5] Turbine bypass control system

The turbine bypass valve is a system which leads the main steam section to the condensator directly bypassing the turbine. The functions of the turbine bypass control system can be summarized as follows:

<1> To operate the turbine bypass valve to ease off the temperature rise of the primary system accompanying to the reduction of load at the transition if the load is reduced over 10 % stepwise or over 5 %/min rampwise. The standard design requires that the turbine bypass valve of the capacity of approximately 40 % of the rated steam flow should be installed making it possible for the valve to follow the 50 % stepwise load reduction without causing reactor trip. Also, some plants have a turbine bypass valve of the capacity of approximately 70 % of the rated steam flow installed to ensure the plant's ability to continue operation by switching to the generator output corresponding to the in-house load (approximately 5 % of the rated power) in case the load is cutoff due to an accident in the grid system, etc.

<2> To operate the turbine bypass valve to suppress excessive rise of the steam pressure on the secondary side and to switch the plant to the hot shutdown condition without operating the main steam safety valve at the transition in case of the plant trip from the power operation.

<3> To remove residual heat from the reactor to maintain the hot shutdown condition and to cool the reactor system during the operation for lowering temperature toward the cold shutdown.

For the control described in <1> and <2> above, the opening angle of the turbine bypass valve is adjusted corresponding to the deviation from the referential temperature to control the steam flow damped into the condensator. For the conditions where the plant has not tripped, the turbine bypass control system has a dead zone wider than that of the control rod control system against a temperature deviation signal, so that any interference with the control rod control system which has the same control signal T_{avg} may be avoided in the process of adjustment between both systems. As a result, once the deviation between T_{avg} and T_{ref} is reduced and enters this zone in the stabilization process after sharp decrease in the load, the turbine bypass valve is closed fully, and T_{avg} is controlled by the control rod control system thereafter. For the conditions after the recovery from the plant trip, T_{avg} is controlled by the turbine bypass control system only, and the proportional control is performed without the dead zone.

For the function described in <3> above, the control signal is switched to the main steam header pressure by the mode selection switch and the steam flow is controlled automatically to maintain the pressure at the set value. If the degree of vacuum of the condensator gets down making it unable to lead the steam to the condensator, the main steam escape valve (atmospheric release valve) installed in the main steam line is operated to control the steam pressure by releasing the steam directly into the air.

An interlock is installed in the reduced load detection circuit to enable the turbine bypass control system to work only when needed, because a wrong operation of the system during the normal plant operation results in increasing unnecessarily the load on the reactor. The turbine bypass valve works if abrupt reduction of load occurs over 10 % step-wise or a variation rate of 5 %/min. A rapid open signal is transmitted if the deviation from the referential temperature extends beyond a specific value to suppress excessive rise of the temperature or pressure of the primary system caused by the reduction of load by a large margin. A pneumatic valve is adopted for the turbine bypass valve, so it is full-opened rapidly bypassing the positioner in case the rapid open signal is transmitted. Figure 8.13 shows the configuration of the turbine bypass control system.

8.3.2 Operation in Normal Conditions

[1] Zero-power reactor physical test

With the PWR, zero-energy reactor physical test is conducted after the criticality operation in the hot shutdown condition (no-load temperature) to check the characteristics of the core after refueling. In this test, critical boron concentration, reactivity value of control rods, and core output distribution, etc., as well as the moderator temperature coefficient and shutdown margin are measured to check the validity of the core design.

The PWR gets heat input from the primary coolant pump even if its core output is zero. So it is necessary to remove heat of the reactor continuously from the turbine bypass valve or the main steam escape valve (atmospheric release valve) to maintain the primary system's temperature at the no-load temperature. Using the turbine bypass valve is preferred from the viewpoint of the consumption of the secondary system's make-up water. Instead of such a valve, the main steam escape valve is used more often with its opening angle fine-adjusted manually for measuring the moderator temperature coefficient, etc., which require relatively sensitive temperature control. To feed water to the steam generator, the water filling control valve is operated by automatic control.

The reactor physical test involves the dilution and concentration of boron, causing difference in the concentration of boron between the fluid phase section of the pressurizer and the primary coolant loop. To maintain this difference in concentration lower than a fixed value, the pressurizer's second reserve heater is put in manually and the pressurizer pressure control system is operated automatically in the mixing mode accompanied by the operation of the spray valve. In this control mode, the differential (D) element of the PID controller shown in Fig. 8.10 is disabled to avoid vibrational change in the opening angle of the spray valve.

[2] Generator integration and output increase

To secure approximately 5 % of initial generator output when the generator is integrated, the reactor output is raised up to approximately 10 % beforehand. The turbine bypass valve is used for the heat removal before the generator integration. The turbine bypass control system is operated in the main steam header mode to control the main steam pressure at a constant level. The turbine revolution speed and phase are fine-adjusted to be synchronized with the frequency of the grid system by the operator referring to the indication of the synchronization tester,

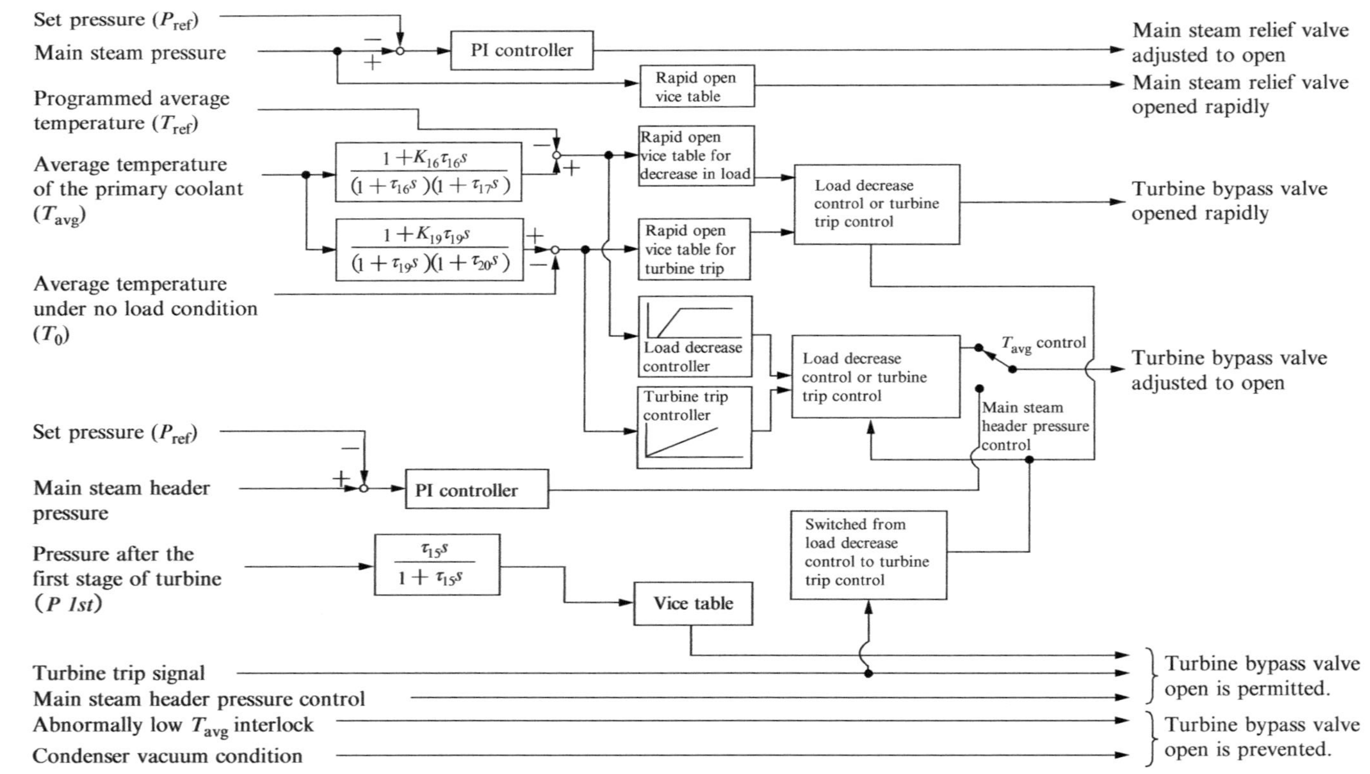

Fig. 8.13 Turbine bypass control system

and the generator integration is attained by putting in the main transformer break switch or the GLBS. Once the break switch is put in, the governor valve opens to maintain the initial load of the generator, causing the automatic operation of the turbine bypass valve in the closing direction, though it is not closed fully but continues controlling the steam pressure.

Continued rise of output is attained by increasing the output instructions from the turbine side at a fixed rate. In this process, the turbine bypass valve is closed fully, so the control rods are operated manually to prevent the average temperature of the primary coolant from dropping excessively and to raise the reactor output.

Once the generator reaches approximately 15 % power, the main water feeding to the steam generator is switched from the bypass control valve to the main feed-water control valve. Some plants can perform the switching of those valves automatically. After the control rod control system is switched to automatic operation at approximately 15 % power, the control rods are extracted automatically following the increase in the turbine load, raising the reactor output. In the process where the load is increased, the output distribution in the axial direction in the core is monitored and boron is diluted as required to maintain the distribution within a predetermined range. Ultimately, the control rods of the control group reach the condition where they are extracted almost fully when the rated power is attained.

[3] Rated power operation

The fuel cost accounts for just a minor portion of the generation cost of a nuclear power plant. So continuing its operation at as high output as possible leads to reduced unit cost of power generation. For this reason, nuclear power plants in Japan are operated at the rated power level regardless of changes in the requirements from the electricity system except for the periodical inspections including refueling. Any fluctuation in electricity requirements during the day or season is absorbed by adjustment of the power generated by other thermal or hydraulic power plants.

The rated power is maintained by the load instructions issued from the turbine controller (load limiter) operated manually by the operator monitoring the generator output or the reactor thermal output. The load instruction changes the opening angle of the governor valve located at the turbine inlet, regulating the steam flow into the turbine. In the past, the plant was operated at the constant rated generator power, requiring the operation for adjusting the opening angle of the governor valve. It is because the change in the turbine efficiency caused by change in degree of vacuum of the condenser, etc., may be compensated for maintaining the constant generator power. Nowadays, however, all the PWR plants in Japan are operated at the constant rated thermal power, requiring almost the same opening angle of the governor valve all through the year.

An operation test is required for the governor valve, though, to check the speed control and excessive speed prevention functions of the turbine on a regular basis. The test is conducted by closing and opening fully in turn each of four governor valves installed in parallel (stem-free test). The test requires the lowering of the plant power down to the extent where no more than three governor valves can support the power. This lowers the reactor output at the same time. No scheduled fluctuation in power occurs besides the stem-free test, and the plant is operated at a rated power all through the cycle in principle.

The entire reactor control system is under the automatically controlled condition in the rated power operation, except the concentration of boron control system which requires manual operation. The concentration of boron in the primary coolant is highest at the beginning of the cycle to absorb any excess reactivity of fuel. Minor discrepancy in output between the primary and secondary systems caused by the reduced reactivity accompanying to the progress of combustion results in the decrease of T_{avg}, which in turn adds the positive reactivity to the core, thus making possible the continued operation of the reactor at constant power. This decrease in T_{avg} caused by the fuel combustion is a very slow change allowing its observation in an order of hours. Any deviation between T_{avg} and the referential temperature (T_{ref}) is checked by an operator on a regular basis. If the deviation gets smaller than the control value, pure water is added by the dilution function of the concentration of boron control system to lower the concentration of boron in the primary coolant and to compensate for the reactivity reduced by the combustion. In general, this control value is far lower than the dead zone set by the control rod control system. So the control rods are never operated so long as the plant continues stable operation at the rated power; the rods are held at the almost fully extracted position all through the cycle.

8.4 Behavior and Operation of the Plant When Load Changes

8.4.1 Behavior When Load Decreases Rapidly

The control performance of the reactor control system against disturbance to the designed load change has been confirmed by the kinetics simulation with a computer, the transition response test of plant conducted during the commissioning of the power plant, and the results of later operation.

With the PWR, the extent of change in the reactor power and the T_{avg} against the same disturbance to the load change differs largely between BOC and EOC by a large margin. In general, the change in the T_{avg} or the pressurizer pressure tends to be larger in BOC when the absolute value of the moderator temperature coefficient is small, because the effect of the negative reactivity feedback caused by the rise in the primary coolant's temperature is small, causing the reactor output subject to less change. The control of the PWR is designed, however, to have the control setting values which allow the plant to follow any designed load changes stably without trip all through the life of the core. Here, the behaviors of major parameters of the reactor system, including the operation of the control system, are outlined by way of illustration of a rapid step-wise load decrease from 100 % to 50 % power (i.e., the case where steam flow from turbine decreases sharply step-wise from the rated

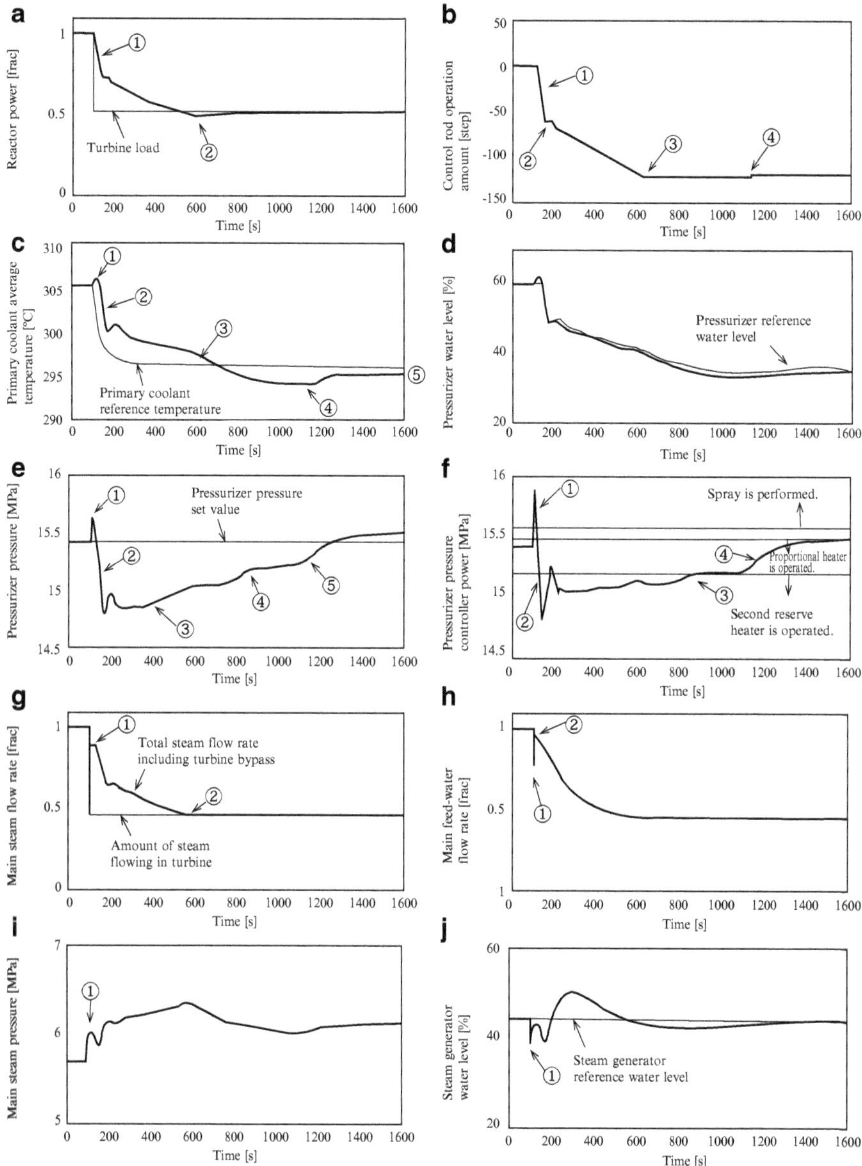

Fig. 8.14 Example of plant parameter behaviors in sharp decrease of load

output down to a level corresponding to 50 % power) in BOC selected from the designed load changes. In the subsequent description of the behavior, superscript notations "(a)①," etc., show the graph and arrow numbers in Fig. 8.14.

[1] Overview of plant behavior

Figure 8.14 shows the change in major parameters of the plant when load decreases sharply.

Abrupt decrease in load causes a sharp decrease of the steam flow into the turbine, making the heat removal from the primary system inadequate, which in turn raises the T_{avg} at first. However, as the turbine bypass control system comes into operation automatically, and contributed by the control rods' operation, the reactor attains the equilibrium value, 50 % power, without trip and stably as shown in the figure.

The reactor power is decreased rapidly by the control rods inserted at full speed immediately after the load decreases rapidly. But the speed becomes lower as the T_{avg} comes closer to the T_{ref}.

As the T_{avg} is raised or lowered, the pressurizer pressure and water level immediately after the rapid decrease of load is raised/lowered transiently; thereafter, they are controlled toward their referential values.

Immediately after the sharp decrease of the load, the water level of the steam generator is dropped abruptly by the reduced boiling area of the secondary side of the steam generator. As the turbine bypass valve opens, however, the main steam pressure turns to decrease, extending the boiling area and thus putting the water level to rapid recovery. After that, the control system is used to decrease the main feed-water flow as the main steam flow is decreased, and the water level of the steam generator is controlled toward its referential level.

The main steam pressure is raised sharply immediately after the sharp decrease of the load caused by the rapidly closed governor valve. As the steam flow is recovered by the rapidly opened turbine bypass valve, however, the rise in the pressure is suppressed. After that, as the turbine bypass valve is closed gradually following the drop in the T_{avg}, the main steam pressure is settled at the level corresponding to a new load.

[2] Behavior of the reactor power and T_{avg}

Sharp decrease in the load causes the pressure after the first stage of turbine (P 1st), which is the index of the turbine load, to drop sharply, which in turn causes the T_{ref} to drop down to the temperature corresponding to 50 % power ultimately (c)⑤. The T_{avg} is raised transiently at first (c)①, but it is controlled toward its referential temperature by the insertion of the control rods and the operation of the turbine bypass valve.

Immediately after the sharp decrease in the load, the output from the output discrepancy channel decided by the deviation between the reactor output and the P 1st increases sharply, causing a large deviation signal for control (T_{err}) transmitted instantly. After that, the output from the T_{avg} channel decided by the deviation between T_{avg} and T_{ref} increases, with the control rods inserted for approximately 60 s at their maximum speed, 72 step/min (b)①.

The reactor output is decreased rapidly (a)① by this full-speed insertion of the control rods, and the T_{avg}, too, by such insertion coupled with the operation of the turbine bypass valve (c)②. The control rods are stopped at the point of approximately 160 s (b)② at first; this is resulted from the set off of the signal of the T_{avg} channel in the inserting direction with the signal in the direction of extraction output by the inexact differential circuit of the output discrepancy channel in the control rod control system, because the sharp decrease in the reactor output has caused the deviation from the turbine load to change in the direction of reduction. After that, as the rate of decrease of the reactor output drops causing the effect of the output discrepancy channel decreased, the control rods are inserted more slowly by the effect of the T_{avg} channel mainly. At the point of approximately 600 s, T_{err} enters the dead zone range at first (c)③, causing the control rods to stop. Here the reactor output is slightly smaller than the turbine load (a)②, causing the T_{avg} to continue dropping until the T_{err} goes over the dead zone (c)④. Thus slight extracting operation is caused at the point of approximately 1,100 s (b)④. Thereafter, the T_{avg} is settled in a range of $T_{ref} \pm$ dead zone.

[3] Behavior of the pressurizer pressure

When the load changes abruptly, the expansion/contraction of the primary coolant due to the change in its temperature causes its inflow or outflow surge to or from the pressurizer, causing increase or decrease of the volume of the fluid phase in the pressurizer, and thus change in the volume of the gas phase, resulting in the fluctuation in the pressurizer pressure. Immediately after sharp decrease in the load, raised T_{avg} (c)① increases the pressurizer pressure (e)①. Then the pressurizer pressure control system causes the pressurizer spray valve to operate (f)①, suppressing the rise in the pressure. When the T_{avg} drops as the reactor output decreases, the pressurizer pressure is decreased down to a level lower than the set pressure, 15.41 MPa (e)②. Here the output (f) from the pressurizer pressure controller (PID controller), whose input signal is the pressurizer pressure, shows the same behavioral trend. It contains differential elements, however, so its fluctuation band represents the enlarged fluctuation in the pressurizer pressure.

The proportional heater output, which has been decreased down to zero as the pressure rises in the beginning, is at the full output level at the point of approximately 130 s and later because of decreased output from the pressurizer pressure controller, with the second reserve heater put in additionally (f)②. At the point of approximately 170 s and later, the T_{avg} continues dropping still; the drop in the pressurizer pressure is suppressed by all the pressurizer heaters put in, though, turning to the recovery trend (e)③. At the point of approximately 900 s and later, the pressure is on an uptrend (e)④ with the proportional heater used only (f)③. At the point of approximately 1,100 s, however, T_{avg} is turned to an uptrend by the extraction of control rods, causing the pressure raised by a large margin (e)⑤. Against such change in the pressure, the output from the proportional heater is decreased sharply (f)④, though the pressure surpasses the set value once. Thereafter, the pressure is turned to downtrend gradually by radiation from the pressurizer because the T_{avg} has been stabilized already, controlled slowly toward the set pressure value.

[4] Behavior of the pressurizer water level

The referential water level of the pressurizer (d) shows the behavior substantially matching with T_{avg} until the plant is stabilized, because it is designed to correspond substantially to the change in the volume of the primary coolant caused by the change in T_{avg}. The pressurizer water level shows the similar behavior in the same way, though slight deviation from the referential water level is caused in the process of transient change. The pressurizer water level control system has a low setting of the control system gain to prevent its abrupt control operation, so a relatively small deviation of the water level may continue for a long time. Such a deviation will be removed over time, though, as the filling flow is increased gradually by the effect of the integral operation of the control system.

[5] Behavior of the steam generator water level

Immediately after the sharp decrease in the load, the deviation from the main feed-water flow is expanded instantly by sharp decrease in the main steam flow. This causes great decrease at first in the control signal for the main feed-water control valve of the steam generator's water level control system, resulting in sharp decrease in the main feed-water flow (h)①. However, the output from the controller turns to increase rapidly and the feed-water flow recovers promptly the level close to that attained during the previous rated operation (h)② so as to compensate for the sharp drop in the steam generator's water level (j)① and because the turbine bypass valve opens rapidly increasing the main steam flow (g)①.

The water level of the steam generator drops sharply (j)① immediately after the rapid decrease of the load due to sharp rise in the main steam pressure (i)① caused by the rapid decrease of the load. It turns toward recovery, however, because the turbine bypass valves, all of which get fully opened, suppress the rise in the main steam pressure and that the main feed-water flow is turned to recovery. After that, the main feed-water flow is controlled stably following the decrease in the main steam flow due to the turbine bypass valve closed gradually so that the steam generator's water level matches the referential water level.

[6] Behavior of the turbine bypass valve

When the load decreases rapidly, the load decrease controller in the turbine bypass control system operates, adjusting the opening angle of the turbine bypass valve corresponding to the deviation between T_{avg} and T_{ref}.

Sharp decrease in the turbine load (P 1st) causes decreased T_{ref}, while the T_{avg} is increased by the inadequate heat removal from the steam generator. A large deviation in temperature between the sharply dropping T_{ref} and the rising T_{avg} increases the input into the load decrease controller rapidly, causing the signal requesting for the turbine bypass valve opening angle to reach at 100 % almost instantly. As a result, all the turbine bypass valves are opened rapidly immediately after the sharp decrease in the load (g)①. At this time, more steam than the rated 40 % is dumped into the main steam flow (dump flow). This is caused by all the turbine bypass valves opening fully, and is added to by the effect of the raised main steam pressure.

After that, T_{avg} drops toward T_{ref} as the control rods are inserted, and the turbine bypass valves are closed in sequence as the T_{avg} drops. After approximately 580 s (g)②, the signal of the deviation between T_{avg} and T_{ref} enters the dead zone of the load decrease controller, and the turbine bypass valves are all closed fully. Thereafter, the T_{avg} is controlled by the control rod control system.

8.4.2 Ability to Switch to In-House Single Operation

The grid system in normal conditions configures a network where plural power plants are linked to each other. In case one power plant is isolated from others for any reason, it is switched to a single system with a partial load. This condition is called the "single operation." In general, if a nuclear power plant is switched to single operation, the isolated system has a remaining load which is smaller than the load before the relevant event. This condition means for the plant that the load is reduced abruptly. If the grid system is cut between the plant and the closest transformer station, the plant is switched to the single operation where only the in-house load remains (in-house single operation). This means for the plant the largest decrease in the load (load cutoff) among all the possible cases.

The ability of the plant to continue operation by switching to the in-house single operation without causing the reactor trip in case of such abrupt decrease of the load is preferable to the reactor shutdown from the viewpoint of safety in securing the power source of the reactor. It will also contribute to a stable supply of electricity because it allows starting the supply of electricity once the electric system is recovered.

As standard, the PWR plant is designed to tolerate the step-wise 50 % decrease in the load by installing the turbine bypass system with the capacity of approximately 40 % of the rated steam flow. Some other plant designs ensure, however, the in-house single operation all through the life of the core by increasing the capacity of the turbine bypass valve up to a value corresponding to approximately 70 % of the rated steam flow. Such plants are called "in-house single plants."

The reactor protection system of the PWR does not allow the reactor to trip directly if the load decreases abruptly, but only if any processing volume, such as the temperature, pressure, water level, of the reactor, goes beyond the tolerated range and reaches at the set point for tripping.

If the load is reduced abruptly, the margin for the tripping factors described below decreases basically. If the trip caused by such factors can be avoided, even a standard plant with the turbine bypass capacity of approximately 40 % can be switched to the in-house single operation.

<1> Over-temperature ΔT is high

This is a trip factor set for protecting the cladding of fuel rods. A raised primary coolant temperature results in a smaller margin for the DNB (transition to film boiling) heat flux, resulting in a lower value set for trip. Thus, the trip may be reached if the T_{avg} is raised to a large degree by the output discrepancy between the primary and secondary systems when the load is reduced sharply. The margin for such a trip setting is smaller in the beginning of the reactor life where the absolute value of the moderator temperature coefficient is smaller and decreases in proportion to the reduced turbine bypass capacity.

<2> Steam generator water level is abnormally low
This is a trip factor set for protecting the reactor from the loss of the heat removal function of the reactor system. The trip may be reached if the water level drops to a large extent due to a change in the steam pressure when the load is reduced sharply. The margin for this trip setting is smaller in proportion to the reduced turbine bypass capacity, but the effect of the moderator temperature coefficient on the margin caused by the life of the core is small. The drop in the steam generator water level immediately after the sharp decrease in the load is caused by raised pressure on the secondary side due to sharp drop in the steam flow. Accordingly, such drop, which occurs at the beginning of the reactor life, can hardly be suppressed by the feed-water control.

<3> High neutron flux in the output range
This is a trip factor set for preventing excess output from the core. The trip may be caused if the output from the reactor increases to a large extent. When the reactor is separated from the grid system, the revolution speed of the turbine generator is raised transiently, and also the frequency of the in-house power source which receives the electricity from the generator, although for a very short time during load cutoff. This is accompanied by the increase in revolution speed of the primary coolant pump motor, causing a temporary increase in the coolant flow in the reactor. Here, the dropped core temperature adds to the positive reactivity, leading to increase in the reactor output. The larger the absolute value of the moderator temperature coefficient in the end of life, the larger the increase in the output is, resulting in reduced margin for the trip. No effect is created by the turbine bypass capacity.

A plant with a turbine bypass capacity of approximately 40 % is not warranted by the design in that it can be switched to the in-house single operation when the load is cutoff at the rated power. It has been confirmed, however, to have a level of ability against load reduction that is well over the designed ability based on the performances of the plants in Japan as shown at the load-cutoff tests in the commissioning and at the actual load-cutoff events after the commencement of operation. The main reasons include that actual moderator temperature coefficient works in the real core instead of a coefficient having an extremely small or large absolute value as assumed by the design, and that the primary coolant referential temperature for the rated operation is set and operated at a lower value than designed in Japan.

8.4.3 Operation After the Single Operation

[1] Determination of events

If the plant is switched to the single operation, many alarms are transmitted and the turbine bypass valve is opened, showing a condition like the plant trip. Unless the first-out alarm is transmitted, there is a possibility that the plant is operating in the single operation. The single operation can be determined by checking the conditions of the generator output, main transformer break switch, generator load switch, and grid system break switch, etc. If the operation of the turbine's governor switch to the "Low" direction reduces the frequency, single operation can be determined.

[2] Check of automatic operation

Immediately after the plant is switched to the single operation, the automatic operation condition of each device is checked as follows:

<1> Operation of the over speed protection control (OPC) closes the governor valves and the intercept valves fully once.
<2> Check the indicators of revolution speed, axial vibration, axial position, and difference in extension to see that the turbine has no abnormalities in the operational condition.
<3> Check the operation indicator of each turbine bypass valve to see that the "Permissive" indicator is lit permitting the valve to be opened and that the valve is open; here, the main steam escape valve may be opened.
<4> Check to see that the control bank control rods have been inserted automatically.
<5> Check the indication of the outlet temperature of the pressurizer escape valve, and the pressure, water level and temperature of the pressurizer escape tank to see that the escape valve is closed securely, if the escape valve has operated transiently.
<6> Check the steam generator water level for its large decrease immediately after sharp decrease in the load and its subsequent recovery.
<7> Check to see that OPC has been recovered and that the opening angles of the governor valve and intercept valve are stabilized.

[3] Operation of the reactor system

Operation of the control rod control system reduces the reactor output slowly towards the output corresponding to the in-house load. The control rod control system is designed to allow stable control until approximately 15 % power is attained. In the actual operation, however, the system is switched to manual operation when the reactor output is reduced down to approximately 25–30 %, keeping the output at a level slightly higher than 15 %, for the following reasons:

<1> A sharp drop in the reactor output causes an increase in xenon with some time lag. Reactor output is kept at a higher level so that it may not be reduced excessively corresponding to the additional negative reactivity due to the additionally created xenon.
<2> Insertion of the control rods to an increased extent may make it difficult to recover the output distribution in the axial direction to a targeted range of values.
<3> Output lower than approximately 15 % makes it difficult for the main feedwater control valve to control the steam generator water level stably, thus requiring more time for settling the water level.

The heat generated by the reactor over the in-house load is removed automatically by the turbine bypass control system. So it is possible to maintain the reactor at a high output condition over 30 % unless the capacity range of the turbine bypass valve is surpassed. It is realistic, though, to switch the control rods operation to manual at approximately 25–30 % output as a rough idea, considering the time required for determining the event against which measures should be taken first, checking the plant conditions, etc., if a load-cutoff event occurs.

[4] Recovery operation

If a cause for failure in the grid system has been removed and the in-house single operation of the power plant is continued stably, the operation for the generator integration can be performed to recover the power plant to the grid system promptly according to the instruction given by the power supply direction center. Here, the operations for the integration and raising the power are the same as those for the normal plant operation.

8.5 Behavior and Operation of the Plant in Case of an Accident

8.5.1 Reactor Protection System and Engineered Safety Features

The reactor protection system is installed to monitor the major plant processing volumes and transmits the reactor trip signal so that the soundness of the fuels or the reactor coolant pressure boundary is not lost even if an abnormal transient change or accident might occur beyond the assumption for the design of the reactor control system. If the reactor trip switch is opened by the reactor trip signal, the power source to the control rod cluster driver is cutoff, and all the control rod clusters are inserted into the core by their own weight, shutting down the reactor safely. The reactor trip signal is transmitted if the output or pressure of the reactor, temperature or flow of the primary coolant, the pressurizer water level, or the water level on the secondary side of the steam generator, etc. becomes abnormal.

The engineered safety features are installed as equipment to cool the core, protect the reactor containment boundary, and to ensure public safety in case assumed loss of the reactor coolant or fracture of the main steam pipe caused by rupture in the reactor coolant pressure boundary might occur. They include the emergency core cooling system (ECCS), containment vessel spray, etc., and are activated automatically if any processing volume such as the reactor pressure, pressurizer water level, main steam line pressure, etc., becomes abnormal.

As an example of the plant behavior and operation in case of an accident, the processes of the automatic operation of the reactor protection system and engineered safety features, as well as those for closing the accident by the operator's operation, in case of the steam generator tube rupture are described in the following.

8.5.2 Overview of a Steam Generator Tube Rupture Event

Rupture of the steam generator tube is an event where the steam generator tube is ruptured for some reason during the power operation of the reactor causing release or the primary coolant outside the reactor container through the secondary coolant system. If any radioactive substance is contained in the primary coolant, it may be released into the environment through the secondary side of the steam generator. Accordingly, if the steam generator tube is ruptured, the main operation required is to close the main steam isolation valve, etc., leading to the ruptured steam generator and operate the main steam escape valve and the pressurize escape valve on the sound steam generator side, so that the primary system may be cooled and its pressure may be reduced at an early stage, stopping the release of the primary coolant to the secondary side.

If the extent of the rupture is minor, the normal plant shutdown operation with the pressurizer water level maintained works well because the make-up water volume from the filling pump is controlled automatically to make up the dropped pressurize water level. If it is major, however, the reactor is tripped by the reactor trip signal, "Low reactor pressure" or "Over-temperature ΔT high," issued by the reactor protection system. The release of the primary coolant continues after the reactor shutdown so long as the pressure of the primary coolant system is high. The ECCS activated automatically by its activation signal, "Low reactor pressure matched with the low pressurizer water level" or "Abnormally low reactor pressure," causes the boron water in the refueling water tank injected into the core, resulting in the maintained cooling ability of the core.

Those automatic operations of the reactor protection system and engineered safety features ensure the reactor shutdown and the maintained cooling function of the core in case of the rupture of the steam generator tube. Appropriate operation by the operator is expected for, though, to close the release of the primary coolant to the secondary side of the ruptured steam generator at an early stage. The operator should close the main steam isolation valve, which leads to the ruptured steam generator and can be operated from the central control room, to avoid diffusion of the radioactive substances to the secondary side. Then the operator should operate

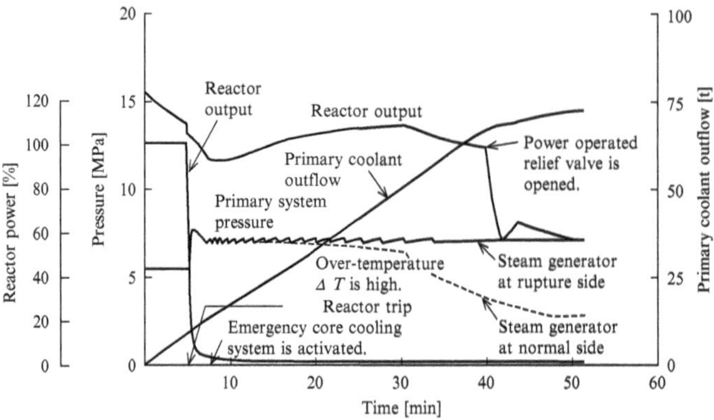

Fig. 8.15 Event sequence after steam generator tube rupture

the turbine bypass valve or the main steam escape valve of the steam generator on the sound side, and then the pressurizer spray valve and escape valve from the central control room, to reduce the pressure of the primary coolant system down to the substantially same level as the secondary side of the steam generator at an early stage. Those operations to equalize pressure will stop the release to the secondary system and close the accident.

8.5.3 Analysis of the Accident Process

Figure 8.15 shows the result of analysis of an accident under conservative assumption of the core conditions, initial plant conditions, and the post accidental power source conditions, time gap before the operation by operator, etc., so that the result significantly stricter than real may be obtained. Changes in major events and the responses of the plant are described below:

<1> One steam generator tube is ruptured instantly at time point 0, and the primary coolant flows out to the secondary side, causing drop in the reactor pressure and the pressurizer water level.

<2> After approximately 5 min, the reactor trip signal, "Over-temperature ΔT high" is issued, causing the reactor to trip. The turbine also is tripped at the same time, causing the pressure on the secondary side to rise, which activates the main steam escape valve.

<3> The release continues from the primary to the secondary side of the steam generator after the reactor trip. Thus the ECCS activation signal, "Low reactor pressure matched with the low pressurizer water level" is issued approximately 7 min later, causing the boron water injected into the core.

<4> Approximately15 min after the accident occurred, the operations to isolate the steam generator on the ruptured side and to stop the make-up water supply

to the steam generator are performed by the operator. Approximately 25 min after the accident occurred, the closing of the main steam isolation valve leading to the steam generator on the ruptured side is completed.

<5> Approximately 25 min after the accident occurred, the operation to open the main steam escape valve is started by the operator, to reduce the reactor pressure.

<6> The cooling operation is not enough to cause a drop in the reactor pressure; the pressurizer escape valve is opened manually by the operator when the primary coolant attains under saturation well to force the reduction in the pressure. When the reactor pressure becomes equal to that of the steam generator on the ruptured side, the pressurizer escape valve is closed by the operator.

<7> When the conditions for stopping the ECCS are met after the pressurizer escape valve is closed, the ECCS is stopped by the operator.

<8> Stopped ECCS, cooling of the primary coolant system by the main steam escape valve of the steam generator on the sound side, and continued operation for pressure reduction reduces the pressure of the primary coolant system, stopping the release of the primary coolant to the secondary cooling system when the equilibrium is reached between the reactor pressure and the steam generator pressure on the ruptured side.

Thus, even if the steam generator tube is ruptured, the reactor is stopped automatically by the reactor protection system. After that, the primary cooling system is cooled and its pressure reduced at an early stage by the proper operation of the operator can lead to isolated steam generator ruptured and closed accident.

8.6 Central Control Panel

8.6.1 New Type of Central Control Panel

The centralized system using the central control panel installed in the central control room is adopted for the surveillance and control of the nuclear plant. The central control panel comprises many controllers, surveillance instruments, switches for the plant operation and its surveillance are involved to a large extent in human system interface (HSI) with an operator. For the purpose of improving operation, surveillance and operability of the plant, the HSI has been improved continuously by utilizing the computer technology, ergonomics and knowledge engineering, resulting in the development of the new type of central control panel which can make the most of the operator's ability.

The new type of central control panel is designed to centralize and sophisticate the conventional information and automatize the operation to reduce largely the potential human error and workload of the operators engaged in a series of work. Also, the conventional individual monitoring instruments and controls have been

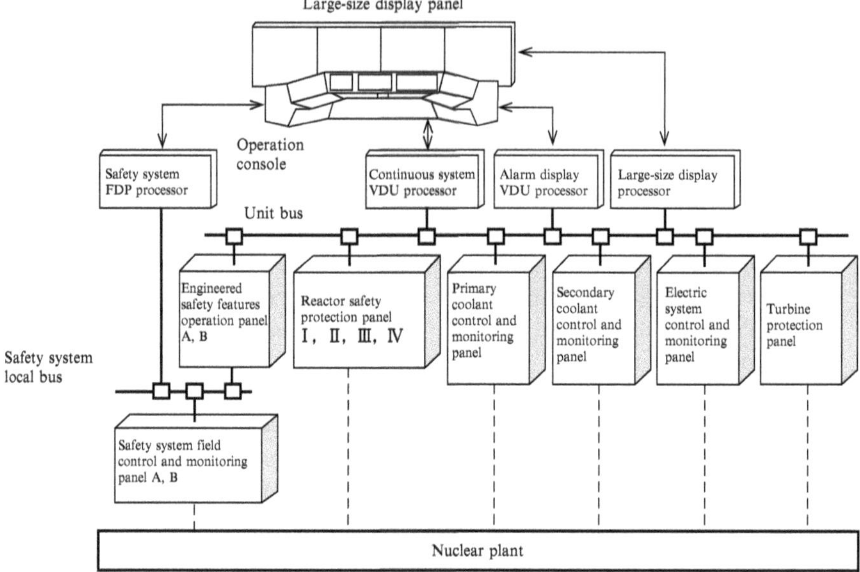

Fig. 8.16 Conceptual diagram of PWR full-digital instrumentation and control system

computerized in principle, allowing their operation by touching the display with a touch panel. As a result, the operators can now operate and surveille the plant just sitting against the centralized screen on a compact console-type panel.

To meet various needs for, and to improve the reliability, operability, and maintainability of the latest plants, the recent measuring control equipment has the entirely digitalized system configuration. The new type of central control panel, as the core of such measuring control system, communicates the surveillance and operation signals by multiplex transmission with various control devices functioning to control and protect the plant. Figure 8.16 shows the system configuration of the new type of central control panel and controllers.

8.6.2 Configuration of New Type of Central Control Panel

The new type of central control panel is designed to meet the conditions listed below from the viewpoint of operation. The panel comprises the operation console, operation instruction console, and large-size display panel as shown in Fig. 8.17.

<1> All the surveillance and operation on the central control panel can be performed from the operation console.

<2> Large-size display panel allows the full-time surveillance of the entire important information and the sharing of the information.

<3> Operation instruction console allows the chief of the personnel on duty to perform his job, including control and supervision for operation, reporting, etc.

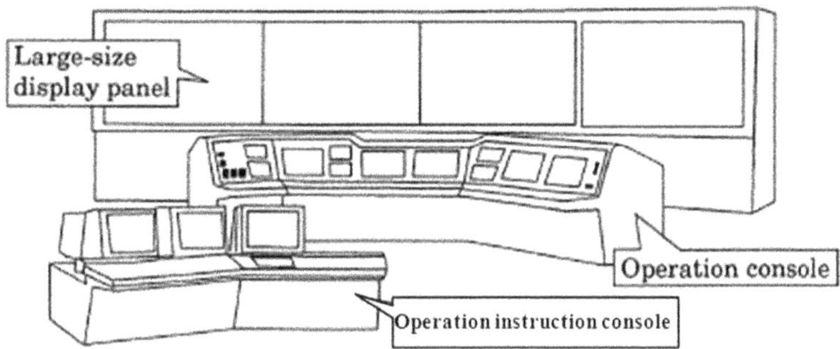

Fig. 8.17 Configuration of new central control panel

[1] Operation consoles

The operation console, which has integrated functions of the conventional central control panel, allows the surveillance and operation in a concentrated and integrated way. Its system is configured as follows:

<1> Visual display unit (VDU) for surveillance and operation:
Displays the information of the regular operation and safety systems and has also the function to operate the devices in the regular operation system.
<2> Flat display panel (FDP) for the safety system:
Has the function to display the safety system's controls and to operate the safety system devices. Also, it displays the readings of the accident surveillance instruments requiring multiplexity, independence and earthquake protection among the accident surveillance parameters.
<3> VDU for alarms:
Include a function to display alarms categorized by dynamic importance.
<4> Hardware devices:
Some system level operation switches, including those for operating manually the reactor trip, engineered safety features, etc., in case of emergency which the operators should respond to instantly, are configured by dedicated hardware controls.

[2] Large-size display panel

The large-size display panel displays the surveillance information of the entire plant at the highest plant level and the alarm information to facilitate grasping the entire plant conditions and to provide the information to be shared by the operation crew. Also, it has a variable display section where any item can be selected for display to support the sharing of the surveillance information to be focused on corresponding to the plant operation. The large-size display panel displays the following items:

<1> Surveillance information of the entire plant and alarms:
The schematic view of the entire plant showing major parameters and the first-out and general alarms allowing the conditions of the entire plant promptly in both normal and accidental conditions to be grasped.

<2> System conditions:
The system conditions based on the system chart and the parameters of major control systems in the plant.

<3> Equipment conditions:
The surveillance information of the conditions of equipment.

[3] Operation instruction consoles

The operation instruction console has the functions to surveille the plant and support the plant operation other than those to operate the plant so that the chief of the personnel on duty can surveille the conditions of the plant operation to the same extent as the operators. Also, it has a function to display the screen shown on the VDU for regular operation console onto that for the operation instruction console, upon request from the chief of the personnel on duty.

8.6.3 Method for Operating Devices (HSI)

[1] Operational conditions

<1> Sit-down operation
The controls on the new type of central control panel have been computerized to make the console more compact, allowing the operators to surveille and operate the plant in the sitting position.

<2> One-man operation
With the new-type central control panel, only one operator is enough for the operation and surveillance described in the operation manual. An assistant operator is supposed to be assigned considering the operability of the plant in case of temporary absence of the operator or an event he cannot respond to, etc. In such a case, two operators share the task as required by the existing operational conditions.

[2] Plant operation

The basic plant operation consists of a flow of a series of surveillance and operations, i.e. “surveillance,” “determination,” “operation,” and “checking.” With the conventional central control panels, the hardware devices such as controls, controllers, indicators, pilot lamps, alarm windows, recorders, etc. are installed at places separate from each other, and the plant is surveilled and operated at a spacious operation area. Whereas with the new type of central panel, touch operation technology has been introduced for the centralized surveillance and operation to enable smooth performance of the surveillance and operation flow. The new type of central panel features the following:

1) Improved surveillance and operationability

<1> Integration of surveillance with operation by the touching operation
<2> Large-size display making possible full-time surveillance of the entire plant and sharing of information

2) More secure determination

<1> Alarm sheet introduced (support in case of alarm issuance)
<2> Operation aid system introduced (support in case of accident)

The controls and controllers are operated according to the following procedures:

<1> Call up a device or controller by touching its symbol on the system chart or controller screen
<2> Release the misoperation protection tag attached to the name of the relevant control or controller
<3> Operate the device by "opening" or "closing" it or adjust the processing volume
<4> Set the misoperation protection tag and erase the control or controller displayed

The controls and controllers for the surveillance and operation of the plant are separated between those for regular operation system and those for safety system. They have a uniformed method for surveillance and operation irrespective of the system, and their functions are integrated by allowing the regular operation system to call the control on the safety system panel. An ON/OFF type valve or an auxiliary machine can be operated by calling the control by touching the rectangular frame in the touching area for the relevant valve or the auxiliary machine shown on the surveillance system screen. Also, the screen for the controllers for continuous adjustment is configured by the combination of the parameters controlled and the controllers, considering the linkage of both.

Chapter 8 Exercises

1. Explain why the reactor following system can be adopted for the power control system for the PWR.
2. Quote two core reactivity control devices of the PWR, and explain the main purpose of their use.
3. Explain why a compromized system harmonizing the constant steam pressure system and the constant T_{avg} system has been adopted for the primary coolant average temperature program of the PWR.
4. Explain when and how the reactor criticality operation is performed in the PWR plant startup operation.
5. Explain the effect of the difference of the core life (beginning or end of life) on the plant behavior when the load fluctuates.

Bibliography

1. Hirota (July 1974) Control of the PWR nuclear plant. Technol Bull Mitsubishi Heavy Ind 11(4):473–481
2. Sano and Hirota (1981) Operational performance of the PWR nuclear plant. IEEJ J 101(4):285–294
3. Thermal and Nuclear Power Engineering Society (1983) Instrumentation control and automation course, Tokyo

Chapter 9
Actual Operation and Control of Fast Reactor

Hidetaka Takahashi and Kiyoshi Tamayama

9.1 Overview of Prototype Reactor for MONJU Fast Breeding

A fast breeder reactor has several features that make it different from a light water reactor. The fuel is plutonium and uranium-mixed oxide, the reactor core enables breeding mainly by reaction of fast neutrons, and it has high power density and burn-up levels. In addition, sodium is used as the coolant and the operating temperature is far below the boiling point (approximately 880 °C at 1 atmospheric pressure). So, the reactor cooling system is designed to operate at low pressure and high temperatures. Sodium is active chemically, and the sodium liquid surface has to be covered with inactive gas.

Heat generated in the prototype fast reactor for MONJU fast breeding is taken out with the sodium cooling system which is composed of three independent loops and is transmitted to the primary sodium system, the secondary sodium system, and then to the water/steam system. The cooling systems are shown in Fig. 9.1.

The primary system sodium is introduced from the lower body of the reactor vessel at approximately 397 °C. It is then heated in the reactor and flows out of the upper body of the reactor vessel at approximately 529 °C. The secondary system sodium exchanges heat with the primary system sodium in the intermediate heat exchanger, and the temperature rises from approximately 325 °C to approximately 505 °C. The secondary system sodium exchanges heat with steam in the helical coil type steam generator (the evaporator and the superheater), and superheated steam at approximately 483 °C and approximately 12.5 MPa is produced and is transmitted to the turbine directly connected to the generator. The thermal output of the reactor is 714 MW, and the electrical output is about 280 MW at the generating end. Since the boiling point of sodium is high, the reactor needs not to pressurize the coolant as in case of a light water reactor and is operated at a pressure close to the atmospheric pressure. The primary system sodium piping connecting equipments is installed at a high location. Device called a guard vessel is attached to the reactor vessel, main pumps, the intermediate heat exchanger, and the piping connecting them. Adopting these measures, the primary system sodium for cooling the reactor core could be

Y. Oka and K. Suzuki (eds.), *Nuclear Reactor Kinetics and Plant Control*,
An Advanced Course in Nuclear Engineering,
DOI 10.1007/978-4-431-54195-0_9, © Springer Japan 2013

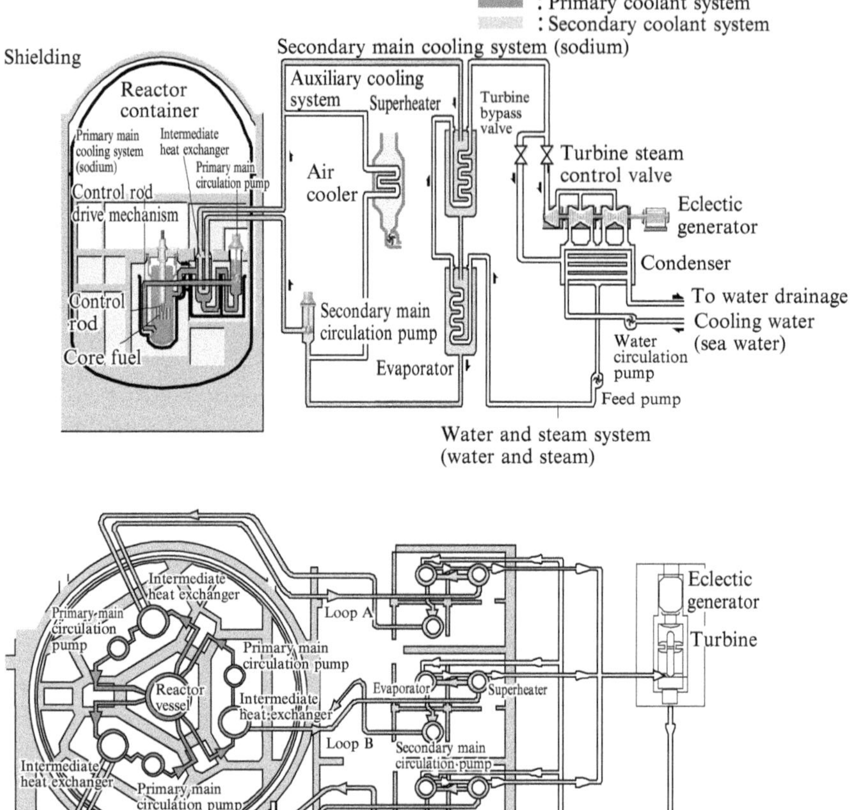

Fig. 9.1 Outline drawing of MONJU reactor

secured even if coolant leaks. Since sodium is active chemically, the internal liquid surface is covered with argon gas atmosphere, and the chamber housing the primary system holding radiosodium is filled with a nitrogen atmosphere.

In addition to these main cooling systems, an auxiliary cooling system is installed to eliminate decay heat of the reactor core at the time of a reactor shutdown due to fuel exchange and in case of emergency. These auxiliary cooling systems are branched from the secondary sodium system and installed with an air cooler in parallel with a steam generator. When these auxiliary cooling systems are operated, the pony motors of the primary system and secondary system main pumps circulate coolant.

The nuclear reactor houses the core and the reactor core internal structure in a steel cylindrical reactor vessel about 7 m across, and a lid called a shielding plug is placed on the top.

Fig. 9.2 Core component diagram

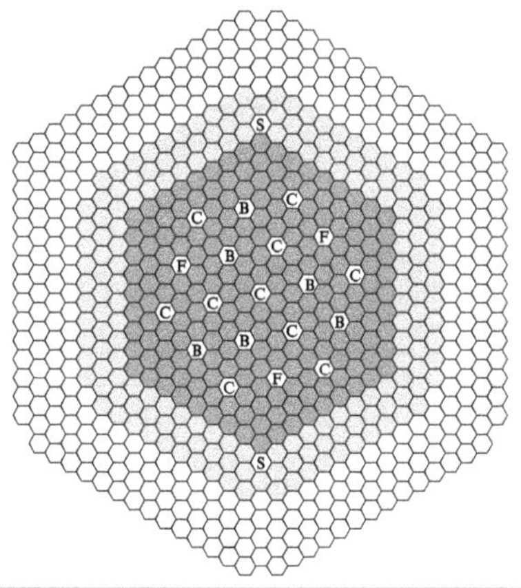

Core elements		Symbols	Quantity
Core fuel assembly		⬢	198
Blanket fuel assembly		⬡	172
Control rod assembly	Fine control rod	Ⓕ	3
	Coarse rod	Ⓒ	10
	Backup shutdown rod	Ⓑ	6
Neutron source assembly		Ⓢ	2
Neutron shield, etc.		⬡	324

Figure 9.2 shows the core component diagram. The core consists of a core fuel assembly, a blanket fuel assembly, a control rod, and a neutron shield. The core fuel assembly houses 169 core fuel elements of 6.5 mm across that seal the core fuel of plutonium and uranium-mixed oxide and the axial blanket fuel of depleted uranium oxide pellet to be placed on and under the core fuel in the wrapper tube hexagonal in section. The blanket fuel assembly houses 61 blanket fuel elements sealing depleted uranium oxide pellets in the same way. The overall length of the fuel assembly is about 4.2 m.

A control rod system is composed of 13 regulating rods and 6 backup shutdown rods. The regulating rods (three fine control rods and ten coarse rods) are used to control the power for ordinary operations of startup and shutdown of the nuclear reactor. The backup shutdown rods are used for emergency scram. The reactor shutdown system is designed to have two independent systems. Table 9.1 shows main characteristics of sodium. Table 9.2 shows main particulars of the plant.

9.2 Plant Control System Equipment

9.2.1 Overview of Control System

Features being considered when building a fast reactor operation control system are described in the following.

Table 9.1 Main characteristics of sodium

Item	Property values	Units
Melting point	97.82	°C
Boiling point	881.4 (at 1 atmospheric pressure)	°C
Density (solid/liquid)	0.968 (at 20 °C)/0.856 (at 400 °C)	g/cm^3
Specific heat (solid/liquid)	0.480 (at 20 °C)/0.305 (at 400 °C)	cal/g
Heat transfer coefficient[a]	60.0 (water thermal conductivity: 0.340)	kcal/m h°C
Heat transfer coefficient[a]	92,400 (water thermal conductivity: 32,340)	$kcal/m^2$ h°C

[a]Thermal conductivity and heat transfer coefficient at 300 °C with sodium at 1 atmospheric pressure and water at 87.61 atmospheric pressures

Table 9.2 Main items of MONJU fast breeder reactor

Type of nuclear reactor	Sodium cooled fast reactor
Thermal output	714 MW
Electrical output	280 MW approx.
Fuel (core/blanket)	Plutonium and uranium mixed oxide/uranium dioxide
Breeding ratio	1.2 approx.
Core fuel average fetch burn-up	80,000 MWD/T approx.
Linear heat generation rate (core average/ core maximum)	210/360 W/cm approx.
Fuel cladding tube material	SUS316
Cladding tube outside diameter/thickness (core/blanket)	6.5 approx./0.47 mm approx./11.6 approx./ 0.5 mm approx.
Core fuel cladding tube maximum temperature	675 °C approx.
Type of reactor vessel	Cylindrical vertical vessel with base mirror
Primary coolant flow rate	15.3×10^6 kg/h
Primary coolant temperature (reactor inlet/ reactor outlet)	397/529 °C
Number of loops	3
Type of intermediate heat exchanger	Vertical non-level parallel counter-flow type
Secondary coolant flow rate	3.7×10^6 kg/h (1 loop)
Secondary coolant temperature (low temperature side/high temperature side)	325/505 °C
Type of steam generator (evaporator, superheater)	Helical coil once-through separate type
Type of steam turbine	Tandem 3-cylinder 4-flow exhaust non-reheat type
Steam temperature (in front of main steam shutoff valve)	483 °C
Steam pressure (in front of main steam shutoff valve)	12.5 MPa
Amount of steam flowing in turbine	1.1×10^3 t/h

<1> Large temperature difference between the reactor vessel outlet and the inlet
The sodium coolant has superb thermal conduction characteristics and a high boiling point (approximately 880 °C at 1 atmospheric pressure). So, at any

Table 9.3 Fast reactor plant control method (Hori (ed) 1993)

Control method		
Coolant temperature at reactor outlet	Coolant flow rate	Features
Variable	Constant	Since control with pump, control valve, etc. is not necessary and the control system is simple, this method is advantageous for an experimental reactor, etc.
Constant	Variable	Since steam conditions can be kept constant even for change of load, this method is advantageous for a power reactor.

time of operation, no phase change occurs and less pressure change occurs in the sodium system. Therefore, it is possible to make large the temperature difference between the rector vessel outlet (the high temperature side) and its inlet (the low temperature side). On the other hand, thermal transient conditions are so severe to avoid large thermal changes even over a short period.

<2> A large amount of dead time due to delay in heat transportation
Since the control system is composed of the primary main cooling system (sodium), the secondary main cooling system (sodium) and the water/steam system, there is a large amount of dead time due to delay in heat transportation.

<3> Adoption of superheated steam turbine
Since it is possible to make the reactor vessel outlet temperature high, a superheated steam turbine is adopted.

These features make two methods available for control of fast reactor plants as shown in Table 9.3.

These control methods have various respective features. The control method adopted for MONJU is to proportionate the coolant flow rate to the plant output, to create a large temperature difference between the reactor vessel outlet and the inlet irrespective of plant output, and to make main steam temperature and pressure constant irrespective of plant output as requested for turbine design. As shown in Figs. 9.3 and 9.4, MONJU is then controlled.

The overview of the plant control system of a MONJU reactor is shown in Fig. 9.5.

The plant control system of MONJU forms a hierarchical system with the output command device on the top. Under the device, there are output command device, reactor output control system, primary main cooling system flow rate control system, secondary main cooling system flow rate control system, feed water flow rate control system and sub-systems for main steam temperature control system and main steam pressure control system, all of which are required by the turbine.

The reactor output control system changes the reactor output with the control rod in accordance with the plant output command showing electrical output from the output command device of the plant control system in percentage form while controlling the temperature of sodium at the reactor vessel output in accordance with the output command. The primary main cooling system flow rate control

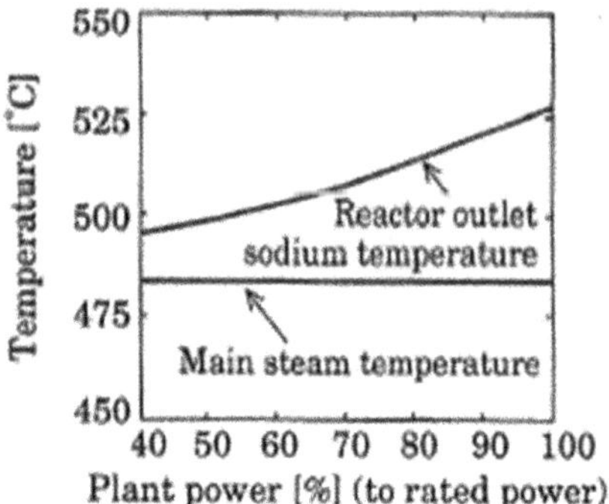

Fig. 9.3 Relationship between plant power and reactor vessel outlet temperature/main steam temperature

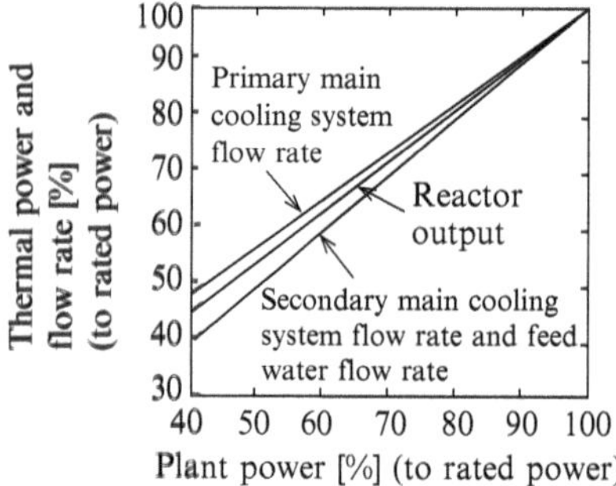

Fig. 9.4 Relationship between plant power and reactor power/coolant flow rate

system, the secondary main cooling system flow rate control system, and the feed water flow rate control system control coolant flow rates so as to satisfy the output command. On the other hand, the control method to make the main steam temperature and the pressure constant is adopted for the main steam temperature control system and the main pressure control system so as to satisfy the superheated steam turbine requirements. In Table 9.4, control and operation items with each automatic control system are summarized.

9.2.2 Basic Functions of Control System Sub-systems

[1] Output command device

Figure 9.6 shows the block diagram of the output command device.

This device is available in two modes, i.e., "Calculator Mode" which receives plant output command target value signals from the central computer and "Manual Mode" for which an operator sets plant output command target value signals to the output command station by manual operation. Step-like plant output command target value signals input with the "Calculator Mode" or "Manual Mode" are

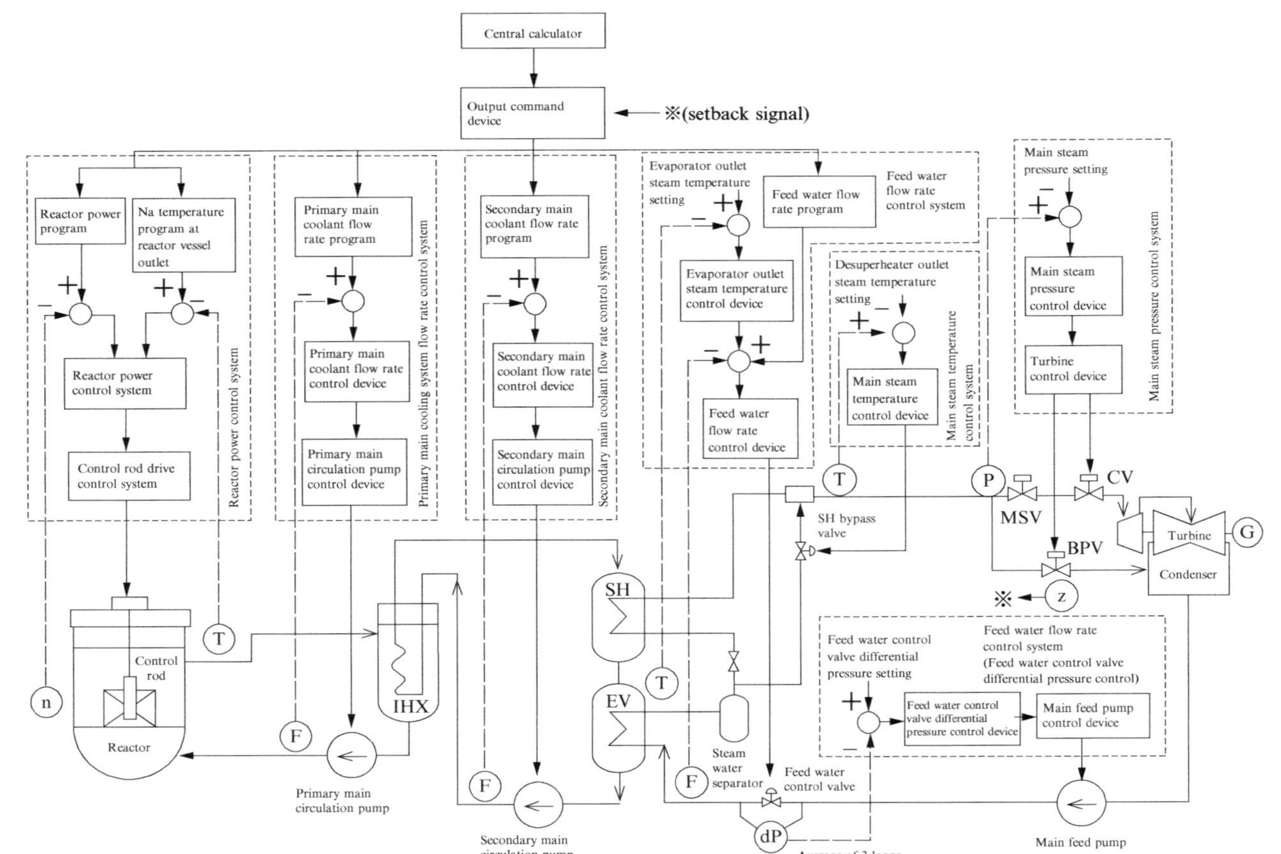

Fig. 9.5 Outline diagram of plant control system

Table 9.4 Control and operation items in automatic control states

Control system names		Control items	Auxiliary signals	Operation items
Reactor power control system		Na temperature at reactor vessel outlet	Neutron flux	Na temperature at reactor vessel outlet
Primary main cooling system flow rate control system		Primary main cooling system Na flow rate	–	With variable speed fluid coupling M-G set Position of rake pipe with fluid coupling
Secondary main cooling system flow rate control system		Secondary main cooling system Na flow rate	–	Thyristor inverter output frequency
Feed water flow rate control system	Evaporator outlet steam temperature control	Evaporator outlet steam temperature	Feed water flow rate	Feed water control valve opening
	Feed water control valve differential pressure control	Differential pressure between feed water control valve outlet and inlet	–	Turbine steam regulator valve for driving feed water pump
Main steam temperature control system		Desuperheater outlet steam temperature	–	Superheater bypass valve opening
Main steam pressure control system		Main steam pressure	–	Steam control valve opening, turbine bypass valve opening

transmitted to the lower control sub-systems at the specified variation rate with an output command variation rate setting device and according to variation rate restricting element (1). The output variation rate is 0.4 %/min. at the time output rises, or 2 %/min. at the time output falls.

When the load on the power generator reduces suddenly, the steam usage and supply are unbalanced resulting in increasing main steam pressure. To maintain constant main steam pressure, the main steam pressure control system, mentioned later in item [7], opens the turbine bypass valve. To eliminate the imbalance between steam usage and supply of the entire plant, the system also has functions to reduce reactor output (setback), to detect opening of the turbine bypass valve, and to transmit the setback target value converted to a previously determined variation rate signal to the lower control systems with the setback variation rate setting device and according to variation rate restricting element (2). The setback signal is transmitted when the turbine bypass valve opens. The setback change-over element is changed over to the setback side according to the setback signal and lowers the plant output command at the lowering rate (5 %/min.) previously set to the setback variation rate setting device. The plant output command continues lowering the plant output command down to 40 % set to the lower limit setting device or until the setback signal is released (the turbine bypass valve shuts). When the setback signal is released, the plant output command holds the value at that time.

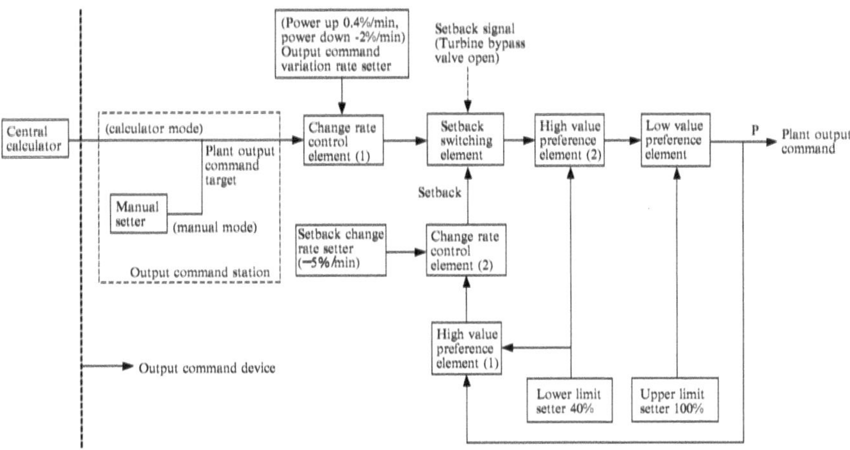

Fig. 9.6 Block diagram of output command device

[2] Reactor power control system

The reactor power control unit transmits fine control rod drive speed control signals to the control rod drive mechanism so that the reactor vessel outlet sodium temperature comes to the value corresponding to the plant output command target value signal.

Figure 9.7 shows the block diagram of the reactor power control system.

The plant output command is input to the output command arithmetic circuit ($f_1(x)$) for reactor vessel outlet sodium temperature control target value signal (T_{REF}) and to the output command arithmetic circuit ($f_2(x)$) for reactor power target value signal (Q_{REF}).

The output command arithmetic circuit ($F_1(x)$) is a function programmed to seek the reactor vessel outlet sodium temperature control target value signal (T_{REF}) that realizes appropriate heat balance at the outlets of each plant for the plant output command (P). This is to be nonlinear function as shown in Fig. 9.8. On the other hand, the output command arithmetic circuit ($f_2(x)$) is a function programmed to seek the reactor power (neutron flux) target value signal (Q_{REF}) at each plant output level for the plant output command (P). This is to be linear function as shown in Fig. 9.8.

Each loop of the primary main cooling system is fitted with a 2-channel reactor vessel outlet sodium thermometer for control. The thermometers create a high value out of these temperature signals to 3-loop average temperature signal (T_{AVG}) as the high sorting signal. In the temperature measurement, such thermal delays occur in the order of extraction/insertion of the fine control rod, output change and change of reactor vessel outlet sodium temperature when the output command changes. This is because the temperature detector fitted on the primary main cooling system piping is a long way from the reactor outlet and the large heat capacity of the coolant causes response delay. So, the control characteristics need to be improved

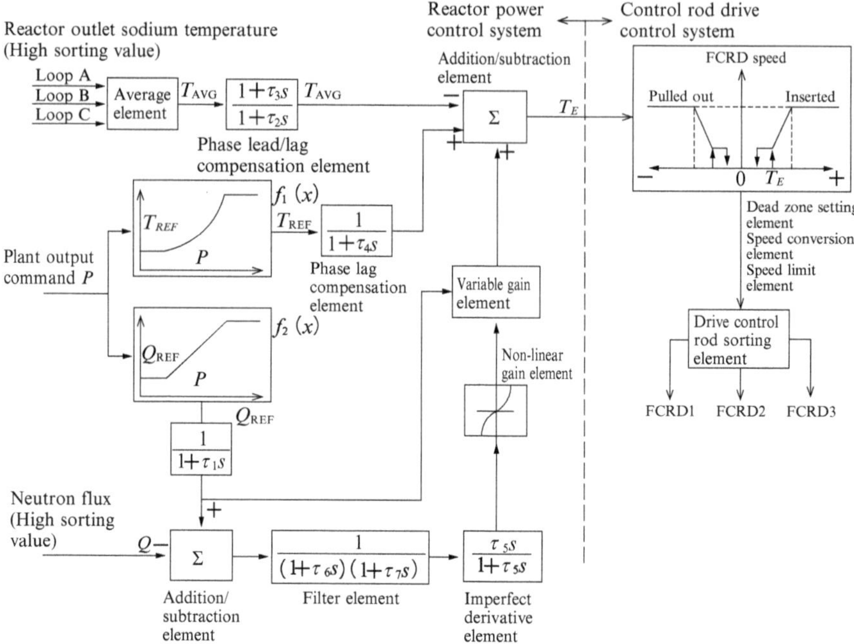

Fig. 9.7 Block diagram of reactor power control system

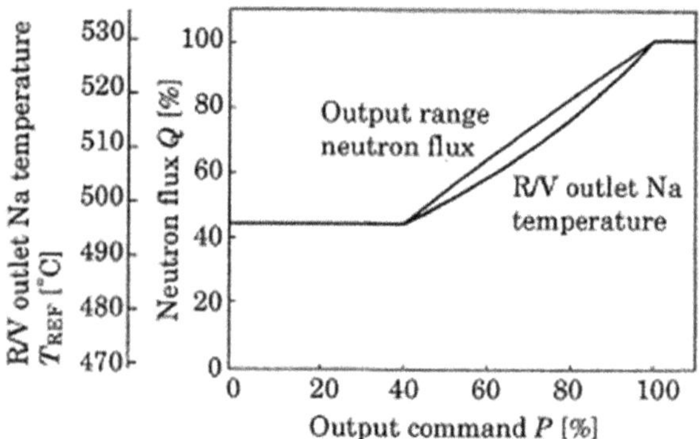

Fig. 9.8 Reactor program and reactor vessel outlet Na temperature program

by providing the after-mentioned auxiliary signal and compensating the delay of the average temperature signal using the phase/lag compensation elements. These lag compensation elements have a function that makes the T_{REF} signal equivalent to thermal transport delay and suppresses overshoot or undershoot.

The auxiliary signal (neutron flux signal) is available to improve quick response and stability when reactor power changes. When an output command is changed, T_{AVG} may excessively deviate from the T_{REF} signal (overshoot). To restrict the

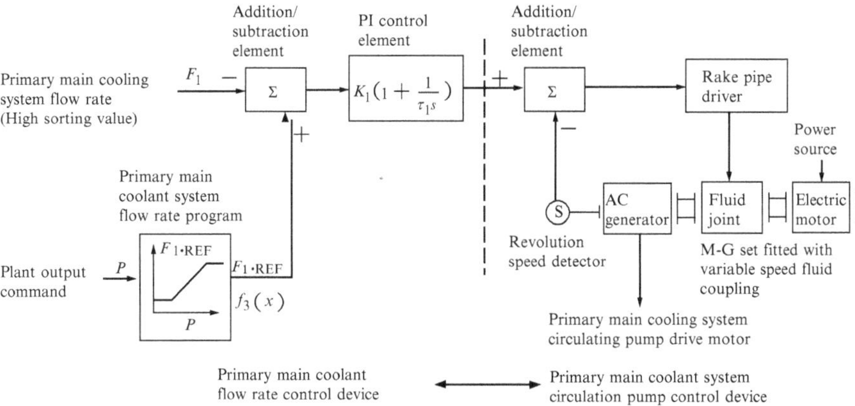

Fig. 9.9 Block diagram of primary main coolant flow rate control system

deviation, stability is obtained by feeding back a neutron flux signal equivalent to the reactor power without causing thermal delay. After being received, the neutron flux signal (Q) is compared with the reactor power target value signal (Q_{REF}). The deviation signal between Q_{REF} and Q passes through filter wave element, imperfect derivative element, nonlinear gain element and variable gain one after another and is added to the main control signal.

In the control rod drive control unit, the control signal (T_E) of the fine control rod drive device (FCRD) passes through dead zoned element, speed change element, speed limit element, drive control rod sorting element, etc. for protecting the FCRD. Then the control rod is inserted or extracted with the FCRD.

[3] Primary main cooling system flow rate control system

This control system includes functions to create a control signal by comparing the primary main cooling system flow rate signal with a flow rate control target value from the output command device to create a control signal, and to control the primary main cooling system flow rate of each loop by adjusting the revolution speed of the primary main cooling system circulating pump so that the flow rate will come to be approximately proportional to the plant output. The revolution speed of the primary cooling system circulating pump is adjusted with the M-G set fitted with a variable speed fluid coupling that has large inertia force in order to secure the primary cooling system flow rate immediately after a reactor scram.

The primary main cooling system flow rate control systems control each loop separately, and Fig. 9.9 shows the block diagram of the primary main cooling system flow rate control system for a loop.

The plant output command (P) is input to the output command arithmetic circuit ($f_2(x)$) for the primary main cooling system flow rate control target value signal ($F_1 \cdot {}_{REF}$). This arithmetic circuit is a function programmed to seek the primary main cooling system flow rate control target value signal ($F_1 \cdot {}_{REF}$) at the output level of each plant, and the function programmed is linear as shown in Fig. 9.10.

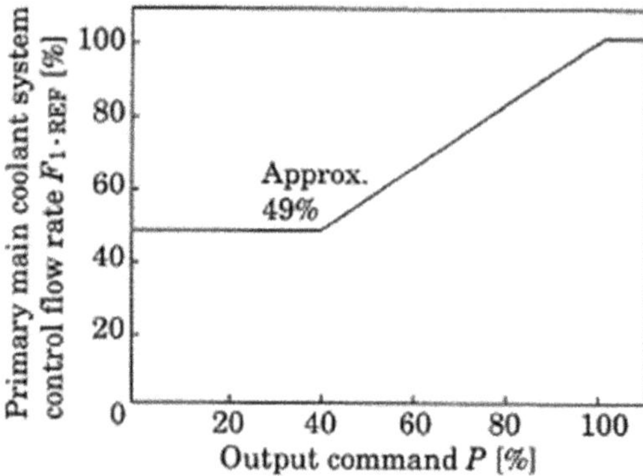

Fig. 9.10 Primary main coolant system flow rate program

A 2-channel primary main cooling system flowmeter for control is fitted to each loop of the primary main cooling system and creates a flow rate signal (F_1) making a high value out of the flow rate signals a high sorting signal. This F_1 and the primary main cooling system flow rate control target value signal ($F_{1 \cdot REF}$) create a flow rate deviation signal. The flow rate deviation is input to the PI control element which creates a revolution speed command signal based on it. And then the signal is sent to the primary main cooling system main circulating pump control unit through the speed limit element so that no excessive revolution speed command signal is output.

In the primary main cooling system main circulating pump, the revolution speed command signal is compared with the generator revolution speed measuring signal of the M-G set fitted with a variable speed fluid coupling, and the deviation signal is input to the speed control element. The speed control element is processed as a signal so that the above deviation signal satisfies the characteristics of the rake pipe driver.

The input power source (6,600 V/60 Hz) makes the voltage to frequency ratio (V/F) of the generator constant by controlling the position of the rake pipe of the variable speed fluid coupling on the M-G set and is converted to the one of 684–3,420 V/11–57 Hz approx. satisfying the plant output command and is supplied to the primary main cooling system main circulating pump drive motor to adjust the revolution speed.

[4] Secondary main cooling system flow rate control system

This control system changes the output frequency and voltage of the static variable frequency power supply unit (VVVF: variable voltage and variable frequency control method) of the secondary main cooling system circulating pump, so that the secondary main cooling system flow rate comes to the value corresponding to the plant output command target signal and controls the revolution speed of the secondary main cooling system circulating pump. Having basically the same configuration as the primary main cooling system flow rate control unit, this control

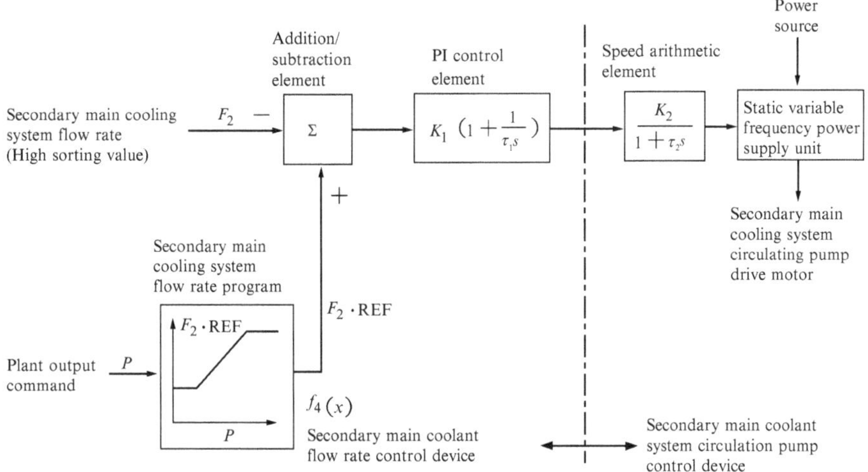

Fig. 9.11 Block diagram of secondary main coolant flow rate control system

system transmits signals corresponding to the plant output command and the output target signals from the master station to the circulating pump control unit. Figure 9.11 shows the block diagram of the secondary cooling system flow rate control system.

The plant output command (P) is input to the output command arithmetic circuit ($f_4(x)$) for the secondary main cooling system flow rate control target value signal ($F_2 \cdot {}_{\mathrm{REF}}$). This circuit is a function programmed to seek the primary main cooling system flow rate control target value signal ($F_2 \cdot {}_{\mathrm{REF}}$) at the output level of each plant for the plant output command (P) and comes to be linear as shown in Fig. 9.12.

A 2-channel secondary main cooling system flowmeter for control is fitted to each loop of the secondary cooling system. This creates a flow rate signal (F_2), which makes a high value coming out of the flow rate signal a high sorting signal. This F_2 and the secondary main cooling system flow rate control target value signal ($F_2 \cdot {}_{\mathrm{REF}}$) create a flow rate deviation signal. The flow rate deviation is input to the PI control element and output as a revolution speed command signal.

In the secondary cooling system main circulating pump control unit, this revolution speed command signal passes through the speed arithmetic element, which has the first-order lag function to match the response characteristic with the M-G set fitted with variable speed fluid coupling, i.e., the primary main cooling system circulating pump drive motor driver, and the revolution speed limit function to limit increase of the secondary main cooling system flow rate due to failure of the secondary main cooling system flow rate control unit, and is output to the VVVF. The input power source (6,600 V/60 Hz) makes the voltage to frequency ratio (V/F) of the generator constant through these controls and is converted to the one of 1,260–3,920 V/18–56 Hz approx. satisfying the plant output command and is supplied to the secondary main cooling system main circulating pump drive motor to adjust the revolution speed.

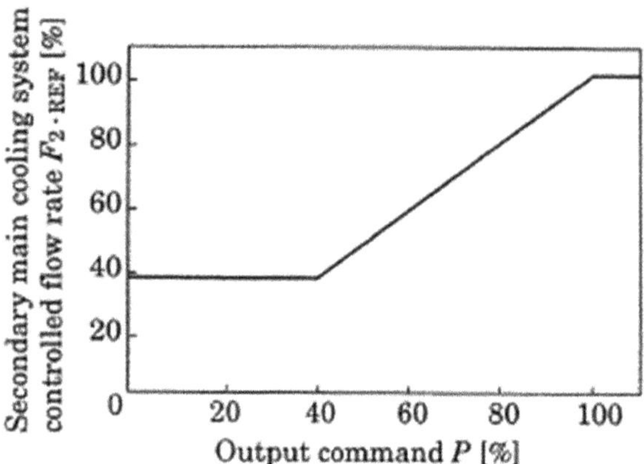

Fig. 9.12 Secondary main coolant system flow rate program

[5] Feed water flow rate control system

This control system has a function to make the deviation between evaporator outlet steam temperature and an evaporator outlet steam temperature control target value the main control signal so that the degree of superheating of the evaporator outlet steam temperature becomes constant. It also makes the deviation between the feed water flow rate of each loop and the feed water flow rate control target value from the output command device, an auxiliary signal in order to improve the response so that it adjusts the opening of the feed water control valve of each loop and controls the feed water flow rate of each loop.

In addition, this control system has a function that detects differential pressure before and after the feed water control valve so that the feed water control valve of each loop can control the valve opening in order to deliver high control performance, to adjust the feed water pump revolution speed so that the average signal makes the differential pressure of the feed water control valve constant, and to control the total feed water flow rate.

Figure 9.13 shows the block diagram of the feed water flow rate control system (evaporator outlet steam temperature control).

A 2-channel evaporator outlet steam thermometer is fitted to each loop of the feed water system and creates a temperature signal (T_W) making a high value out of the temperature signal the high sorting signal. This Tw signal and the evaporator outlet steam temperature control target value signal ($T_{w} \cdot {}_{REF}$) to create a temperature deviation signal. In the temperature measurement, the temperature detector fitted to the feed water system piping causes a response delay, and a temperature change causes a thermal delay. Therefore, the auxiliary signal mentioned later is provided and the temperature deviation signal is passed through the phase/lag compensation element to compensate for the delay and to improve the control characteristics. The temperature deviation signal is input to the PI control element, changes to a flow rate command signal, and is output to adjust the feed water control valve.

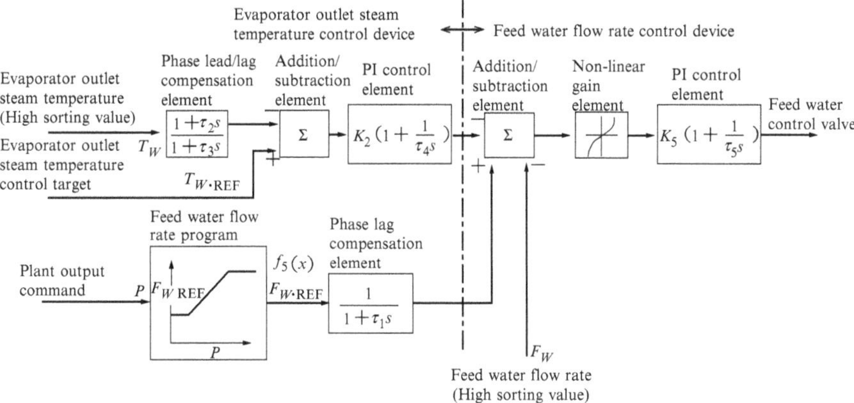

Fig. 9.13 Block diagram of feed water flow rate control system

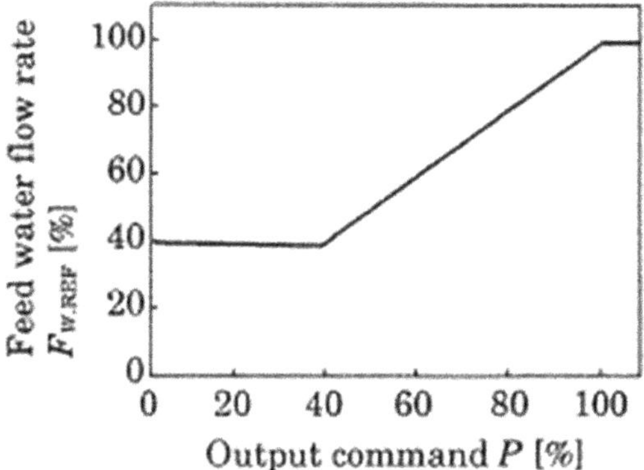

Fig. 9.14 Feed water flow rate program

The auxiliary signal (feed water flow rate signal) is available to improve quick response and stability at the time of a plant output command change. A 2-channel feed water flow meter for control is fitted to each loop of the feed water system, and the feed water flow rate signal creates a feed flow rate signal (F_w) making a high value out of the flow rate signal a high sorting signal. The feed water flow rate control target value signal ($F_w \cdot {}_{\mathrm{REF}}$) obtained from this F_w and the plant output command (P) and the feed water flow rate program $f_5(x)$ shown in Fig. 9.14 is added to or reduced from the above-mentioned flow rate command signal. The output of the feed water flow rate program has a function to delay the $F_w \cdot {}_{\mathrm{REF}}$ signal equivalent to temperature response delay of the temperature signal to improve the controllability.

Figure 9.15 shows the block diagram of the feed water flow rate control system (feed water control valve differential pressure control).

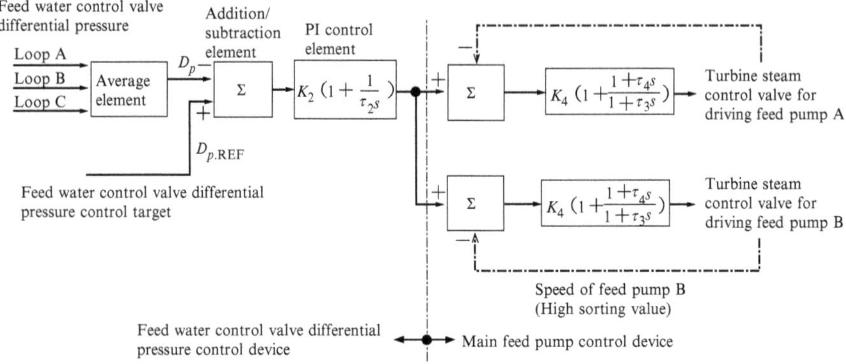

Fig. 9.15 Block diagram of feed water flow rate control system (feed water control valve differential pressure control)

A 2-channel feed water control valve differential pressure gauge for control purposes is fitted to each loop of the feed water system. This creates an average differential pressure signal (D_{p}) of differential pressure signals of the three loops by making a high value out of the differential signals a high sorting signal. This D_{p} signal and the feed water control valve differential pressure control target value ($D_{\mathrm{p*REF}}$) creates a differential pressure deviation signal.

This differential pressure deviation signal is input to the PI control element, changes to a differential pressure command signal, output to the turbine revolution speed control unit for driving feed water pump and adjusts the opening of the turbine steam regulator valve for driving the main feed water pump.

[6] Main steam temperature control system

This control system has a function to detect rise of desuperheater outlet steam temperature to protect the turbine, to adjust opening of the superheater bypass valve of each loop to bypass super heater inlet steam to the super heater outlet side when this temperature rises over the control target value and to maintain the main steam temperature at the predetermined value.

Figure 9.16 shows the block diagram of the main steam temperature control system.

A deviation signal is created from steam temperature at the outlet of desuperheater fitted at the outlet of the superheater of each loop and the same control target value and input to the PI control element. An opening command signal of the superheater bypass valve is output from the PI control element.

[7] Main steam pressure control system

This control system has a function to detect main steam pressure fluctuation resulted from output change on the reactor side and to control opening of the steam control vale so that the main steam pressure become constant. Also, the control system has a function to detect rises of main steam pressure, to open the turbine

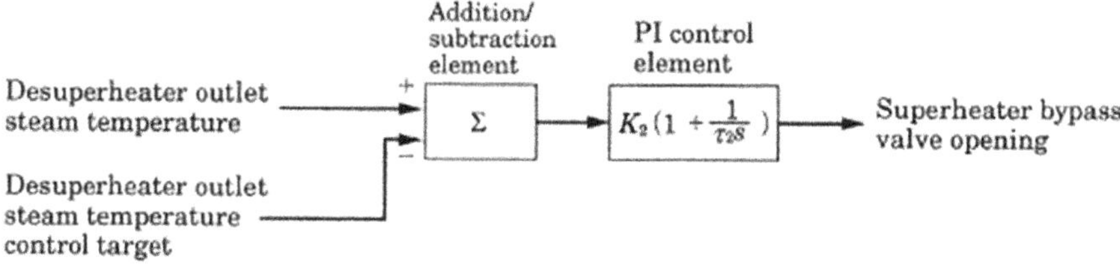

Fig. 9.16 Block diagram of main steam temperature control system

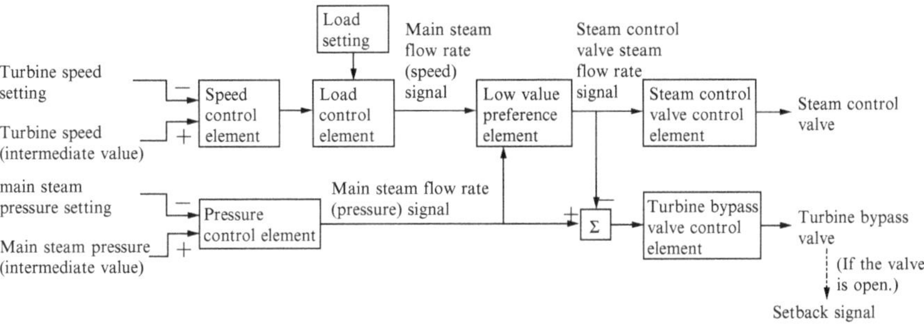

Fig. 9.17 Block diagram of main steam pressure control system

bypass valve, and to reduce main steam pressure in order to maintain the main steam pressure at the predetermined value when the load reduces and the main steam pressure rises because the steam control valve is shut.

Figure 9.17 shows the block diagram of the main steam pressure control system.

The turbine is fitted with a 3-channel revolution speed detector, and the medium value of the revolution speed signal is regarded as the turbine revolution speed signal, and the signal is input to the speed control element together with the speed setting signal. In addition, an output from the speed control element is input to the load limit control element together with load setting signal, and an output is regarded as a main steam flow rate (speed) signal and is input to the low value preference element. On the other hand, the main steam pressure control system is fitted with a 3-channel pressure detector, and the medium value of the pressure signal is regarded as a main steam pressure signal and is input to the pressure control element together with the main steam pressure setting signal, and an output is input to the low value preference element together with the main steam flow rate (pressure) signal.

The steam control valve steam flow rate signal output from the low value preference element is input to the steam control valve control element and is output as the steam control valve opening command signal to the steam control valve. Also, the steam control valve steam flow rate signal is compared with the main steam flow rate (pressure) signal output from the pressure control element, and the deviation, which is regarded as the turbine bypass valve steam flow rate, is input to the turbine bypass valve control element and output as the turbine bypass valve opening command signal to the turbine bypass valve.

A turbogenerator is generally synchronized with an electric power system, and there is little deviation between its revolution speed set point and its turbine revolution speed (actual speed). Therefore, its main steam pressure set point and the above set point are mostly equal, and the steam control valve is controlled by "pressure control element → low value preference element → steam control valve control element." This means that the steam control valve opening command signal and the main steam flow rate (pressure) signal are equal, and the turbine bypass valve steam flow rate reaches zero and the turbine bypass valve is completely shut. However, when the turbogenerator revolution speed (the electric power system frequency) rises, the deviation from the speed set point increases, the main steam flow rate (speed) signal becomes smaller than the main steam flow rate (pressure) signal because the revolution speed is tried to be lowered, and the steam control valve steam flow rate signal output from the low value preference element becomes the main steam flow rate (speed) signal. Therefore, a deviation occurs between the steam control valve opening command signal and the main steam flow rate (pressure) signal, and the turbine valve opens as large as the value, and a setback signal is transmitted.

9.3 Operation Control of MONJU

9.3.1 Concept of Operation Control at the Time of Plant Output Change

Table 9.5 shows processing volume in plant output changes (rise and fall).

MONJU uses sodium as coolant and can raise sodium temperatures of the primary system and the secondary system and also main steam system steam temperature, contributing to an improvement in heat efficiency. On the other hand, high temperature and sodium heat transfer several times larger than that of water makes severer the thermal transient conditions when temperature changes. So, a limit is set in the temperature change of sodium caused at the time of plant output change in order to secure structural integrity.

Specifically, the reactor outlet the temperature change of sodium rate is 15 °C/h in the period from the time of reactor startup to the time of 40 % plant output, and the plant output change rate is limited to 0.4 %/min. in the period from 40 % to 100 % of plant output in order to raise the output for security purposes to limit the temperature change rate. At the time of a fall in output, the sodium temperature change is limited in order to secure structural integrity. The reactor output change rate is limited to 2 %/min in the period when plant output falls from 100 % to 40 %, and the outlet sodium temperature change rate is limited to 15 °C/h in the period from the time when plant output reduces from 40 % to the reactor shutdown. In addition, even when the plant load reduces suddenly to 50 % as a result of abnormal plant operating conditions during operation with 100 % output, output falls at the

Table 9.5 Main processing volumes for various plant output states

Processing volume	Planet output (%)			
	At shutdown	At criticality	40	100
Reactor power (%)	–	0	45	100
Reactor outlet sodium temperature (°C)	200	200	493	529
Intermediate heat exchanger outlet secondary system sodium temperature (°C)	200	200	489	505
Main steam temperature (°C)	–	–	483	483
Feed water temperature (°C)	–	195	195	240

rate of 5 %/min in order to secure structural health in a similar way as to normal conditions, and to limit the temperature change of sodium.

9.3.2 Operation Control in Plant Startup Operation

[1] Operation control until after parallel to generator power system (rated plant output: 40 %)

This plant starts the primary main cooling system main pump, the secondary main cooling system main pump, the feed water system condensate pump, starting feed water pump, etc., while the steam generator and the turbine does not contain any water/steam before startup, and they secure flow rates equivalent to 40 % of the rated plant output. Then, the control rod is pulled out, nuclear heating starts, and the plant reaches a state where plant output is 40 % with operation control of the water and steam systems.

In the water·steam system, there shall be phase changes, such as water in single phase, air, and liquid in double phases and superheated steam. The piping shall not contain an electric preheater or similar device in the sodium system. Since an evaporator generates superheated steam to prevent moisture from entering the superheater, and the superheater is aerated, the operation control is different from that of the sodium system. Therefore, the piping is blown by water and is initially warmed and also water is passed through the evaporator. The systems are changed over by using the startup bypass system to the flush tank and then the superheater and the generator turbine are started. The plant is prepared for operation of the output control system in this way.

At this time, the plant comes to the state of 40 % output from the shutdown condition by raising the temperature, flow rate and pressure, and the control system is also used to maintain this state. To increase the feed water flow rate, the flow rate setting is gradually raised by hand and the PI control system works to open the feed water control valve to the setting. Also, to change other plant parameters by operating other control valve, etc., the plant state changes and the control system works to maintain the set point. For example, to raise the flush tank pressure while

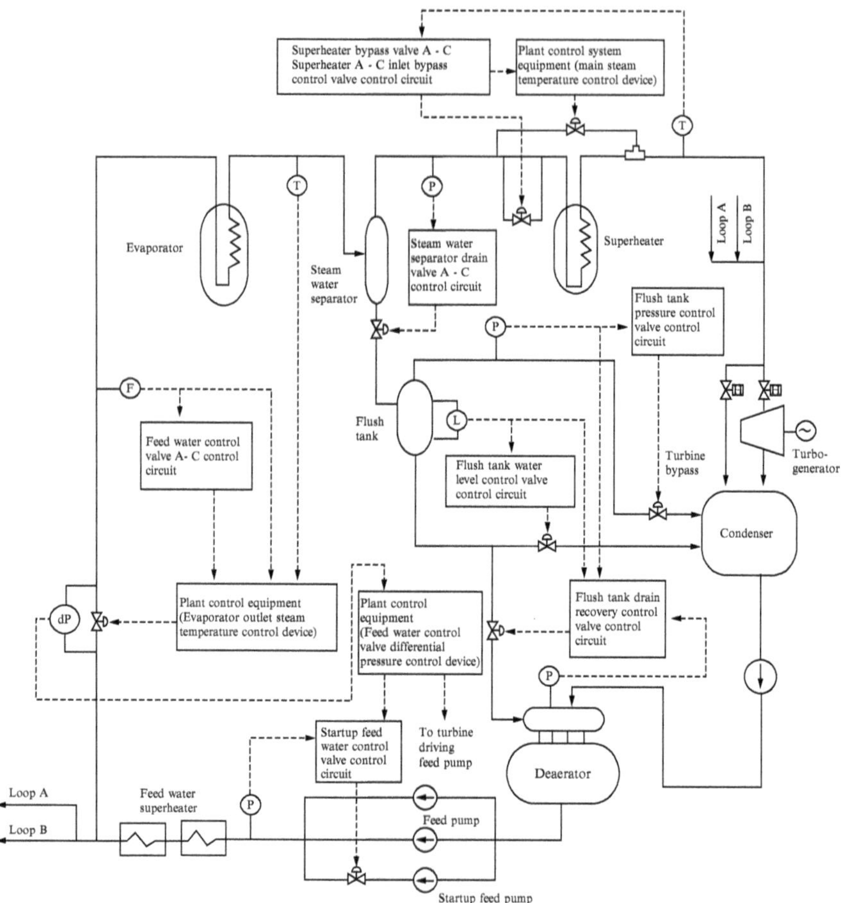

Fig. 9.18 Diagram of water/steam systems around the steam generator

maintaining a constant flow rate, the flush tank pressure control valve is apt to shut but the feed water flow rate control system works and the feed water control valve comes to open automatically. In case of lowering the pressure, the operation in reverse is carried out.

The operation control steps at the time of startup are explained below. In addition, Fig. 9.18 shows the diagram of the water*steam system around the steam generator, and Table 9.6 shows control items and operation items at the time of startup of the water and steam systems.

<1> Operation control from blowing to warming and flowing water through the evaporator

Just before the evaporator is started, the sodium system is at a temperature of approximately 200 °C and flows on the sodium side of the steam generator. The water system feeds water to piping for the purpose of filling water to

Table 9.6 Control and operation items at startup of water-steam system

Control system names		Control items	Auxiliary signals	Operation items
Feed water flow rate control system	Evaporator outlet steam temperature control	Evaporator outlet steam temperature	Feed water flow rate	Feed water control valve opening
	Feed water control valve differential pressure control	Feed water control valve outlet/inlet differential pressure	–	Opening of turbine steam control valve for driving main feed pump
Main steam temperature control system		Desuperheater outlet steam temperature	–	Superheater bypass valve opening
Startup feed water control system		Feed pump outlet header pressure	–	Feed pump outlet pressure control valve opening
Steam water separator drain valve control system (Loop A to Loop C)		Steam water separator outlet pressure	Superheater inlet sodium temperature	Steam water separator drain valve opening
Flush tank pressure control valve control system		Flush tank outlet pressure	–	Flush tank pressure control valve opening
Flush tank water level control valve control system		Flush tank water level	–	Flush tank water level control valve opening
Flush tank drain recovery control valve control system		Flush tank outlet pressure, deaerator pressure, flush tank water level	–	Flush tank drain recovery valve opening

blow, and starts the feed pump for startup by using water from the deaerator. The feed water control valve controls the feed pump outlet header pressure and feeds water. Auxiliary steam is used to clean piping of the feed water system and the water-steam system around the steam generator by hot water. After the cleaning, the evaporator is filled with water of 195 °C approx by feeding. At this time, the feed water control valves starts control of feed water flow rate to the evaporator, and the steam water separator drain valve starts control of the steam water separator outlet pressure.

<2> Operation control from control system automatic operation to the superheater bypass operation, turbine startup, and superheater aeration

Water/steam fills the evaporator separates from the steam water separator and flows in the startup bypass line to the flush tank. The output is raised by nuclear heating from the reactor side and changes over to the evaporator outlet steam temperature control of the plant control system at the rate of approximately 38 %.

Then, the flow route to the flush tank is changed over to the superheater bypass, and the steam water separator drain valve is throttled, increasing the

flow rate of the superheater bypass. The generator turbine is started by using steam though the superheater bypass, and then the flow route changes over from the superheater bypass valve to superheater inlet bypass valve to complete aeration to the superheater.
At this time, the startup feed pump outlet header pressure control of the startup feed pump leading feed water to the high pressure feed water heater·evaporator, the feed water flow rate control and the evaporator outlet steam temperature control work. Also, the steam water separator outlet pressure is controlled by the steam water separator drain valve on the startup bypass line to the flush tank before change-over, and steam from the evaporator is led to the flush tank. In the flush tank, the liquid phase is drained, heat recovery to the deaerator is made by regulating and controlling the flush tank drain recovery valve, the pressure in the deaerator is controlled and excess drain is led to the condenser by the water level control valve to control the water level in the flush tank constant. Steam generated from the flush tank is used for heat recovery to the deaerator and the feed water heater, and excess steam is led to the condenser by controlling the flush tank pressure control valve.

<3> Operation control of generator parallel to power system and introduction of plant control system with 40 % output
Then, the generator is paralleled to the electric power system with the rated plant output of approximately 38 %. Since the startup feed pump driven by the electric motor is operating at this stage, it is changed over to two main feed pumps driven by turbine. Also, the revolution speed of the main feed pump is controlled by controlling differential pressure of the feed water control valve. Main steam pressure is approximately 11.9 MPa, main steam flow rate 465 t/h, and main steam temperature 483 °C become steam conditions for the turbine, and then the operation control moves to the plant control system by means of the plant output command device.

[2] Operation control from reactor output 45 % (plant output 40 %) to plant output 100 %

After the reactor power reaches 45 %, the electrical output comes to 40 %. At this stage, sub-systems of the rector power control system and other plant control systems determine set points by themselves. Then the "Demand Mode" for control based on these set points changes over to the "Cascade Mode" for control based on set points of command value from upper systems, and the control based on demand signal from the output command device starts. The operation control from plant output 40 % to plant output 100 % is as follows:

<1> The output command target value (100 %) from the output command device is set with the central calculator from the automation system or by manual operation on the output command station. When new fuel is loaded, the output command target value is set to 89 % to secure fuel soundness by accurately proceeding with tissue change of new fuel in the plant output raising process, and the output raising method described in the following <3> is adopted.

<2> The output command device outputs target value signals to each lower control system at the change rate of 0.4 %/min up to the plant output 89 % through the command change rate setting device and the change rate limit value circuit. Each lower control system raises the rector power, the primary main cooling system flow rate, the secondary main cooling system flow rate, and feed water flow rate according to the target value signals.

<3> When new fuel is loaded, the following output raising method is taken from the plant output 89 % (reactor power 90 %) to the plant output 100 % (reactor power 100 %):

- The output rise stops once at the plant output 89 %, and the output is held for 24 h.
- After the output was held for 24 h, the output command target value is set to 100 % from the output command device and the plant output rises up to the plant output 100 % at the change rage of 0.4 %/min.
- After the plant output reaches 100 %, the output command target value from the output command device is fixed to 100 % and the output is held for 24 h. In other words, the combustion compensation operation by the fine control rod is not carried out.

9.3.3 *Operation Control During Steady Operation*

In the steady operation with the rated plant output of 100 %, the output command target value from the output command device is 100 %, and each lower control system regulates the reactor power, the primary main cooling system flow rate, the secondary main cooling system flow rate, and the feed water flaw rate according to the target signal.

As the steady operation continues, fuel loaded in the rector core burns on, and the fine control rod is gradually pulled out to maintain the reactor power at 100 %. At this time, the three fine control rods are alternately pulled out in the fixed sequence so that the neutron flux in the reactor core does not strain.

When the fine control rods are pulled out up to the determined position, the coarse rod is inserted by hand, and the fine control rods are inserted to the predetermined position just like to compensate it. After the combustion compensation operation finished following the position adjustment with these fine control rods and the coarse rod, the reactor power is regulated again with the fine control rods.

9.3.4 *Operation Control During Plant Shutdown Operation*

Even when the plant output is lowered, the plant output is automatically controlled from the rated plant output 100 % to 40 %. At this time, the operation is as follows:

<1> The automation system sets the output command target value (40 %) for the output command device from the central calculator or by manual operation on the output command station.
<2> The output command device outputs the target value signals to each lower control system at the plant output change rate of 2 %/min through the command change rate setting device and the change rate limit value circuit.
<3> Each lower control system reduces the reactor power, the primary main cooling system flow rate, the secondary main cooling system flow rate, and feed water flow rate according to the target value signals.

From the rated plant output 40 %, the reactor power is controlled to insert the control rods in accordance with the predetermined procedures to reduce the reactor power, the primary main cooling system flow rate, the secondary main cooling system flow rate, and the feed water flow rate while keeping the reactor vessel sodium temperature drop rate below 15 °C/h by hand.

9.3.5 Control When Abnormal Condition Occurs

A 50 % load quick reduction event at the plant output of 100 % is explained hereunder as an abnormal condition, and Fig. 9.19 shows the outline of the event progress.

<1> When the load reduces quickly due to a trouble of the electric power system during operation at the plant output 100 %, the electric power system frequency rises. Since the turbogenerator is synchronized to the electric power system, the revolution speed changes as the system frequency changes, but when the system frequency increases to 60 + 3 Hz or more, the main steam control valve closes to restrict increase of the frequency.
<2> Following to the closing operation of the main steam control valve, the turbine bypass valve for restring rise of main steam pressure opens. The plant control system detects that the turbine bypass valve opened and sends a setback signal to the output command device.
<3> The output command device, which received the setback signal, outputs a target value signal to each lower control system at the plant output change rate of 5 %/min through the command change rate setting device and the change rate control value circuit, and each lower control system reduces the reactor power, the primary main cooling system flow rate, the secondary main cooling system flow rate, and the feed water flow rate according to the target value signals.
<4> The setback signal is output until the target value signal of the output command device comes to 40 % or until the setback signal disappears (the turbine bypass valve closes), and the target value signal keeps decreasing. After the setback completes, the stabilization of the electric power system is awaited.

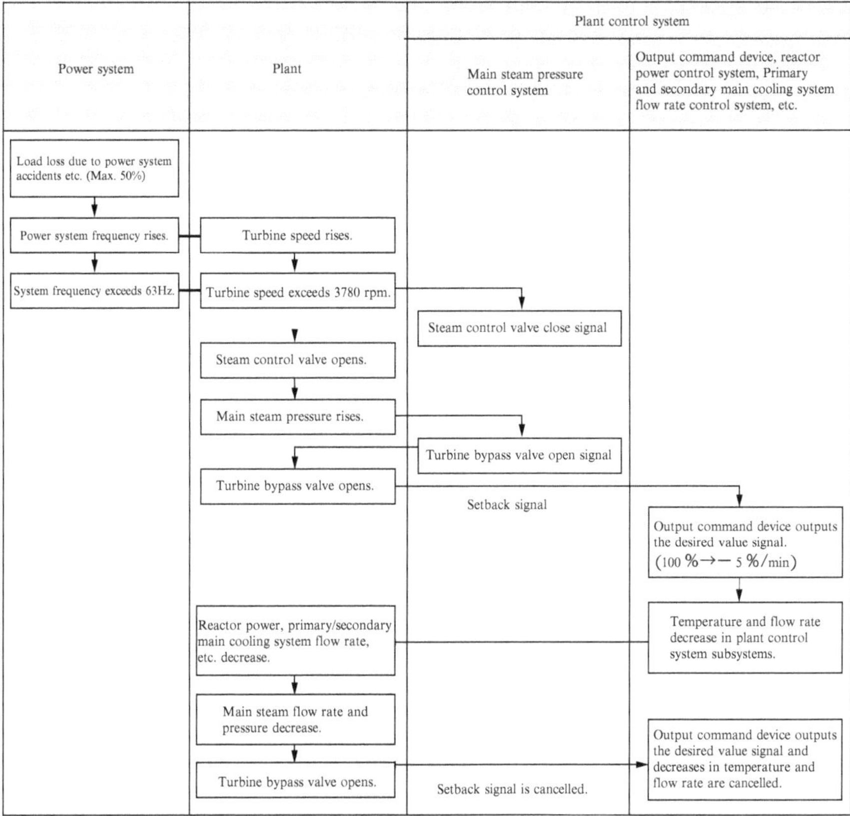

Fig. 9.19 Event sequence diagram when abrupt reduction of 50 % load occurs

When the load reduces more than 50 %, the capacity of the turbine bypass valve to be operated in <2> above becomes insufficient, and the rise of the main steam pressure cannot be restricted. So, the interlock of "main steam pressure amount" works, and a reactor trip occurs due to the predetermined interlock.

9.4 Central Control Panel

9.4.1 Functions Required for the Central Control Monitoring System

In a nuclear power station, there are plant component systems and instruments, which are linked complicatedly, and many parameters are displayed in the central control room. So, operators are requested to make appropriate judgment and operation depending on plant conditions. To operate the plant safely and certainly, the man–machine interface has to be improved from the following viewpoints:

<1> Operators can easily and certainly judge and operate.
<2> Abnormal events can be discovered early and can be treated promptly and appropriately.
<3> Operators' misjudgments and operation mistakes can be prevented.

In addition, sodium is used as coolant in MONJU, and preheating equipment for keeping sodium equipment work and sodium leak detecting device are installed to prevent sodium from being frozen during plant shutdown. As a result, many monitoring parameters and operations are required, and the central control monitoring system improving the man–machine interface is installed.

Regarding design of the central control monitoring system, the following targets are set:

<1> To improve monitoring performance and operability of operation
<2> To improve plant safety
<3> To reduce burden of operator
<4> To simplify instrumentation and control system

To realize these targets, target achievement policy and concrete measures are taken as summarized in Table 9.7.

9.4.2 Overview of Central Monitoring Control System

The central control monitoring panel, the central control panel, and the central auxiliary panel are arranged in the central control room as the central monitoring control system, and they share after-mentioned functions and achieve the above-mentioned design target and policy. Panel shapes and dimensions are designed using human engineering and referring to NUREG-0700, "Guidelines for control room design." The central control panel is a bench type, the central monitoring panel is a desk type (The panel is as high as the vertical portion of the central control panel can be seen in sitting posture.), and the central auxiliary panel is an erect type. Also, these panels are arranged considering the point to shorten flow lines of operators and are designed considering easy access to central control panel from the central monitoring panel in abnormal cases and surveillance performance from the central monitoring panel to the central control panel. Figure 9.20 shows the panel layout in the central control room.

[1] Central monitoring panel

Main functions of the central monitoring panel are <1> overall plant surveillance to display plant conditions, abnormal states, and operation guide (operation to charge and drain sodium in and from the central monitoring panel at the times of normal startup and shutdown), <2> plant operation automation including

Table 9.7 Relations among targets, target achievement policy, and concrete measures

Targets	Target achievement policy	Concrete measures
To improve monitoring performance and operability of operation	To make monitoring operation and function appropriate	• Functional partition of panel • Panel structure in view of human engineering • Panel layout in view of shortening flow line
	Centralization of plant status surveillance	• Centralization of surveillance with CRT
To improve plant safety	Measure to prevent operation mistake and misjudgment	• Panel layout in view of human engineering • Identification of equipment by color and shape • Mixing and grouping
	Early detection of abnormal condition	• Trouble surveillance with calculator
	Measures to be taken in abnormal cases	• Measures like double instrumentation in cases of plant-related problems and calculator system error
To reduce burden on the operator	To facilitate plant operation.	• Automatic control of plant control system and automation of water/steam and turbine/generator equipments • Operation guide with CRT
To simplify instrumentation and control system	To eliminate redundant functions	• To eliminate overlapping instrumentation
	Effective use of CRT-calculator	• Replacement of parts of existing instruments to CRT

automation of startup bypass system and setting plant output target value to the output command device at a plant output operation (49–100 %), <3> operation of control rods from the panels to be operated for output change and combustion compensation and <4> preheating the sodium system, charging sodium, and drain operational management.

[2] Central control panel

The central control panel is composed of the engineered safety feature panel (CRT: 1 unit), auxiliary cooling and reactor auxiliary feature panel, primary and secondary main cooling system equipment panel, reactor equipment panel (CRT: 1 unit), Water·steam system equipment panel, turbine·generator equipment panel (CRT: 1 unit), and the in-plant electric equipment panel. An alarm window to indicate abnormal conditions to operator, a CRT to indicate processing volume, an indicating instrument, a recorder, remote control switches of valves and pumps, and an indicator light of these states (open/close state of valve and run/stop state of pump).

[3] Central auxiliary panel

An alarm window, an indicating instrument and a recorder for equipment and devices not operated frequently and not operated in emergency, a remote control

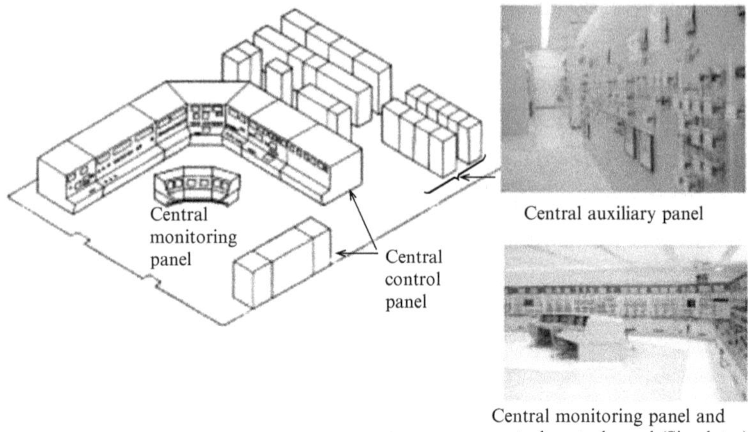

Fig. 9.20 Panel arrangements in the central control room

switch of valves and pumps and an indicator light of these states are arranged from the top on the central auxiliary panel installed on the rear side of the central control panel as shown in Fig. 9.20. When equipment and devices arranged on the central auxiliary panel get in an abnormal condition, an alarm is sounded to the central auxiliary penal as well as a general alarm for them is sent to the central control panel at the same time. So, the operator comes to the central auxiliary panel, checks the abnormal condition, and takes appropriate measures.

Chapter 9 Exercises

1. Explain why a neutron flux signal is used as an auxiliary signal in a reactor power control system.
2. Some input signals are used in a plant control system other than a main steam pressure control system, and all of them are "high sorting values" that use the highest value for later controls. Explain why "high sorting value" is used.
3. Explain why a hierarchical system with an output command device at the top is adopted in the configuration of MONJU plant control system.
4. Explain features of the fast reactor plant control system. Explain what the MONJU control system is.
5. Describe differences in steam conditions for MONJU, BWR, and PWR.
6. Describe the steam generator type in MONJU.

Bibliography

1. Hori M (ed) (1993) Basic fast reactor engineering. Nikkan Kogyo Shimbun Ltd

Chapter 10
Actual Operation and Control of High-Temperature Engineering Test Reactor

Shigeaki Nakagawa

10.1 Overview of High-Temperature Engineering Test Reactor

The high-temperature engineering test reactor (HTTR) (Fig. 10.1) is the first high-temperature gas-cooled reactor in Japan. It first attained criticality in November 1998, and the reactor outlet temperature reached 850 °C in December 2001, and 950 °C in June 2004, both world firsts.

The schematic diagram of the HTTR cooling system is shown in Fig. 10.2.

The HTTR is a graphite moderation/helium cooling-type reactor with a thermal output of 30 MW. Through the primary pressurized water cooler which is a helium-to-pressurized water heat exchanger and the intermediate heat exchanger which is a helium-to-helium heat exchanger (heat is removed for the secondary side high-temperature helium by the secondary pressurized water cooling), the heat generated in the reactor is ultimately diffused into atmosphere through the pressurized water air cooler.

The HTTR has two operation modes. One is a rated operation in which the reactor is operated at the rated power operation of 30 MW and the outlet temperature reaches 850 °C under the conditions of reactor inlet temperature 395 °C and primary coolant flow rate 12.4 kg/s. The other operation mode is a high-temperature test operation mode in which the reactor outlet temperature reaches 950 °C under the conditions of reactor inlet temperature 395 °C and the primary coolant flow rate 10.1 kg/s. Concerning cooling system composing of the primary pressurized water cooler and the intermediate heat exchanger with secondary pressurized water cooler, there are also two configurations. One is called single operation mode in which only the primary pressurized water cooler is operated. The other is parallel operation mode in which the primary pressurized water cooler and the intermediate heat exchanger are operated in parallel. Each operation mode of the reactor is operated with one of these cooling system configurations.

Y. Oka and K. Suzuki (eds.), *Nuclear Reactor Kinetics and Plant Control*, An Advanced Course in Nuclear Engineering,
DOI 10.1007/978-4-431-54195-0_10,

Fig. 10.1 Appearance of HTTR

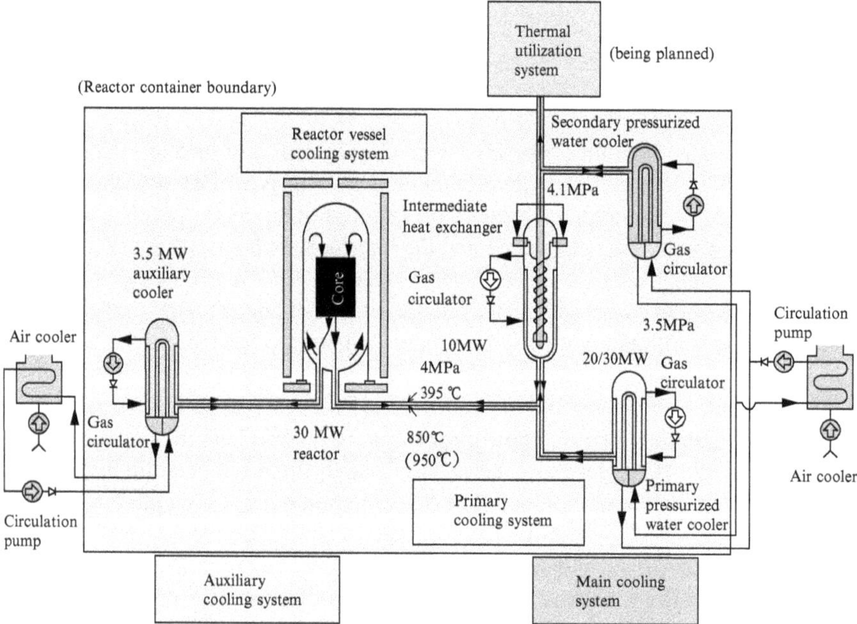

Fig. 10.2 Coolant system geometry of HTTR

10.2 Reactor Core Management

Yearly inspection is performed for the HTTR. Based on the measurements, the control rod reactivity worth, including "reactivity shutdown margin," "maximum reactivity addition rate," "excess reactivity," and "reactivity restraining effect," is evaluated and made sure to satisfy the limit values that are clearly determined in the application form for construction of reactor facility. Checks are performed before the reactor is started.

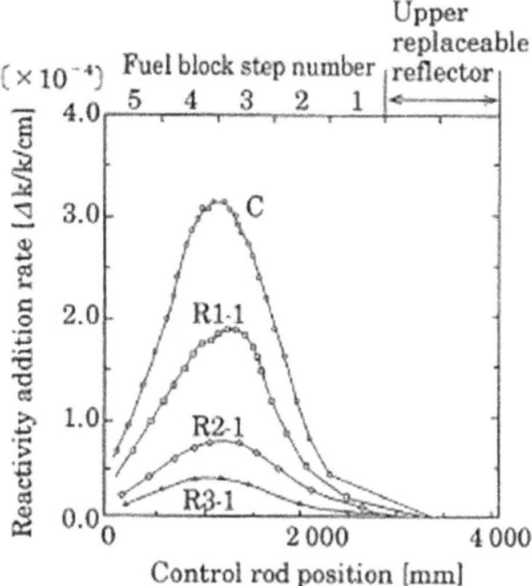

Fig. 10.3 Measurement results of reactivity change rate curve

10.2.1 Measurement of Control Rod Worth

Measurement of the control rod worth is also called as control rod calibration. The control rod reactivity change rate is obtained by dividing the reactivity change with reference to the positional change of the control rod by the positional change. And the reactivity changes in relation to the control rod positional changes are cumulated to obtain the control rod reactivity worth. The inverse kinetics method is applied to obtain the reactivity change in relation to the positional change of the control rod, by measuring the time change of the neutron flux when the control rod is pulled out, and by integrating the kinetic equation of one-point core dynamics approximation to calculate the reactivity. The inverse kinetics method is different from the rod drop method or period method. It can measure the reactivity by successively changing the neutron flux.

Figures 10.3 and 10.4 show the measurement results of the reactivity change rate curve and reactivity worth curve of the HTTR critical property test.

The HTTR includes 16 pairs of control rods. To measure the control rod worth for the control rod group (Fig. 10.5) of R1 (6 pairs), R2 (6 pairs), and R3 (1 pair), in which multiple control rods having the same performance are arranged on the same circumference of circle of the rod, the reactivity change rate curve and the reactivity worth curve are obtained for a single pair of representative control rods.

Since the C control rod located at the center of the core has only one pair, the C control rod is used as the representative control rod. Control rod worth is measured under a constant operating condition with the coolant temperature of approximately 120 °C (the core is isothermal at approximately 120 °C). Heat is input from the helium circulator and the pressurized water circulation pump.

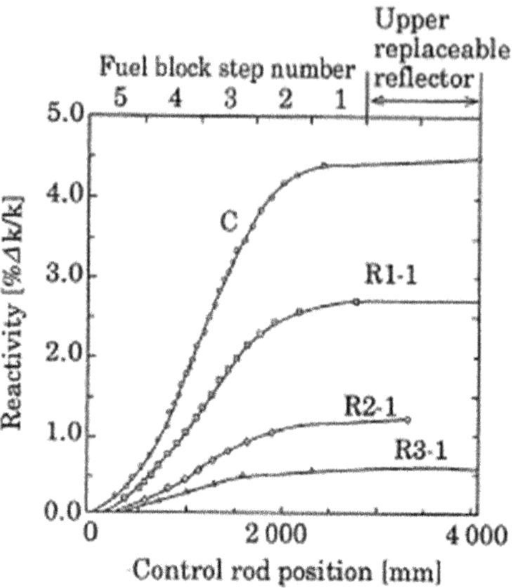

Fig. 10.4 Measurement results of reactivity worth curve

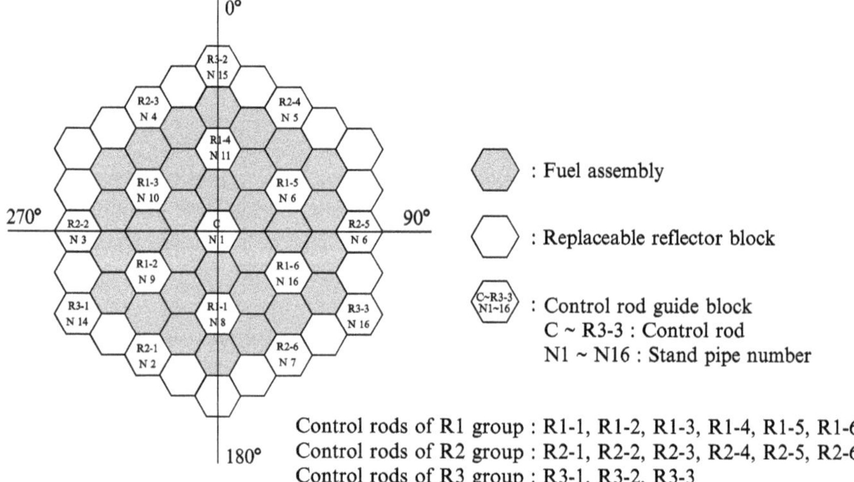

Fig. 10.5 Control rod arrangement of HTTR

10.2.2 Core Management Parameters and Method

<1> Reactivity shutdown margin

The reactivity shutdown margin is checked to ensure 0.01 $\Delta k/k$ or more in the reference condition (temperature 27 °C).

The reactivity shutdown margin is obtained by calculating the subcritical reactivity when all of the 16 pairs of control rods are inserted based on the reactivity worth curve and the critical control rod position, and by subtracting the reactivity worth of one pair of control rods with the maximum reactivity worth. The reactivity shutdown margin is obtained according to the Eq. (10.1) with correction factors for the temperature effect and the interference effect of the control rods.

$$\rho_{SD} = C \times \left(\sum \rho_i - \rho_{max}\right) - A \tag{10.1}$$

where

ρ_{SD}: Reactivity shutdown margin [$\Delta k/k$];
ρ_i: Reactivity worth [$\Delta k/k$] of one pair of i control rods that integrate over the interval from critical control rod position to all control rods inserted position;
ρ_{max}: Reactivity worth [$\Delta k/k$] of one pair of i control rod having the maximum reactivity worth;
C: Correction factor [−] for removing interference effect of the control rod;
A: Reactivity compensation [$\Delta k/k$] for converting the core temperature to 27 °C.

<2> Maximum reactivity addition rate

The reactivity addition rate is checked to ensure that the maximum reactivity addition rate among all driving strokes of the control rod is 2.4×10^{-4} $\Delta k/k$/s or less. The maximum reactivity addition rate is obtained using the Eq. (10.2) based on the reactivity change rate curve (Fig. 10.3) and the driving speed of the control rod.

$$\beta = \gamma \times V \tag{10.2}$$

where

β: Maximum reactivity addition rate [$\Delta k/k$];
γ: Maximum reactivity change rate [$\Delta k/k$/mm];
V: Control rod driving speed [mm/s] when the control rod is pulled out at high speed.

<3> Excess reactivity

The excess reactivity is checked to ensure 0.165 $\Delta k/k$ or less in the reference condition (core temperature 27 °C). The excess reactivity is obtained by calculating the reactivity of all of the 16 pairs of control rods over the interval from the critical control rod position to the all control rods pulled-out position, based on the reactivity worth curve (Fig. 10.4) and the critical control rod position. The calculation is carried out by Eq. (10.3) with correction factors for the temperature effect of the core and the interference effect of the control rods.

$$\rho_{EX} = C \times \sum \rho_k + A \tag{10.3}$$

where

ρ_{EX}: Excess reactivity [$\Delta k/k$];
ρ_k: Reactivity worth [$\Delta k/k$] of one pair of k control rods that integrates over the interval from the critical control rod position to the all control rods pulled-out position;
C: Correction factor [−] for removing interference effect of the control rod;
A: Reactivity compensation [$\Delta k/k$] for converting the core temperature to 27 °C.

<4> Reactivity restraining effect

The reactivity restraining effect is checked to ensure that the reactivity restraining capability is 0.18 $\Delta k/k$/s or more. The reactivity restraining capability is obtained using Eq. (10.4) based on the excess reactivity value and the reactivity shutdown margin value obtained in advance.

$$\rho_{CON} = \rho_{EX} + \rho_{SD} \tag{10.4}$$

where ρ_{CON}: Reactivity restraining capability [$\Delta k/k$].

10.3 Operation Control When Starting/Stopping

When starting and raising the power of the HTTR, the reactor power is raised so that the reactor outlet temperature is increased from the normal temperature to 950 °C. Therefore, care must be taken to high-temperature creep of the components such as the heat exchanger and pipings that are used under high-temperature conditions. For this reason, the rate of temperature rise is limited to 35 °C/h in the reactor outlet temperature range below 650 °C, and 15 °C/h in the reactor outlet temperature range above 650 °C. The limited of 35 °C/h is set for SUS 321TB as the materials used in such as the heat exchanger tube of the primary pressurized water cooler and Hastelloy XR as the materials used in such as the heat exchanger tube of the intermediate heat exchanger. The limit of 15 °C/h is set for Hastelloy XR as the materials used in components such as the heat exchanger tube of the intermediate heat exchanger, which is used in particular in the high-temperature range. In addition, to prevent brittle fracture of the major components that constitute the primary coolant pressure boundary, the minimum operating temperature (operating limit temperature) is specified. For the HTTR, the operating limit temperature for the entire plant is specified as 52 °C, which temperature is designed to meet the requirement for the reactor pressure vessel that is the severest among the major components.

Figure 10.6 shows the transition of the parameters up to the rated power operation both in the high-temperature test mode with the single operation mode.

The operating state of the reactor is divided into the manual operation of up to 30 % power and the automatic control operation subsequent to it.

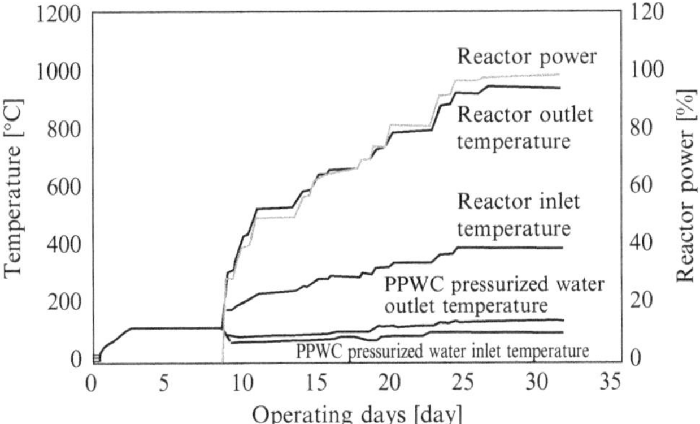

Fig. 10.6 Variation in parameters during power rise in high-temperature test in single operation mode

To satisfy the operating limit temperature of 52 °C of the plant, the components that constitute the main cooling system such as the reactor pressure vessel and the heat exchanger are heated up to approximately 120 °C before the reactor is started (the control rods are pulled out). Heat is input from the helium circulator and the pressurized water circulating pump via the primary helium coolant, secondary helium coolant, and pressurized water as bypassing almost all of the pressurized water flow in the air cooler of the final heatsink. Then, gradually increasing the pressurized water flow of the air cooler, pulling out the control rods starts with the power increase rate to satisfy the limit of temperature rise rate of 35 °C/h. The operator manually operates the reactor up to this point. Then, when the reactor power reaches 30 % and the parameter variations slow down, the reactor control system, pressurized water temperature control system, and the reactor inlet temperature control system are activated to start the automatic operation. At the 30 % power, the reactor inlet temperature is approximately 180 °C, and the reactor outlet temperature is approximately 346 °C (315 °C in the rated operation mode). From the 30 % power, the automatic control system raise to the 100 % power under the following control requirements.

<1> The power must be raised while maintaining the reactor outlet temperature at a predetermined value.
<2> The temperature rise rate of the coolant at the reactor outlet must be limited to 15 °C/h.
<3> The primary coolant (helium) pressure must be maintained higher than the pressurized water of the primary pressuried water cooler.
<4> The primary coolant (helium) pressure must be maintained lower than the secondary coolant (helium) of the intermediate heat exchanger.

To satisfy requirement <1>, the reactor power, the reactor inlet temperature, and the primary coolant flow rate are automatically controlled. The reactor inlet temperature is automatically controlled to the control target value that is proportional to the control target of the reactor power as shown in Fig. 10.7.

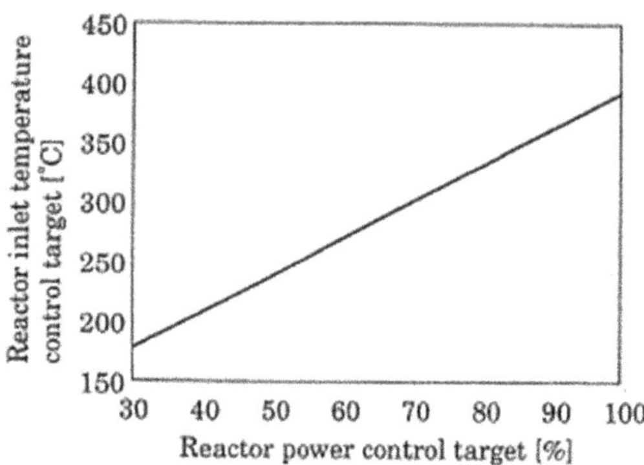

Fig. 10.7 Control target of reactor inlet temperature

The primary coolant flow rate is controlled at a constant rate for all time. Using this control method, the reactor outlet temperature is raised while controlling to a value appropriate to the reactor power. Consequently, to satisfy the requirement <2>, it is only necessary to control the power change rate. The reactor power control system sets the power change rate to 1.5 %/h to satisfy the temperature rise rate limit of 15 °C/h in the reactor outlet temperature range of 650 °C and above, after the reactor outlet temperature reaches approximately 300 °C. The requirement <3> is to prevent water entry into the core causing the addition of reactivity and the oxidation of graphite structure in core from the case the heat-transfer pipe of the primary pressurized water cooler is broken. The requirement <4> is to prevent the primary radioactive coolant from diffusing into the secondary coolant through breakage in case the heat-transfer pipe of the intermediate heat exchanger is broken. To satisfy the requirements <3> and <4>, the primary coolant-to-pressurized water differential pressure control system and the primary-to-secondary helium differential pressure control system control the differential pressure between the respective coolants to a predetermined value.

Table 10.1 shows the important control systems on power rise of the HTTR.

Also, Fig. 10.8 shows the whole configuration of the reactor power control system including the reactor power controller, and Fig. 10.9 shows the whole configuration of the plant controller.

[1] Reactor power control system

The reactor power control system is used after the reactor power reaches 30 % of full power. When the reactor power is between 30 % and 100 %, the control rod position instruction signal is converted to a drive signal and forwarded to the control rod drive mechanism so that the reactor power follows the set value given by the drive mode selector unit, and therefore the control rod position is adjusted.

[2] Reactor inlet temperature control system and pressurized water temperature control system

The reactor inlet temperature control system adjusts the pressurized water temperature to maintain the reactor inlet temperature of the primary coolant to a predetermined value. It is used when the reactor power reaches 30 %. The set value

Table 10.1 Parameters and components to be controlled by control system on power rise

Name of the control system	Control parameter	Components to be controlled
Reactor power control system	Neutron flux	Control rod position
Reactor inlet temperature control system	Reactor inlet temperature	Pressurized water temperature
Pressurized water temperature control system	Pressurized water temperature	Opening angles of air cooler outlet flow rate control valve and bypass flow rate control valve
Primary coolant flow rate control system	Primary coolant flow rate	Primary helium circulator speed
Primary-to-secondary helium differential pressure control system	Differential pressure between primary coolant and secondary coolant (helium)	Supply valve and drain valve of secondary helium storage and supply system
Primary coolant-to-pressurized water differential pressure control system	Differential pressure between primary coolant and pressurized water	Supply valve and drain valve of nitrogen supply system

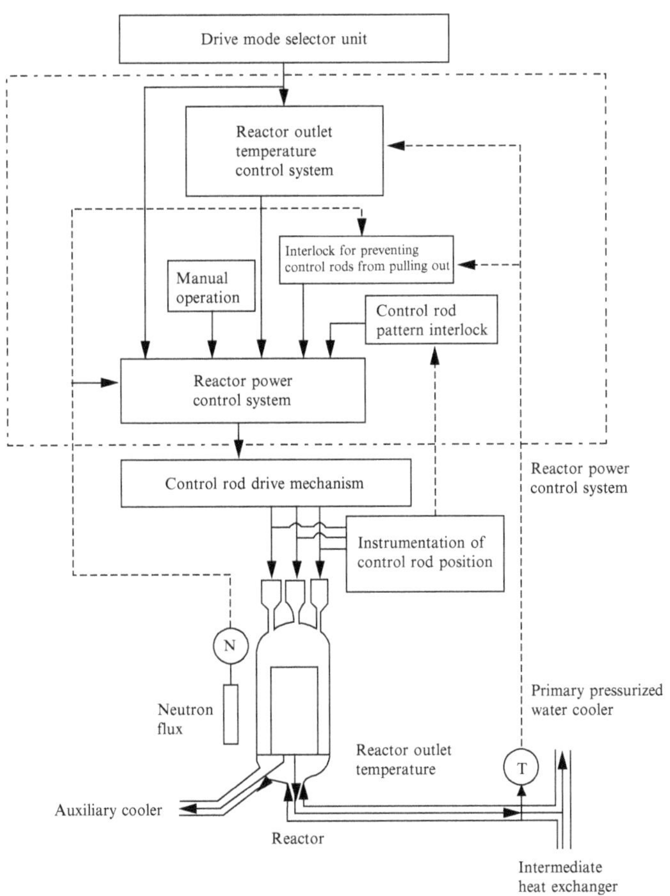

Fig. 10.8 Whole configuration of reactor power control device

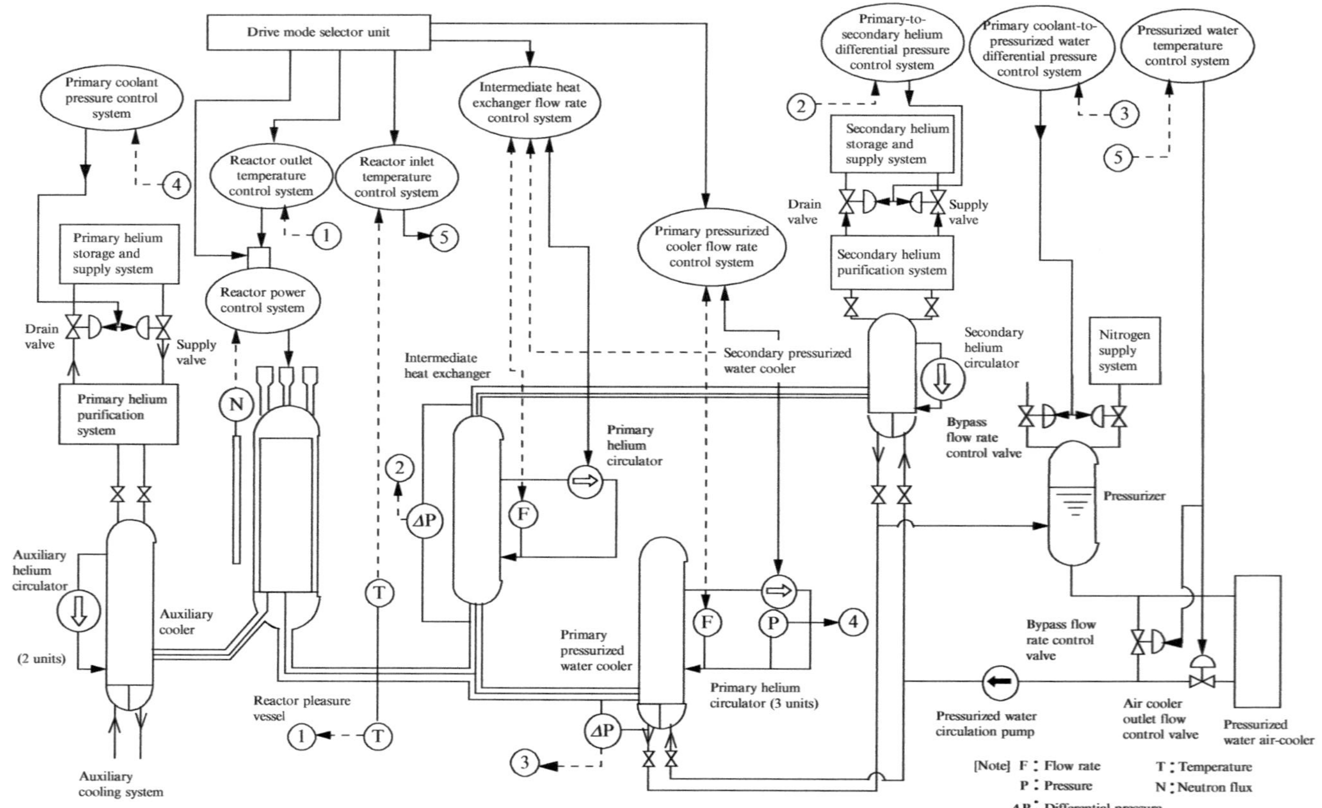

Fig. 10.9 Basic configuration of plant control device

of the pressurized water temperature is given to the pressurized water temperature control system so that the reactor inlet temperature follows the set value specified by the drive mode selector unit. The pressurized water temperature control system maintains the reactor inlet temperature of the primary coolant to a predetermined value. It adjusts opening angles of the air cooler outlet flow rate control valve and bypass flow rate control valve, which adjusts the flow rate of the pressurized water in the air cooler so that the pressurized water temperature at the inlets of the primary pressurized water cooler and secondary pressurized water cooler is controlled at a temperature given by the reactor inlet temperature control system.

[3] Primary coolant flow rate control system

The primary coolant flow rate control system consists of the primary pressurized water cooler flow rate control system and the intermediate heat exchanger flow rate control system. These control systems adjust the flow rate by adjusting the speed of the primary helium circulators installed for the primary pressurized water cooler and the intermediate heat exchanger. The primary pressurized cooler flow rate control system controls the primary coolant flow rate of the primary pressurized cooler to be maintained constant. The intermediate heat exchanger flow rate control system controls the primary coolant flow rate of the intermediate heat exchanger so that it remains constant.

[4] Primary-to-secondary helium differential pressure control system and primary coolant-to-pressurized water differential pressure control system

To prevent the fission products retained in the primary coolant from entering into the secondary coolant (helium), the primary-to-secondary helium differential pressure control system opens and closes the supply valve and drain valve of the secondary helium storage and supply system so that the secondary coolant (helium) pressure is controlled at a predetermined value that is higher than the primary coolant pressure. To prevent water entry into the primary cooling system which will cause nuclear effect (addition of activity) to the core and integrity degradation (oxidation) of graphite structure inside the reactor, the primary coolant-to-pressurized water differential pressure control system opens and closes the supply valve and drain valve of the nitrogen supply system so that the pressurized water pressure is controlled at a predetermined value that is lower than the primary coolant pressure.

The automatic control system of the HTTR is capable of raising the reactor power from 30 % to 100 % successively. However, in actual operation, the power rise per day is set to about 10 % (3 MW) of the rated power as shown in Fig. 10.6. This is because the heat capacity of the core in the high-temperature gas-cooled reactor is quite large compared to the power density, and it needs several hours before parameters are settled. Therefore, the reactor power rise is retained every day and the power is carefully raised while the parameters of the intermediate output stage are checked.

When lowering the power from the rated power to the shutdown power, operations are performed in the reverse sequence of starting and power rising. The automatic control system operates to lower the reactor power from 100 % to 30 %, and the operator manually operates the reactor from 30 % till the reactor stops

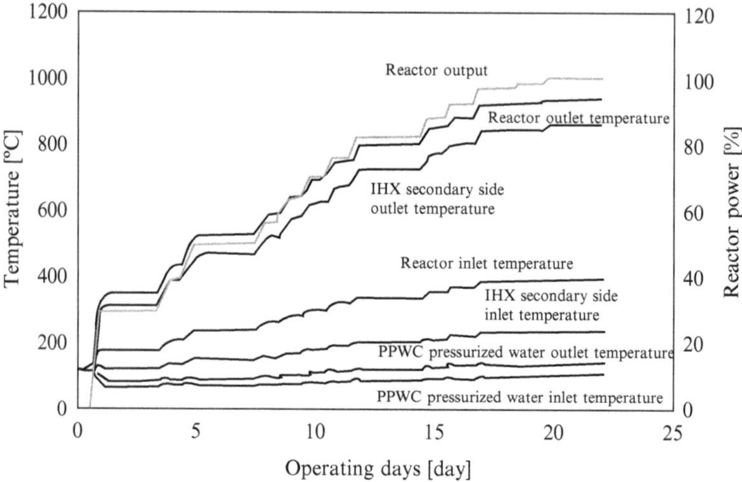

Fig. 10.10 Variation in parameters during power rise in high-temperature test in parallel operation mode

(checking that all of the control rods are inserted). After all of the control rods are inserted, the reactor is switched to the residual heat removal operation. Then, when it is confirmed that the reactor is cooled down to a sufficient level, the operator stops the helium circulator and the pressurized water circulating pump.

Figure 10.10 shows the variation of the parameters up to the rated power operation both in the high-temperature test and parallel operation mode.

In the parallel operation mode, both of the primary pressurized water cooler and the intermediate heat exchanger are operated in parallel, whereas in the single operation mode, only the primary pressurized water cooler is operated. The difference in the reactor operation (e.g., starting and stopping of the reactor) in the parallel operation mode from the single operation mode is that the number of parameters to be monitored is increased, but there is no difference in the operation control. Therefore, the reactor is operated in the same sequence as the single operation mode.

10.4 Operation Control During Steady Operation

The reactor control during steady operation of 30 MW must satisfy the following requirements.

<1> The reactor outlet temperature must be controlled to a predetermined value (850 °C or 950 °C) for stable nuclear heat supply.

<2> The primary coolant pressure must be controlled to a predetermined value (4 MPa).

<3> The primary coolant pressure must be maintained higher than the pressurized water.

<4> The primary coolant pressure must be maintained lower than the secondary coolant (helium) pressure.

Table 10.2 Parameters and components to be controlled by reactor outlet temperature control system and primary coolant pressure control system

Name of the control system	Control parameter	Components to be controlled
Reactor outlet temperature control system	Reactor outlet temperature	Neutron flux
Primary coolant pressure control system	Primary coolant Pressure	Supply valve and drain valve of primary helium storage and supply system

To satisfy the requirement <1>, the HTTR controls the reactor power to 30 MW, the reactor inlet temperature to 395 °C, and the primary coolant flow rate to a predetermined value, and therefore attains the reactor outlet temperature of 850 °C or 950 °C. Also, to control the reactor outlet temperature constant against disturbance, the reactor outlet temperature control system can be activated. The requirement <2> is designed to ensure the stable flow rate of the primarily coolant. To satisfy this requirement, the primary coolant pressure control system is provided. To satisfy the requirements <3> and <4>, in the same manner as when the power is raised or lowered, the primary coolant-to-pressurized water differential pressure control system and the primary-to-secondary helium differential pressure control system control the differential pressure between the respective coolants to a predetermined value. Table 10.2 shows the parameters and components to be controlled by the reactor outlet temperature control system and the primary coolant pressure control system. The values in Table 10.1 are applied for parameters and components to be controlled by other control systems.

[1] Reactor outlet temperature control system

Figure 10.8 shows the whole configuration of the reactor power controller including the reactor outlet temperature system. The reactor outlet temperature control system can be used during normal operation with nearly 100 % of the reactor power. It controls the reactor outlet temperature of the primary coolant to 850 °C during the rated power operation and 950 °C during the high-temperature test operation. When deviation occurs between the reactor outlet temperature of the primary coolant and the set value, the system calculates the deviation signal and gives the reactor power control system a set value so that the temperature deviation of the reactor power becomes zero.

[2] Primary coolant pressure control system

The primary coolant pressure control system ensures the stable flow rate of the primarily coolant. It opens and closes the supply valve and drain valve of the primary helium storage and supply system so that the primary coolant pressure is controlled at a predetermined value.

Figure 10.11 shows the variations of the reactor power, reactor outlet, and inlet temperatures, and the pressurized water temperature throughout one day of steady operation in which the reactor power is controlled to be constant.

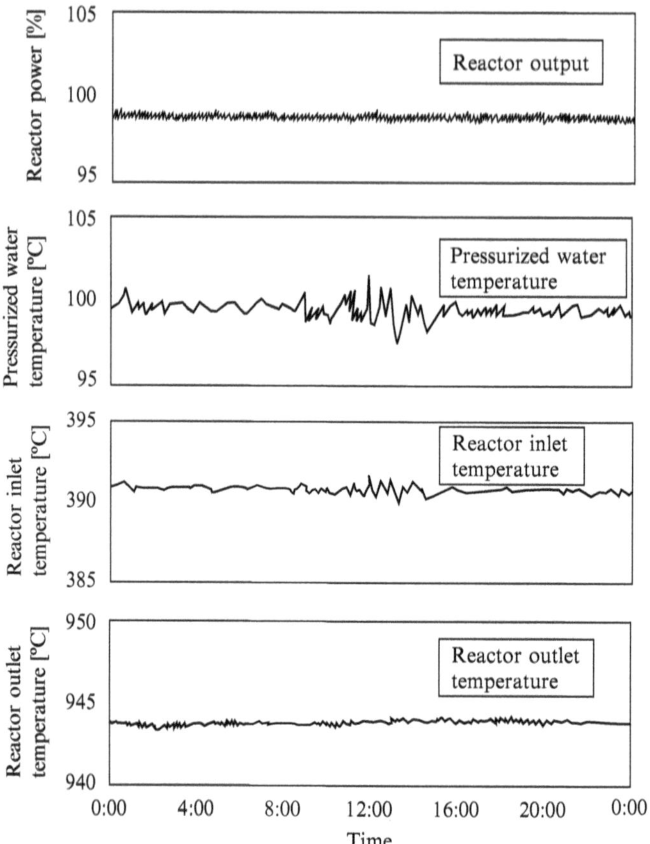

Fig. 10.11 Variation in parameters during constant power operation mode

Daily variation of the outside air temperature and other factors act as disturbances on the reactor inlet temperature. Due to the disturbances, the pressurized water temperature is varied and the reactor inlet temperature control system works to suppress the effects of disturbances on the reactor inlet temperature. So, little effect of disturbances on the reactor inlet temperature is observed. Furthermore, because of the large heat capacity of the core, the variation of reactor inlet temperature has little impact on the reactor outlet temperature. Then, little effect of disturbances on the reactor outlet temperature is observed. From the above, it is clear that the stable nuclear heat supply (constant reactor outlet temperature operation) is possible without the reactor outlet temperature control system.

10.5 Operation Control During Testing Condition

To demonstrate the inherent safety of the high-temperature gas-cooled reactor, the HTTR can be used for the safety demonstration tests such as the control rod pull-out test. In such safety demonstration test, the automatic control system is set to operate

Table 10.3 Control system conditions during control rod pull-out test

Name of the control system	Condition	Description
Reactor outlet temperature control system	OFF	To check feedback effect by addition of reactivity with pulling out of the control rod
Reactor power control system	OFF	To check feedback effect by addition of reactivity with pulling out of the control rod
Reactor inlet temperature control system	ON	To eliminate feedback effect due to variation of the reactor inlet temperature
Pressurized water temperature control system	ON	To eliminate feedback effect due to variation of the reactor inlet temperature
Primary coolant flow rate control system	ON	To eliminate feedback effect due to variation of the primary coolant flow rate
Primary coolant pressure control system	OFF	This control system is used only when the reactor operates at 30 MW
Primary-to-secondary helium differential pressure control system	ON	To prevent the primary radioactive coolant from diffusing, in case that the heat-transfer pipe of the intermediate heat exchanger is broken
Primary coolant-to-pressurized water differential pressure control system	ON	To prevent water entry into the core, in case that the heat-transfer pipe of the primary pressurized water cooler is broken

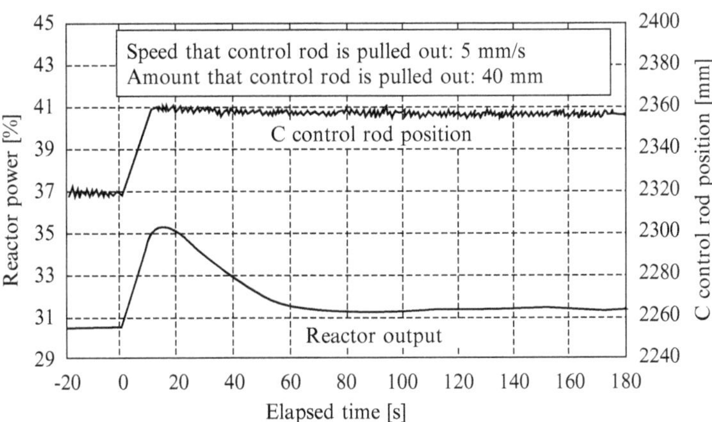

Fig. 10.12 Results of control rod pull-out test with initial power at 30%

under a condition as shown in Table 10.3. The initial reactor power is set to 30–80 % for the control rod pull-out test.

Figure 10.12 shows the result of the control rod pull-out test on the initial power at 30 %.

Analyzing the reactor power variation, the Doppler effect can be evaluated due to variation of the fuel temperature and the moderator temperature effect due to variation of the moderator temperature. This can improve and sophisticate the evaluation techniques of the various temperature coefficients.

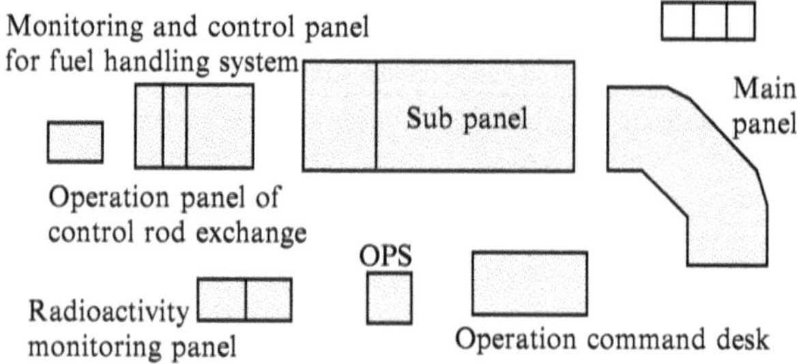

Fig. 10.13 Panel arrangements in the central control room

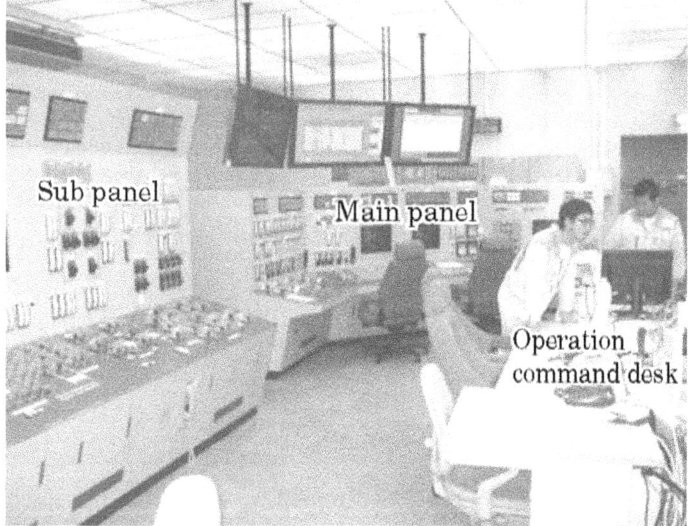

Fig. 10.14 Picture of central control room

10.6 Central Control Panel

The main panel, sub panel, etc. are provided in the central control room. The panels are fitted with the instrumentation and control components for the systems including the reactor control system required to operate the reactor and other major associated systems, the process instrumentation system, the reactor protection system, and the engineered safety features operation system. Central monitoring and control of the nuclear plant can be performed by the panel operation both in the normal operation and abnormal state. Figure 10.13 shows the panel arrangement in the central control room. The picture of the control room is given in Fig. 10.14.

The reactor is operated based on a three-shift system; each operation team consists of five operators (total of five teams work in shift). One person is stationed in the operation command desk as the team leader and provides overall directions regarding the operation. Two people are in the main panel and operate the control rod and the reactor control system. The other two people operate the subpanel, control panels, and components located in the actual devices.

Chapter 10 Exercises

1. During the steady operation of the HTTR, 13 pairs of the control rods use for reactor power under the condition of the control rod positions being flat (relative deviation = 20 mm). Explain the reason.
2. During the steady operation of the HTTR, the primary coolant pressure is controlled to be higher than the secondary coolant pressure. Explain the reason.

Bibliography

1. Yamashita K et al (2000) Critical test of the high-temperature engineering test reactor (HTTR). J Nucl Sci Technol 42(1):30–42
2. Fujikawa S et al (2002) Power up test of the high-temperature engineering test reactor (HTTR). J Nucl Sci Technol (Japanese Series) 1(4):361–372
3. Fujikawa S et al (2004) Achievement of reactor-outlet coolant temperature of 950 °C in HTTR. J Nucl Sci Technol 41(12):1245–1254
4. Tachibana Y et al (2003) Plan for first phase of safety demonstration tests of the High Temperature Engineering Test Reactor (HTTR). Nucl Eng Des 224(2):179–197

Chapter 11
New Control Theory and Its Application

Katsuo Suzuki and Kunihiko Nabeshima

11.1 Expressing a System with State Equations

The PID control system design method based on transfer functions was established in the 1950s and has since made contributions to automation processes in various industrial fields. Transfer functions have features that make you aware of rough characteristics of input–output responses based on the positioning of the poles and zero point and also discuss frequency responses and the stability of the control system using their Bode diagram. Now we will once again review the restrictions required to express dynamic characteristics of the control target with transfer functions.

<1> Transfer functions are defined only for the linear constant system. They are not therefore available for time-varying or nonlinear systems.
<2> No influence is considered for the initial conditions of the system. This means that the influence of the past inputs to the system is ignored.
<3> A transfer function is basically used to describe response characteristics of a single-input–single-output system, so it is not suitable for collectively handling a multi-input–multi-output system (hereinafter referred to as a multiple input–output system).
<4> A transfer function is used to describe an input–output response relationship in the system; however, it does not cover all dynamic states in the system.

A transfer function is not available in a more complicated control target such as a multiple input–output system or nonlinear system. Therefore we need to consider a differential equation that describes dynamic characteristics of a control target, that is, a set of equations that express state equations and output equations with state variables. The state variable refers to a set of variables that indicate the state of the internal system. Using this state variable, we can consider the influence of the initial output state of the system and also understand the dynamic behaviors in all the internal states. We can also handle time-varying coefficients or nonlinear systems as well as linear constant coefficient systems.

Y. Oka and K. Suzuki (eds.), *Nuclear Reactor Kinetics and Plant Control*, An Advanced Course in Nuclear Engineering,
DOI 10.1007/978-4-431-54195-0_11,

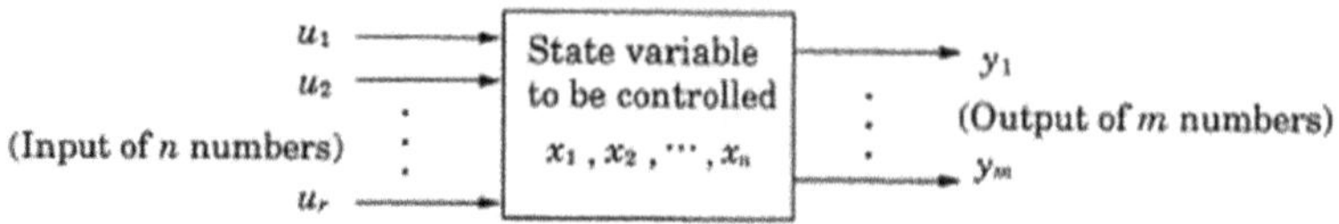

Fig. 11.1 Expression of control target using inputs, outputs, and state variables

A differential equation that describes dynamic characteristics of the system provides various formats; however, the state equation is an order 1 simultaneous ordinary differential equation regarding state variables, which does not include differentials of order 2 or higher. Now, assume a control system that has r inputs $u_1(t)$, $u_2(t)$, ..., $u_r(t)$ and n outputs $y_1(t)$, $y_2(t)$, ..., $y_m(t)$ as shown in Fig. 11.1.

When the cause–effect relationships between these inputs and outputs are expressed by the order 1 simultaneous ordinary differential equations and simultaneous algebraic equations as shown below:

$$\frac{dx_i}{dt} = f_i(x_1, x_2, \ldots x_n, u_1, u_2, \ldots, u_r), \quad i = 1, 2, \ldots, n \tag{11.1}$$

$$y_j = g_j(x_1, x_2, \ldots x_n, u_1, u_2, \ldots, u_r), \quad j = 1, 2, \ldots, m \tag{11.2}$$

Equations (11.1) and (11.2) are referred to as "state equation model" of the control target. Equation (11.1) is referred to as "state equation" and Eq. (11.2) as "output equation." Variables x_1, x_2, ..., x_n are state variables, which are internal variables for connecting inputs and outputs with an order 1 differential equation. Inputs and outputs are variables that represent the interaction between the system and the outside world. No physical obligations, excluding Eq. (11.1), are between n state variables x_i and r inputs u_k.

If vector $\mathbf{x} = (x_1, x_2, \ldots, x_n)^T$ that has state variables as components is introduced, Eqs. (11.1) and (11.2) can clearly be expressed by a vector symbol, enabling us to understand a space with state variables as a geometric image. Therefore, vector $\mathbf{x} = (x_1, x_2, \ldots, x_n)^T$ is referred to as "state vector," which corresponds to a geometric vector drawn from the origin to a coordinate $(x_1, x_2, \ldots, x_n)$ in an n-dimensional space with n coordinate axes. The n-dimensional space, which contains state variables, is referred to as "state space." In a state space, a locus that is drawn by point $(x_1(t), x_2(t), \ldots, x_n(t))$ as time t goes by is referred to as "state trajectory." Figure 11.2 shows the conceptual diagram of the state space and locus in three dimensions.

The following shows a system that is given by the finite-dimensional, linear, and time-invariant state equations and output equations.

(State equation)

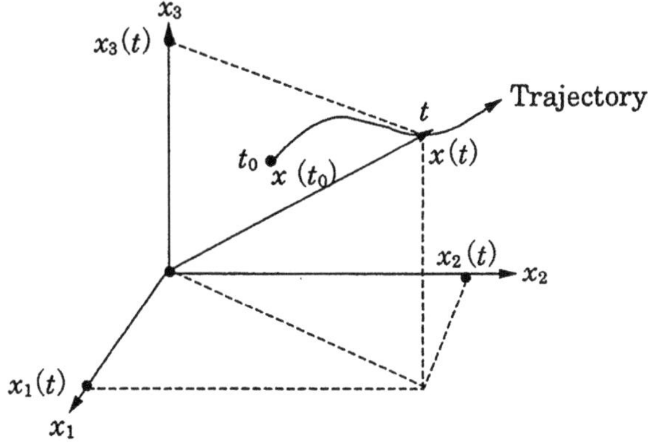

Fig. 11.2 Concept of state space and state trajectory in three dimensions

$$\frac{dx_1(t)}{dt} = \alpha_{11}x_1(t) + \alpha_{12}x_2(t) + \cdots + \alpha_{1n}x_n(t) + b_{11}u_1(t) + b_{12}u_2(t) + \cdots + b_{1r}u_r(t)$$
$$\vdots$$
$$\frac{dx_n(t)}{dt} = \alpha_{n1}x_1(t) + \alpha_{n2}x_2(t) + \cdots + \alpha_{nn}x_n(t) + b_{n1}u_1(t) + b_{n2}u_2(t) + \cdots + b_{nr}u_r(t) \tag{11.3}$$

(Output equation)

$$y_1(t) = c_{11}x_1(t) + c_{12}x_2(t) + \cdots + c_{1n}x_n(t) + d_{11}u_1(t) + d_{12}u_2(t) + \cdots + d_{1r}u_r(t)$$
$$\vdots$$
$$y_m(t) = c_{m1}x_1(t) + c_{m2}x_2(t) + \cdots + c_{mn}x_n(t) + d_{m1}u_1(t) + d_{m2}u_2(t) + \cdots + d_{mr}u_r(t) \tag{11.4}$$

If matrix $\boldsymbol{A}$, $\boldsymbol{B}$, $\boldsymbol{C}$, $\boldsymbol{D}$ is defined using vectors $\mathbf{x}(t)$, $\mathbf{u}(t)$, $\mathbf{y}(t)$ and coefficients as elements

$$\boldsymbol{x}(t) = \begin{bmatrix} x_1(t) \\ \vdots \\ x_n(t) \end{bmatrix}, \quad \boldsymbol{u}(t) = \begin{bmatrix} u_1(t) \\ \vdots \\ u_r(t) \end{bmatrix}, \quad \boldsymbol{y}(t) = \begin{bmatrix} y_1(t) \\ \vdots \\ y_m(t) \end{bmatrix} \tag{11.5}$$

$$A = (a_{ij}), \ \ (i = 1, \ldots n; \, j = 1, \ldots, n), \ \ B = (b_{ik}), \ \ (i = 1, \ldots, n; \, k = 1, \ldots, r) \tag{11.6}$$

$$C = (c_{li}), (l = 1, \ldots m;\ i = 1, \ldots, n), \quad D = (d_{lk}),\ (l = 1, \ldots, m; k = 1, \ldots, r) \tag{11.7}$$

Equations (11.3) and (11.4) are expressed clearly as shown below:

$$\frac{\mathrm{d}\boldsymbol{x}(t)}{\mathrm{d}t} = A\boldsymbol{x}(t) + B\boldsymbol{u}(t) \tag{11.8}$$

$$y(t) = C\boldsymbol{x}(t) + D\boldsymbol{u}(t) \tag{11.9}$$

Vectors $\mathbf{x}(t)$, $\mathbf{u}(t)$, and $\mathbf{y}(t)$ are referred to as "state vector," "input vector," and "output vector," respectively.

The system model based on the state equation of the control target has a feature that analyzes the time transition, explicitly using the internal state of the system, "**x**." This feature enables high-quality control by grasping the interaction between internal variables (state variables) and external variables (inputs and outputs) in the target system, for example, whether it is possible to control state variables as expected by handling inputs (controllability) or whether it is possible to check the internal state by observing outputs (observability). For the basic concept of the modern control theory such as the controllability and observability, refer to the relevant technical book about the control theory [1, 2].

11.2 Optimal Regulator

11.2.1 Formulation of Control Problems and Evaluation Functions

[1] Formulation

The theory of the optimal regulator means a control system design method based on the state equation of the system to be controlled. The optimal regulator is also referred to as "LQ control," taking into account that the target system is a linear system and the evaluation function of optimality is expressed in the quadratic form. The problems about the optimal regulator are intended for the systems expressed by the following state equation.

$$\frac{\mathrm{d}\boldsymbol{x}(t)}{\mathrm{d}t} = A\boldsymbol{x}(t) + B\boldsymbol{u}(t) \tag{11.10}$$

where set $\mathbf{x}(t)$ to an n-dimensional state vector and $\mathbf{u}(t)$ to an m-dimensional input vector (operation amount). These variables must be defined as a fluctuation obtained from an equilibrium position of the system. Therefore, **x** and **u** are set to

0 at the equilibrium position (of course, the equilibrium position does not need to be set to 0 physically).

In this example, the control purpose is to hold the target system to the equilibrium position, that is, to return state **x** near 0 immediately when a deviation from 0 occurs because the value of state x fluctuates due to a disturbance. The optimal regulator is obtained by "defining coefficient matrix **K** of the following state feedback controller for the control target expressed by Eq. (11.10)

$$\boldsymbol{u}(t) = K\boldsymbol{x}(t) \tag{11.11}$$

so that the following evaluation function $J(x_0, u)$ is minimized in the quadratic form."

$$J(\boldsymbol{x}_0, \boldsymbol{u}(t)) = \int_{t0}^{t\infty} (\boldsymbol{x}^{\mathrm{T}}(\tau)Q\boldsymbol{x}(\tau) + \boldsymbol{u}^{\mathrm{T}}(\tau)R\boldsymbol{u}(\tau)\mathrm{d}\tau) \tag{11.12}$$

where in $J(\mathbf{x_0}, \mathbf{u})$, "$\mathbf{x_0}$" indicates the initial state of the system and means that the value of this evaluation function is determined by control inputs $u(t)$ and $\mathbf{x_0}$, considering that the behavior of system state $\mathbf{x}(t)$ depends on $\mathbf{x_0}$ and $u(t)$. Weight matrix **Q** is a positive semidefinite symmetrical matrix, and weight matrix **R** is a positive definite symmetrical matrix. In the theory of the optimal regulator, all the state variables are observed as controlled variables; therefore, an output equation is not necessary.

[2] Meaning of evaluation function

Before proceeding to the next subject, we will discuss evaluation function $J(\mathbf{x_0}, \mathbf{u})$. The evaluation function that consists of the sum of quadratic forms defines a constraint condition to trade off the behavior of $\mathbf{x}(t)$ for control performance and the cost of control input $\mathbf{u}(t)$. For example, this function eliminates controls that improve the system behavior $\mathbf{x}(t)$ using a too great control input $\mathbf{u}(t)$. It also eliminates controls that causes a significant fluctuation of $\mathbf{x}(t)$ using a low input $\mathbf{u}$ (t). Next, we will discuss the meaning of matrixes **Q** and **R** in Eq. (11.12). **R** is a positive definite matrix; therefore, $\mathbf{u}^{\mathrm{T}}(t)\mathbf{R}\mathbf{u}(t) > 0$ is applied to input vectors other than 0, and all inputs $\mathbf{u}(t)$ are to be evaluated with an evaluation function. **Q** may be defined as a positive semidefinite matrix; it does not need to be a positive definite matrix. Therefore, $\mathbf{x}^{\mathrm{T}}(t)\mathbf{Q}\mathbf{x}(t) > 0$ is applied, allowing the quadratic form to be set to 0 even when $\mathbf{x}(t)$ is not 0. This generates state variables that are not constrained by an evaluation function, resulting in an insufficient evaluation function $J(\mathbf{x_0}, \mathbf{u})$. This conclusion is applied when each state variable value $x_i(t)$ is independently defined in the system. Each $x_i(t)$ is actually constrained by a state equation while holding an interaction as a time function, not acting independently. Therefore, if an adequate condition is satisfied even if **Q** is a positive semidefinite matrix, the behavior of all state variable values $x_i(t)$ are evaluated by quadratic form $\mathbf{x}^{\mathrm{T}}(t)\mathbf{Q}\mathbf{x}(t)$. The adequate condition means that pair (***C***, ***A***) is detectable when a positive semidefinite matrix **Q** is equal to $\mathbf{C}^{\mathrm{T}}\mathbf{C}$. Equation (11.12) is then assumed to be a complete evaluation function. When pair (***C***, ***A***) is detectable, it means that the following is satisfied for all the eigenvalues λ_i in the nonnegative real part of matrix **A**.

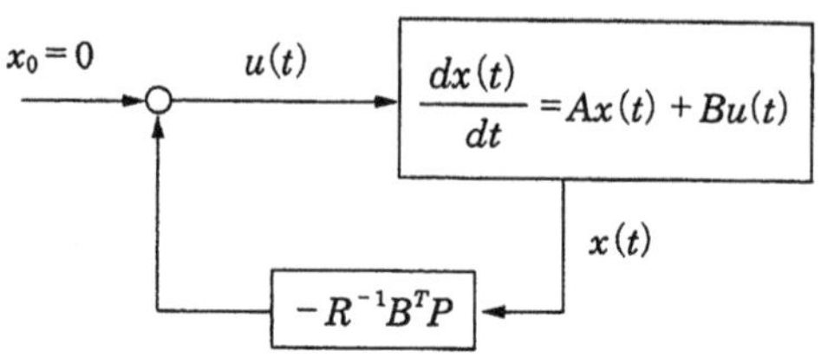

Fig. 11.3 Block diagram of optimal regulator

$$\text{rank}\begin{bmatrix} \lambda_i I - A \\ C \end{bmatrix} = n$$

11.2.2 Optimal Feedback Controller

Problems in the optimal regulator are resolved by applying an input $\mathbf{u}(t)$ in the state feedback form as shown below. However, if $\mathbf{Q} = \mathbf{C}^{\mathrm{T}}\mathbf{C}$, $(\mathbf{C}, \mathbf{A})$ is assumed to be detectable, that is, it assumes that there is no vector $\zeta \neq 0$ which satisfies $\begin{bmatrix} \lambda I - A \\ C \end{bmatrix}\xi = 0.$

$$u(t) = Kx(t) = -R^{-1}B^{\mathrm{T}}Px(t) \tag{11.13}$$

The minimum value of the evaluation function is then $J(x_0, u) = x_0^{\mathrm{T}}Px_0$. Here $\mathbf{P}$ is a positive semidefinite solution of the following equation that is referred to as "Riccati equation."

$$A^{\mathrm{T}}P + PA - PBR^{-1}B^{\mathrm{T}}P + Q = 0 \tag{11.14}$$

This equation is a quadratic (nonlinear) equation of $\mathbf{P}$. In general, it is difficult to obtain a solution, so, usually, we use a computer to perform numeric calculations.

The formulation for resolving problems in the optimal regulator does not include any conditions that generate a control input $u(t)$ using a state feedback controller; therefore, the control input search is performed in the function space including all time functions. In theory, we can conclude that the optimal solution is obtained by the state feedback controller of the gain constant as shown in Eq. (11.13). Figure 11.3 shows the block diagram of the optimal regulator.

11.2.3 Easy Design Example

The following shows a design example of the optimal regulator for a single-input control target.

$$\frac{\mathrm{d}x(t)}{\mathrm{d}t} = \begin{bmatrix} 0 & 1 \\ 0 & -2 \end{bmatrix} x(t) + \begin{bmatrix} 0 \\ 1 \end{bmatrix} u(t)$$

Let us design an optimal regulator that has the following evaluation function "J" as the minimum.

$$J = \int_0^\infty \left\{ x^{\mathrm{T}}(\tau) \begin{bmatrix} 2 & 0 \\ 0 & 1 \end{bmatrix} x(\tau) + u^2(\tau) \right\} \mathrm{d}\tau$$

First, obtain the following Riccati equation of expression (11.14).

$$\boldsymbol{P} \begin{bmatrix} 0 & 1 \\ 0 & -2 \end{bmatrix} + \begin{bmatrix} 0 & 0 \\ 1 & -2 \end{bmatrix} \boldsymbol{P} - \boldsymbol{P} \begin{bmatrix} 0 \\ 1 \end{bmatrix} 1 \begin{bmatrix} 0 \\ 1 \end{bmatrix}^{\mathrm{T}} \boldsymbol{P} + \begin{bmatrix} 2 & 0 \\ 0 & 1 \end{bmatrix} = 0$$

The positive definite solution of this equation is $P = \begin{bmatrix} 3.957 & 1.414 \\ 1.414 & 0.798 \end{bmatrix}$.
Using this **P**, the feedback gain **K** in expression (11.13) is obtained as follows:

$$\boldsymbol{K} = -R^{-1} B^{\mathrm{T}} P = [-1.414 \quad -0.798]$$

Therefore the optimal regulator is obtained as shown below:

$$\frac{\mathrm{d}x(t)}{\mathrm{d}t} = \begin{bmatrix} 0 & 10 \\ -1.414 & -2.798 \end{bmatrix} x(t)$$

Here the eigenvalues are set to -0.662 and -2.136, resulting in the asymptotic stabilization.

11.3 H_∞ Control

The actual control-target system ordinarily has nonlinear characteristics and distribution-based features. It also provides physical parameter values that frequently vary with time. In the modern control theory, a finite-dimensional state equation model is obtained by performing the linearization around the equilibrium position of the system or the approximation of a low-dimensional model. The approximation is handled in various methods; therefore, the obtained state equation model is no more than one of the approximate models of the control target. From this viewpoint, we can consider a method to design a controller for not only one approximate model but also a "set of models," taking into account a modeling error caused due to an approximation.

This is the basic concept of robust control. This term "robust" is an English term that has a meaning "tough." It is used to indicate that the control system has a feature (tough character), which causes its stabilization not to be deteriorated so much even if a modeling error occurs in the control-target system. A design problem in the controller that ensures the robust stabilization is referred to as a

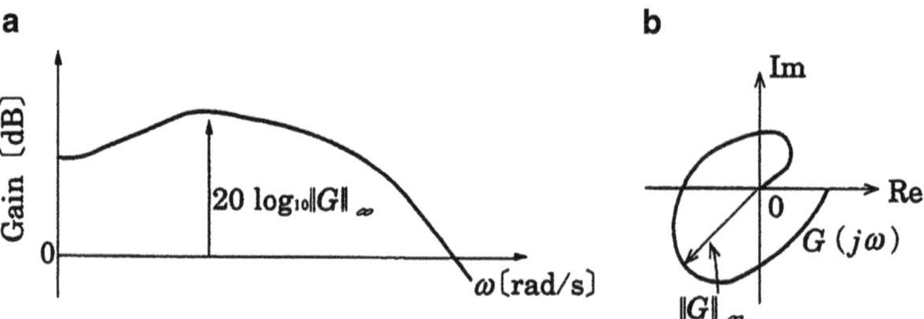

Fig. 11.4 H_{∞} norm of transfer function $G(s)$ in single input–output system (**a**) bode diagram, (**b**) Nyguist vetor locus

robust stabilization problem. H_{∞} control is one of the important design theories that are considered to resolve the problem.

11.3.1 $\mathbf{H}_{\infty}$ *Norm*

H_{∞} control is a feedback control that sets the magnitude (H_{∞} norm) of a transfer function **G(s)** ranging from the specified input point to the specified output point in the closed loop system, below a specific value. In a multiple input–output system, **G(s)** acts as a transfer function of a vector or matrix, so the H_{∞} norm is defined as shown below.

$$||G(s)||_{\infty} = \sup_{0 \leq \omega \leq \infty} \bar{\sigma}\{G(j\omega)\} \tag{11.15}$$

$$\bar{\sigma}\{G(j\omega)\} = \{\lambda_{\max}(G^{*}(j\omega)G(j\omega))\}^{1/2}$$

where, $G^{*}(j\omega) \equiv G^{\mathrm{T}}(-j\omega)$ is the conjugate transpose in the complex matrix. $\lambda_{\max}$ indicates the maximum eigenvalue.

In a single closed input–output loop system, the H_{∞} norm of the transfer function $G(s)$ of a closed loop system is the absolute maximum value of the gain in a Bode diagram shown in Fig. 11.4a, i.e. it indicates the maximum distance from the origin on the vector locus shown in Fig. 11.4b. It is then concluded as follows:

$$||G(s)||_{\infty} = \sup_{0 \leq \omega \leq \infty} |G(j\omega)| \tag{11.16}$$

11.3.2 Formulation and Solution of $\mathbf{H}_{\infty}$ *Control*

[1] Formulation

H_{∞} control has features that use a transfer function for evaluation of control characteristics and use a state equation description for design calculation of a

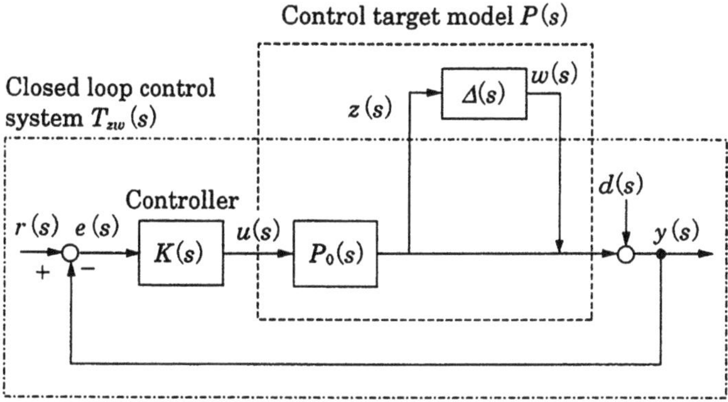

Fig. 11.5 Feedback control system for control-target model with fluctuation $\Delta(s)$

controller, that is, it is based on a theoretical configuration with the frequency response method and state equation method combined. H_∞ control is formulated to design a controller $K(s)$ that satisfies the following four general design conditions for the feedback control system shown in Fig. 11.5.

<1> The closed loop system must be stable (internal stability condition).

<2> The deviation $e(s)$ caused by the target input $r(s)$ or disturbance $d(s)$ of the feedback control system for the nominal model $P_0(s)$ of actual control object must be small (performance characteristic).

<3> The controller $K(s)$ designed based on the nominal model $P_0(s)$ must hold the stability of the feedback control system for the control-target model $P(s)$ (robust stability).

<4> The response characteristics of the feedback control system for the control-target model $P(s)$ must not be deteriorated (robust performance characteristic).

This design problem is handled as follows. First, we will discuss condition <1>. In this condition, the internal stability means that no unstable pole and zero point offset occurs and a specific input–output transfer function is stable. This is the most basic condition in the design process of a control system, where H_∞ control is based on a theoretical configuration to internally stabilize the feedback control system.

Next, we will consider condition <2>. The transfer characteristic of the deviation $e(s)$ caused by the target input $r(s)$ or disturbance $d(s)$ is determined by the sensitivity function in expression (5.22) in Chap. 5. Condition <2> is therefore satisfied by designing so that the controller $K(s)$ of the feedback control system in the nominal model $P_0(s)$ satisfies the following equation.

$$||W_1(s)S_0(s)||_\infty = ||W_1(s)(I + P_0(s)K(s))^{-1}||_\infty < \frac{1}{\rho} \tag{11.17}$$

where ρ is a positive number and $W_1(s)$ is a weight to reflect the values of the target input $r(s)$ and disturbance $d(s)$. $W_1(j\omega)$ is ordinarily set to a rational function that is assigned to a greater value in a low frequency bandwidth in which $r(j\omega)$ or $d(j\omega)$ has a greater value, and that is assigned to a lower value in the other bandwidth. In the actual design, "ρ" should be set to greater as much as possible, fixing $W_1(s)$.

Condition <3> has the following result. Now, assume that the modeling error $\Delta(s)$ of the nominal model $P_0(s)$ satisfies $\{\Delta(s) : ||\Delta(s)W_2^{-1}(s)||_\infty < 1\}$ and the number of unstable poles is equal between the control-target model $P(s) = (1 + \Delta(s))P_0(s)$ and nominal model $P_0(s)$. Condition <3> is satisfied by designing the controller $K(s)$ that satisfies the following Eq. (11.18). (Small gain theorem)

$$||W_2(s)T_{zw}(s)||_\infty = ||W_2(s)P_0(s)K(s)(I + P_0(s)K(s))^{-1}||_\infty < 1 \tag{11.18}$$

Based on this expression (11.18), it is found that a greater modeling error $\Delta(s)$ is allowed as $T_{zw}(s)$ (complementary sensitivity) is lower.

Finally we will consider condition <4>. When the complementary sensitivity function of the feedback control system for the nominal model $P_0(s)$ is set to $T_0(s)$ and that for the control-target model $P(s)$ with the modeling error $\Delta(s)$ from $P_0(s)$ is set to $T(s)$, the following equation is obtained:

$$\frac{T(s) - T_0(s)}{T(s)} = S_0(s)\frac{\Delta(s)}{P(s)} \tag{11.19}$$

Equation (11.19) indicates that the fluctuation $\Delta(s)$ of the control target is multiplied by the sensitivity function $S_0(s)$ and the result is reflected on the complementary sensitivity function $T(s)$ that represents the performance characteristic of the entire control system. To reduce the deterioration of the performance characteristic by an influence of fluctuation $\Delta(s)$, we should design a controller $K(s)$ that decreases the sensitivity function $S_0(s)$. Therefore, it is concluded that this is expressed in the same format as the conditional equation (11.17) in condition <2>.

The discussions above show that the problem is to obtain a controller $\mathbf{K}(s)$ that satisfies Eqs. (11.17) and (11.18) and also realizes the internal stabilization of the feedback control system. In H_∞ control, this can be handled as a mixed sensitivity problem to design a controller $\mathbf{K}(s)$ under the following norm condition with the two equations above collected.

$$\left\| \begin{matrix} \rho W_1(s)(I + P_0(s)K(s))^{-1} \\ W_2(s)P_0(s)K(s)(I + P_0(s)K(s))^{-1} \end{matrix} \right\|_\infty < 1 \tag{11.20}$$

where if $\rho = 0$, the norm condition in Eq. (11.20) becomes available only for a design that realizes the robust stability. Setting ρ to a greater value improves the low sensitivity characteristic or quick-response performance while guaranteeing the robust stability. However, if the value of ρ is too greater, a controller $\mathbf{K}(s)$ may

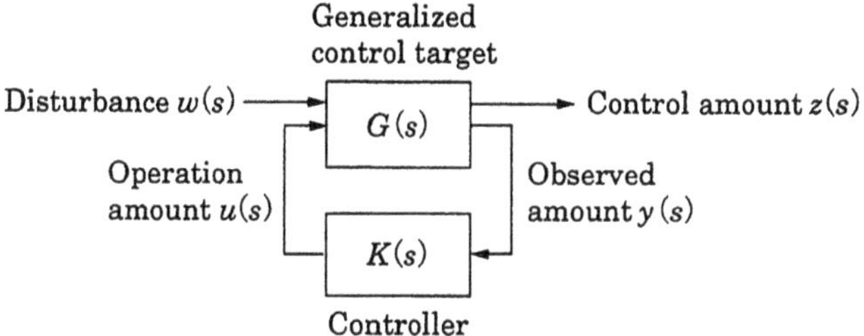

Fig. 11.6 Block diagram of feedback control

not be obtained. If $\mathbf{K}(s)$ is not obtained even by setting $\rho = 0$, it means that the robust stability of this design problem is impossible.

Now, we have explained the formulation of the H_∞ control, using the direct feedback control as an example. In Fig. 11.5, $w(s)$, $d(s)$, and $r(s)$ are described collectively as disturbance $\mathbf{w}(s)$. The input signal $\mathbf{z}(s)$ to a modeling error $\Delta(s)$ is specified as the controlled variable. The observation quantity and operation amount are specified as $\mathbf{y}(s)$ and $\mathbf{u}(s)$, respectively. Using these variables, Fig. 11.5 is represented by the feedback control block diagram shown in Fig. 11.6.

This block diagram is often used for H_∞ control. In this figure, $\mathbf{K}(s)$ and $\mathbf{G}(s)$ are transfer functions of the controller and generalized control target. The relationship between these variables is as follows:

$$\begin{bmatrix} z(s) \\ y(s) \end{bmatrix} = \begin{bmatrix} G_{11}(s) & G_{12}(s) \\ G_{21}(s) & G_{22}(s) \end{bmatrix} \begin{bmatrix} w(s) \\ u(s) \end{bmatrix} \tag{11.21}$$

The Eq. (11.20) of the mixed sensitivity problem is represented in the diagram shown in Fig. 11.7.

Based on Fig. 11.7, Eq. (11.21) is obtained as follows:

$$\begin{bmatrix} z_1(s) \\ z_2(s) \\ y(s) \end{bmatrix} = \begin{bmatrix} \rho W_1(s) & \rho W_1(s)P_0(s) \\ 0 & W_2(s)P_0(s) \\ I & P_0(s) \end{bmatrix} \begin{bmatrix} w(s) \\ u(s) \end{bmatrix} \tag{11.22}$$

This results in the following:

$$G_{11}(s) = \begin{bmatrix} \rho W_1(s) \\ 0 \end{bmatrix}, \quad G_{12}(s) = \begin{bmatrix} \rho W_1(s)P_0(s) \\ W_2(s)P_0(s) \end{bmatrix}, \quad G_{21}(s) = I, \quad G_{22}(s) = P_0(s)$$

Therefore, the mixed sensitivity problem is represented by the following H_∞ norm condition concerning the closed loop transfer function ranging from disturbance $w(s)$ to control quantities $z_1(s)$ and $z_2(s)$.

$$||G_{zw}(s)||_\infty = ||G_{11}(s) + G_{12}(s)K(s)(I - G_{22}(s)K(s))^{-1}G_{21}(s)||_\infty < 1 \tag{11.23}$$

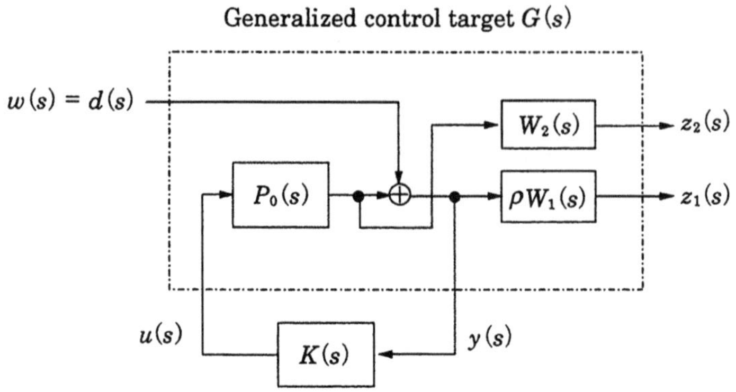

Fig. 11.7 Block diagram for Eq. (11.2) of mixed sensitivity problem

[2] Solution of H_∞ control

Now, we will explain how to obtain the constant gain **K** that satisfies Eq. (11.23) and realizes the internal stabilization of the feedback control system. This is, however, applied only when Eq. (11.22) is expressed by the following state equation, that is, when z has no direct feedthrough term from w or u and y has no direct feedthrough term from u. For general cases, refer to the relevant technical book [2].

$$\begin{aligned}\frac{\mathrm{d}x}{\mathrm{d}t} &= Ax + B_1 w + u \\ z &= C_1 x \\ y &= C_2 x + D_{21} w\end{aligned} \tag{11.24}$$

Here $B_1 D_{21}^{\mathrm{T}} = 0$ (orthogonality) and $D_{21} D_{21}^{\mathrm{T}} = I$ (normalization) must be satisfied.

The necessary and sufficient condition for setting $\mathbf{A} + \mathbf{KC}$ to asymptotic stabilization and assigning the feedback $u = K(s)y$, which causes $||G_{zw}(s)||_\infty = || C_1(sI - A - KC)^{-1}(B_1 + KD_{21})||_\infty < \gamma$, is that the positive definite symmetric solution **P**, which satisfies the following inequality, exists.

$$AP + PA^{\mathrm{T}} + P(\gamma^{-2} C_1^{\mathrm{T}} C_1 - C_2^{\mathrm{T}} C_2)P + B_1 B_1^{\mathrm{T}} < 0 \tag{11.25}$$

Then, **K** is given by $K = -PC_2^{\mathrm{T}}$, and $K(s)$ is obtained by Glover-Doyle solution method [2].

11.3.3 Application Example to Reactor Control

Now we will introduce an H_∞ control system design example for the March-Leuba model that was developed to analyze the unstable oscillation of the BWR core. This model is obtained by the following nonlinear equation system that has a combination of the one-point reactor approximate-kinetic equation and void reactivity ρ_α represented by order of 2 differential equation.

$$\frac{\mathrm{d}n(t)}{\mathrm{d}t} = \frac{\rho(t) - \beta}{\Lambda} n(t) + \lambda c(t) + \frac{\rho(t)}{\Lambda} \tag{11.26a}$$

$$\frac{\mathrm{d}n(t)}{\mathrm{d}t} = \frac{\beta}{\Lambda} n(t) - \lambda c(t) \tag{11.26b}$$

$$\frac{\mathrm{d}T(t)}{\mathrm{d}t} = a_1 n(t) - a_2 T(t) \tag{11.26c}$$

$$\frac{d^2 \rho_\alpha(t)}{\mathrm{d}t^2} + a_3 \frac{d\rho_\alpha(t)}{\mathrm{d}t} + a_4 \rho_\alpha(t) = \kappa k_0 T(t) \tag{11.26d}$$

$$\rho(t) = \rho_\alpha(t) + DT(t) + \rho_c(t) \tag{11.26e}$$

where variables $n(t)$ and $c(t)$ are converted by the following equation as a fluctuation caused from the values N_0 and C_0 of the steady state

$$n(t) = \frac{N(t) - N_0}{N_0}, \quad c(t) = \frac{C(t) - C_0}{C_0}$$

Temperature T indicates a deviation from the temperature in the steady state. Coefficient D indicates a temperature feedback constant, and ρ_c indicates the reactivity applied by the control rod. The parameters of this model are obtained as shown in Table 11.1 for the Vermont Yankee reactor in the USA.

First, we will check the reactor output response when parameters $k = 0.7$ and $k = 1.5$ are set in Eq. (11.26d). Figures 11.8 and 11.9 show a response when the reactivity 0.1$ is applied stepwise. It is then found that the reactor is stable when $k = 0.7$, but unstable when $k = 1.5$.

Therefore we should design an H_∞ controller for a linearization system obtained by Eqs. (11.26a)–(11.26e) when $k = 0.7$. Then let us apply the resulted controller the reactor system with $k = 0.7$ and $k = 1.5$ and check the stability of the controller by response simulation.

Now we will show the design of the H_∞ controller. The following shows a transfer function $P_0(s)$ that has a combination of the linearized reactor system and the control rod drive characteristics (primary delayed system: time constant 1 s).

$$P_0(s) = \frac{2.5 \times 10^4 s^4 + 6.4 \times 10^4 s^3 + 1.9 \times 10^5 s^2 + 5.4 \times 10^4 s + 3.1 \times 10^3}{s^6 + 1.4 \times 10^2 s^5 + 5.1 \times 10^2 s^4 + 1.4 \times 10^3 s^3 + 3.0 \times 10^3 s^2 + 2.1 \times 10^3 s + 1.4 \times 10^2}$$

Figure 11.10 shows the Bode diagram of this formula.

Table 11.1 Parameter values of March-Leuba model [6]

Model coefficient	Value	Unit
α_1	25.04	K·s^{-1}
α_2	0.23	s^{-1}
α_3	2.25	s^{-1}
α_4	6.82	s^{-2}
κ_0	-3.70×10^{-3}	K^{-1}·s^{-2}
D	-2.52×10^{-5}	K^{-1}
β	0.0056	–
Λ	4.00×10^{-5}	s
λ	0.08	s^{-1}

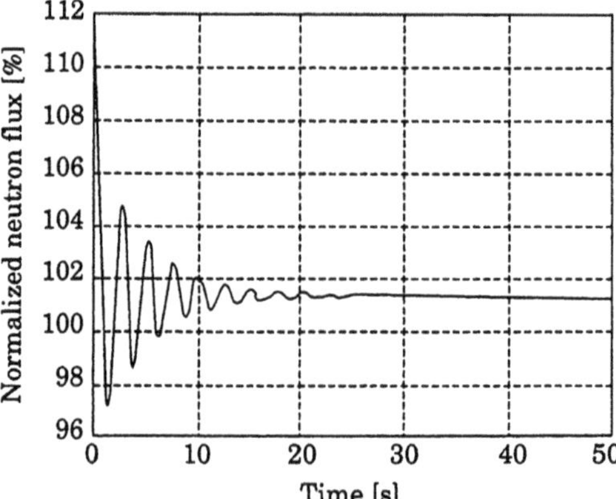

Fig. 11.8 Step response of March-Leuba model ($\kappa = 0.7$) applied reactivity 0.1$

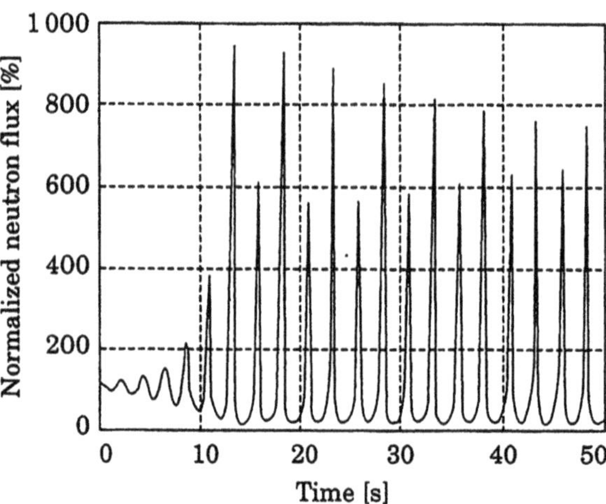

Fig. 11.9 Step response of March-Leuba model ($\kappa = 1.5$) applied reactivity 0.1$

The peak of the gain is placed near 2.4 rad/s. This is a factor that causes the reactor output oscillation shown in Figs. 11.8 and 11.9. It is actually the oscillation frequency caused by the void feedback reactivity.

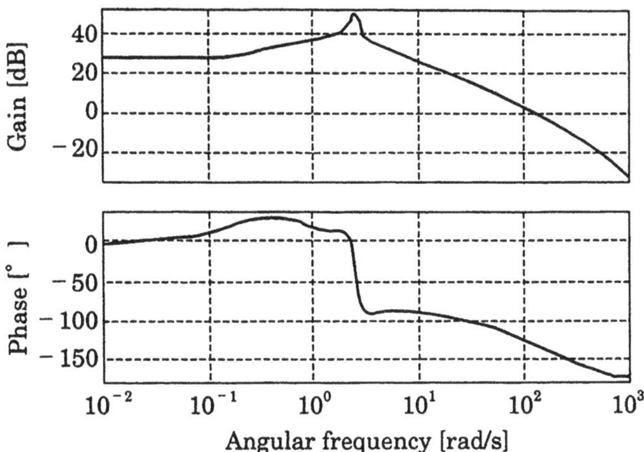

Fig. 11.10 Bode diagram of the controlled object

To obtain the controller, we will resolve a mixed sensitivity problem for the control target $P_0(s)$. Here, specify a weighting function in the following equation.

$$W_1(s) = \rho \frac{\omega_1}{s+\omega_1}, (\rho = 400, \ \omega_1 = 0.0025)$$

$$W_2(s) = K_\mathrm{T}\left(\frac{s+\omega_2}{\omega_2}\right)\left(\frac{s+\omega_3}{\omega_3}\right), (K_T = 10^{-4}, \ \ \omega_2 = 1, \ \ \omega_3 = 10)$$

The following seventh degree controller is obtained by the Glover–Doyle solution [2]. It is a feature of H_∞ controller that the degree of the controller becomes identical to that of the generalization control target $G(s)$.

$$K(s) = \frac{4.0s^6 + 5.8 \times 10^2 s^5 + 2.1 \times 10^3 s^4 + 5.8 \times 10^3 s^3 + 1.2 \times 10^4 s^2 + 8.4 \times 10^3 s + 550}{s^7 + 96.1s^6 + 4.6 \times 10^3 s^5 + 1.2 \times 10^4 s^4 + 3.3 \times 10^4 s^3 + 9.4 \times 10^3 s^2 + 5.7 \times 10^2 s + 1.4}$$

Figure 11.11 shows a Bode diagram of the open-loop transfer function $K(s)P_0(s)$ for the feedback control system.

As shown in Fig. 11.11, the gain margin is approximately 12 dB and the phase margin is approximately 61°; therefore, it is found that this feedback control system is stable.

Finally we will evaluate the response characteristics of this control system. Figure 11.12 shows a response obtained when the target output value is increased 10 % stepwise for a nonlinear reactor system when the void feedback coefficient k is 0.7.

The output overshoots nearly 10 %; however, the output oscillation detected in Fig. 11.8 is reduced, ensuring an appropriate control response: startup time = approximately 0.12 s. and static time = approximately 0.3 s. Figure 11.13 shows the target value response when the characteristics of the reactor system changes to $k = 1.5$.

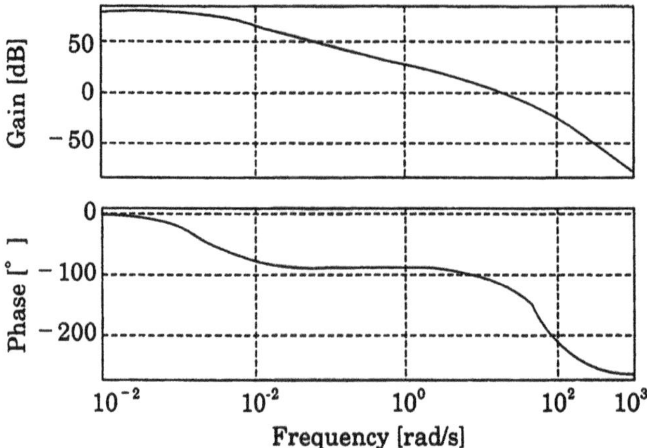

Fig. 11.11 Bode diagram of open-loop transfer function for feedback control system by H_∞ control

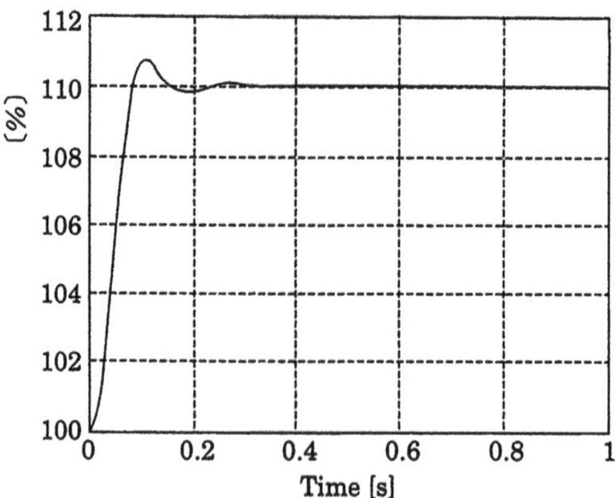

Fig. 11.12 Target response of feedback control system by H_∞ control ($\kappa = 0.7$)

In this case, a void feedback frequency in approximately 2.6 s cycles, which was not detected when $k = 0.7$, occurs in the target value response; however, the divergent oscillation detected in Fig. 11.9 is inhibited without reducing the stability of the control system.

11.4 Application of Artificial Intelligence

11.4.1 Expert System

As advancement and diversification requests for control functions increase recently, research work is proceeding with the control theory that realizes the introduction of the expert's experienced knowledge and the installation of the human skilled and

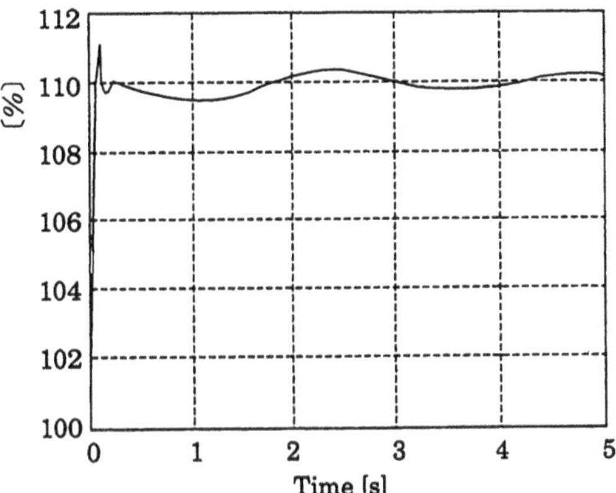

Fig. 11.13 Target response of feedback control system by H_∞ control ($\kappa = 1.5$)

flexible intelligent activity model compared with the conventional and stereotypical automatic control method.

The expert system, which covers a field in the artificial intelligence world, is a modeling system developed to carry out the human expert's problem solution process. Feigenbaum defines "the expert system is an intelligent computer program that makes full use of the knowledge or inference procedure for resolving difficult problems, which require the skilled expert's experiences and inference. The information included in the expert system consists of facts and heuristic knowledge." To configure an expert system, it is necessary to analyze the knowledge of various experts and save the procedure used at decision-making in the system. This knowledge is then used to resolve problems.

In the expert system, "knowledge" is classified into the following three categories:

<1> "Intelligence related to fact": Defines whether or not a specific event is true or false.
<2> "Conditional-relationship knowledge": Defines that a specific conclusion is obtained if a specific condition is satisfied.
<3> "Correlational knowledge": Defines the positional or role-based relationships between events.

Knowledge items <2> and <3> are referred to as "heuristics" in general; it is difficult to strictly distinguish them.

The expert system provides the most widely used methods: a method using a production system with items <1> and <2> combined and a method using a semantic network or frame with items <1> and <3> combined.

In addition to the knowledge items above, the expert system requires the following elements to make inferences and draw a useful conclusion using the knowledge.

<1> Inference mechanism:
Function that extracts event <1> related to the problem to be resolved and combines the extracted event with the inference rules defined in items <2> and <3> to proceed with the heuristic inference.

<2> Knowledge base mechanism:
Function that specifies a frame to save knowledge.
<3> Explanation mechanism:
Function that explains the inference process to derive the conclusion.
<4> Man–machine interface: Function that manages a communication between human and expert system.

In general, the production system in the expert system makes inferences with a set of production rule, "condition $\rightarrow$ conclusion." It is expressed by the following.

if A then B

Here "A" is referred to as a condition part, and "B" as an action part.

The typical methods for concretely executing an inference are classified into two types: forward inference and backward inference. The forward inference starts from the specified fact "A" and advances as follows:

$$A \Rightarrow B_1 \Rightarrow B_2 \Rightarrow \ldots \Rightarrow B_n$$

It is available in a design-type expert system.

The backward inference provides a set of fact and production rule that are identical to the forward inference; however, the objective to be defined as a conclusion is assumed unlike the forward inference. The objective is specified by a question, that is, whether or not a specific event "C" is correct or it is satisfied in this system, following the route shown below.

If B_n then C?
If B_{n-1} then B_n?
:
If A then B_1?
A = true

Here it is assumed that to satisfy "C," "B_n" must be satisfied, and to satisfy "B_n," "B_{n-1}" must be satisfied. Therefore, resolving the problem about "C" means cyclically solving problems "B_n," "B_{n-1}," ..., "B_1."

Now we will introduce the refuel planning expert system of "Fugen" reactor in Japan as an example for applying the expert system to a reactor plant. When refueling at execution of the periodic inspection or stop of planning, it is necessary to remove the dying-down fuel and load new fuel, and also perform shuffling to efficiently burn the fuel. To realize the effective use of fuel, engineers who are familiar with the management of the reactor core have taken a few months and energy to create a plan, taking into account the features of the "Fugen" reactor core. This has developed an expert system with the expert's knowledge arranged and computerized. The "Fugen" reactor core realizes the planarized output distribution, resulting in few differences in use between the MOX and UO_2 fuels. Therefore, the zoned refueling system, which has the reactor core divided circumferentially, is available for expecting the efficient use of fuel, compared with the batch system used in a light-water reactor, etc. This system has advantages that enable flexible responses to refueling at irregular intervals or changing the number of fuel rods and also enables the efficient use of fuel that minimizes shuffling of fuel rods. Based on this knowledge, an expert system embedded in the existing AI tool has been configured. In addition, it has been verified that this system can generate refueling patterns that satisfy core design conditions.

11.4.2 Neural Network

The investigation of the neural network is to mathematically treat the learning abilities that occupy the most part of biotic information processing. The theoretical investigation was triggered by the perceptron offered by Rosenblatt in the early 1960s, and after this, the convergence theorem of the learning algorithm was proved in various forms. However, it is assumed that the full-scale application based on the theory of the learning machine was started by Connectionist Model advocated by Rumelhart and other researchers in 1986. Their large-scale computational experiment led to the belief that the network structure, in which a large number of elements with only simple computational capabilities and that were combined evenly, was generalized by acquiring external information through "learning" carried out while locally transferring information between elements, resulting in the real issues being solved efficiently. In particular, the error back-propagation learning in the multilayer perceptron introduced the high-level latent ability of the learning machine to the public, and its ease encouraged its application into various fields.

[1] Neuron operation model

The brains of living organisms have various functions in a complicated structure; however, nerve cells in the brain are combined in a simple form as shown in Fig. 11.14a. We should, therefore, design a model to which outputs from other cells are input through synaptic connections, assuming that each neuron cell is a multi-input single-output element. When the sum of inputs exceeds the threshold of neuron cells, neuron cells are activated to apply the current; otherwise, they do not output anything. A system created by simplifying and modeling these neuron cell activities is referred to as an engineering model, artificial neural network as shown in Fig. 11.14b.

We will first explain each neuron (unit) as shown in Fig. 11.14. On a simple hierarchical neural network with only a one-way stream from the top to the bottom, in the jth unit in layer k, an input is represented as the sum of weighted outputs from a unit in layer $k-1$ as shown in Eq. (11.27).

$$I_j^k = \sum_i W_{i,j}^{k-1,k} \cdot O_i^{k-1} \tag{11.27}$$

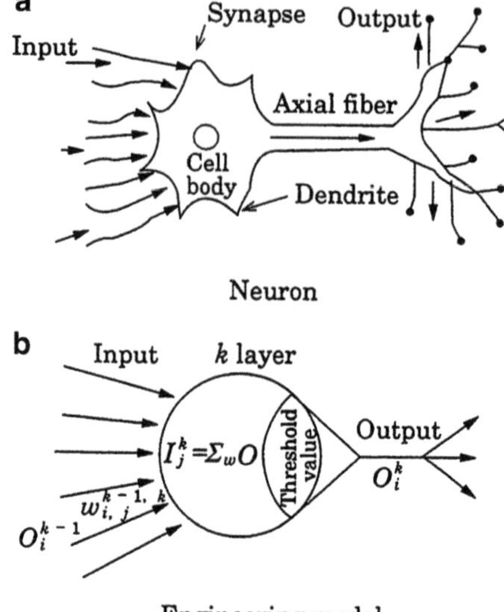

Fig. 11.14 Neuron model (**a**) image of actual neuron in the brain, (**b**) engineering model of the neuron

This input is then converted to an output by the input–output function $f(I)$ in this unit. The neuron of living organisms is basically a δ-function (binary model). If the threshold in the unit is θ, the nonlinear function f is expressed by a unit step function.

$$f(x) = \begin{cases} 1 & (I_j^k \geq \theta) \\ 0 & \left(I_j^k < \theta\right) \end{cases} \tag{11.28}$$

However, using a nonlinear function as the input–output function f enables modeling a nonlinear system in the neural network. When using the neural network to model a nuclear plant, the following sigmoid function is often used as an input–output function, considering that the input–output signal is a continuous value and that the differential value f' of the input–output function required in the learning process in item [3] below can be obtained easily.

$$f(x) = \frac{1}{1 + \mathrm{e}^{-(x-\theta)}} \tag{11.29}$$

Therefore the output signal is expressed as follows:

$$O_j^k = f\left(I_j^k\right) = \frac{1}{1 + \mathrm{e}^{-\left(I_j^k - \theta_j^k\right)}} \tag{11.30}$$

[2] Network form

The neural network structures are classified into two types: hierarchical and interconnected structures. The hierarchical neural network consists of three types of layers: input layer, a moderate number of hidden (intermediate) layers, and output layer. The input layer acts as the first layer to input information, and the output layer acts as the last layer to output information. The hidden layers that intermediate between the input and output layers can be determined arbitrarily depending on the problem to be applied. The hierarchical neuron network ordinarily covers only the connection between layers, ignoring the connection between units in the same layer. In this network, information is flowed once in one direction, resulting in the state change being ended. The most frequently used hierarchical one is a network using the error back-propagation learning algorithm as the learning rule, which requires a teacher signal without a feedback connection.

The interconnected neural network consists of interconnected adjacent units regardless of a hierarchical structure; therefore, input and output units are not distinguished clearly. The network that started from a specific initial state reaches a specific stable equilibrium state while repeating the state change for each unit. This type of network contains a self-organization in the network structure or requires no teacher signals during learning. As the typical interconnected networks, there are Kohonen network, Hopfield-style network, and Boltzmann machine, which are used in image- or character-recognition systems.

[3] Error back-propagation learning algorithm

Now we will discuss the error back-propagation learning algorithm that is generally used as the learning rule of the hierarchical neural network. This algorithm has a feature that changes the weight between units to minimize a square error between the output value in the network and the actual correct answer value (teacher signal). Its name is derived from the fact that, for learning, signals are propagated from the output layer to the input layer with the sum weighted backward, using an error (deviation) between the correct answer value and output value as an input.

In the error back-propagation learning algorithm, the weight $W_{i,j}^{k-1,k}$ of the connection between ith unit in layer $k-1$ and jth unit in layer k in Fig. 11.15 is corrected at the tth learning time as shown below:

$$W_{i,j}^{k-1,k}(t) = W_{i,j}^{k-1,k}(t-1) + \Delta W_{i,j}^{k-1,k}(t) \tag{11.31}$$

$$\Delta W_{i,j}^{k-1,k}(t) = \eta \cdot \delta_j^k \cdot O_i^{k-1} + \alpha \cdot \Delta W_{i,j}^{k-1,k}(t-1) \tag{11.32}$$

Here, η, which indicates the learning ratio in the standard error back-propagation learning algorithm, is a parameter used to determine the weight corrected in one learning process. In Eq. (11.32), the second item in the right part is referred to as a momentum item, preventing learning from being concluded with a minimal value,

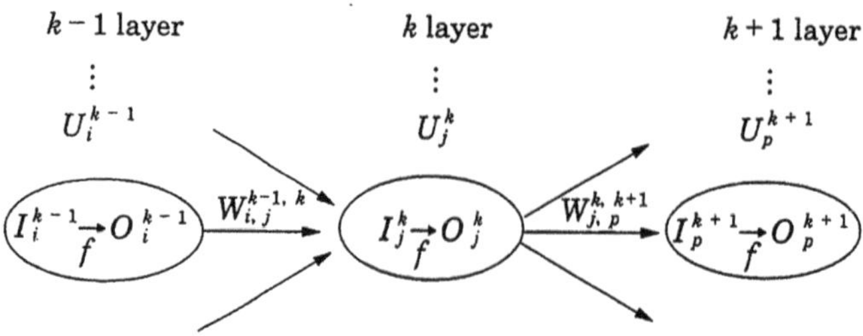

U_j^k : jth unit (neuron) in k layer
I_j^k : Sum of input
O_j^k : Output
$W_{i,j}^{k-1,k}$: Weight of bonding from ith unit of k - 1 layer to jth unit of k layer
f : Input/output function

Fig. 11.15 Neuron bond inside network

not the target minimum value, while accelerating the learning pace with the learning oscillation reduced. "α" is a parameter in the momentum item. "δ_j^k" indicates an error signal for the jth unit in layer k. In the output layer M, it is obtained as follows:

$$\delta_j^M = (Y_j - O_j^M) \cdot f'(I_j^M) = O_j^M \cdot (1 - O_j^M) \cdot (Y_j - O_j^M) \qquad (11.33)$$

In other intermediate layers, it is obtained as follows:

$$\delta_j^k = f'(I_j^k) \cdot \sum_p W_{j,p}^{k,k+1} \cdot \delta_p^{k+1} = O_j^k \cdot (1 - O_j^k) \cdot \sum_p W_{j,p}^{k,k+1} \cdot \delta_p^{h+1}, (k = M - 1, \ldots, 2) \qquad (11.34)$$

Here, δ_p^{h+1} indicates an error signal in layer $k + 1$, and $W_{j,p}^{k,k+1}$ indicates the weight of the connection between the jth unit in layer k and the pth unit in layer $k + 1$, In Eq. (11.33), "Yj" indicates a teacher signal and "f" indicates a differential parameter of the input–output function. This equation contains a sigmoid function like Eq. (11.30); therefore, finally it can be expressed in a simple form in the same way as Eq. (11.33) or (11.34). It is generally said that learning is concluded earlier as the learning ratio and momentum factor are greater in the initial stage of learning, or they are lower if an error is reduced because the learning process advances to a certain degree.

[4] Application to control field

To apply the neural network to the control field, we will acquire an inverse model. In many cases, the learning purpose for control is to find a control signal that matches the actual output result with the expected one. The direct method shown in

Fig. 11.16 Direct method

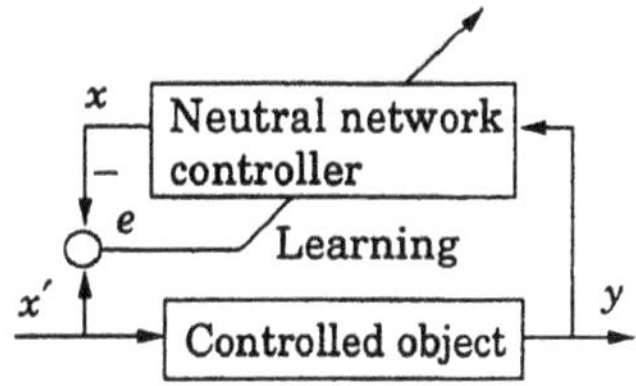

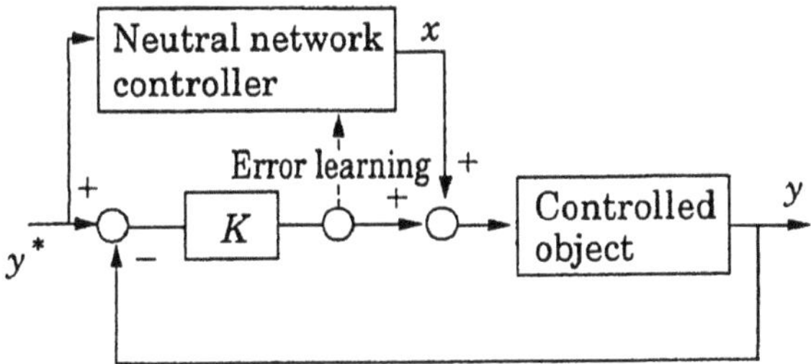

Fig. 11.17 Learning method of feedback error

Fig. 11.16 is available when directly acquiring an inverse model based on the error back-propagation learning algorithm so that the difference between the output and test vectors of the neural network controller is cleared to 0.

The feedback error learning method applies a feedback control to clear the difference between the input vector (target value of output: y^*) of the neural network controller and the actual output vector (y) of the plant to 0. In particular, as shown in Fig. 11.17, the relation $(\partial y/\partial x)^T$ between the control-target minimal input "x" and control-target minimal output "y" is linearized and approximated as a constant value. Here "K" indicates a feedback gain. In addition, there is an inverse kinematics learning method that learns and uses a forward system model to be controlled.

11.4.3 Fuzzy Control

The fuzzy theory is advocated by Zadeh in 1965, triggering the investigation with the fuzzy set introduced. The fuzzy set is an extended one of normal set (crisp set). The crisp set has elements that are clearly bounded, while the fuzzy set has elements that are ambiguously bounded. The crisp set explicitly defines whether or not each element belongs to the set. The fuzzy set allows an intermediate value between 0 and 1 to indicate the level at which element belongs to the fuzzy set. For example, the fuzzy set allows the state where the ratio between the levels at which each element belongs and does not belong to the set is fifty–fifty. The value between 0 and 1 is referred to as a grade. A function that assigns a grade to each element is

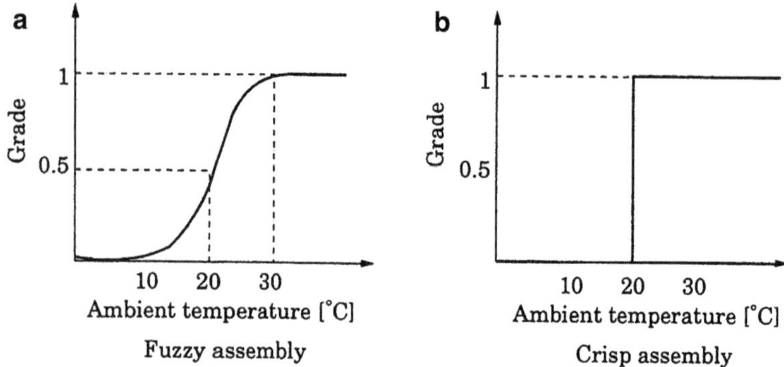

Fig. 11.18 Membership function that represents "hot" (**a**) Fuzzy set, (**b**) Crisp set

referred to as a membership function. To define a membership function is almost equivalent to define a fuzzy set.

Using the fuzzy set, we can represent the meaning of words or the fuzziness of definitions. For example, we can suppose the membership function shown in Fig. 11.18a to express a word "hot." Here, the fuzzy set is defined by assigning a grade to the numeric value that indicates the temperature. Using the crisp set, however, results in the temperature being expressed as shown in Fig. 11.18b. In the crisp set, it is discovered that the temperature is suddenly set to the "hot" state on reaching a specific value, causing an unnatural state to occur. The fuzzy set can represent a more natural temperature state compared with the crisp set. Of course, the form obtained by a membership function varies depending on individuals or conditions. In general, it is not easy to define the meaning of each word and represent it with a membership function. If meaning or definition can be represented by a membership function, the subjective fuzziness can be handled quantitatively. The fuzzy control uses the feedback function to determine a membership function, which gives the better result, by "tuning."

As described above, the fuzzy theory is created from the engineered idea "fuzziness" separately from the conventional theory. For example, when controlling a specific target, as is often the case, we cannot obtain better results using the conventional precise control system, but we can obtain appropriate results by operator's manual control. Therefore, we tried to describe the operator's control method in theory; however, the method includes fuzziness, which we cannot ignore. Fuzzy theory is available to solve this problem.

Since approximately 1975, fuzzy theory has applied to actual systems, for example, control systems such as heat exchanger plant and cement kiln. These control systems have high nonlinear performance, so it was difficult to carry out adjustments or operations based on the conventional PID control. It was, however, verified that it was possible to perform operations with an expert's control knowledge using fuzzy control. As a good example of a fuzzy control system, we can introduce the automatic train operation system, based on the predictable fuzzy control practically applied to the Sendai City's subway in Japan in 1987, that runs a train at a limited speed after it leaves a station instead of requiring a train

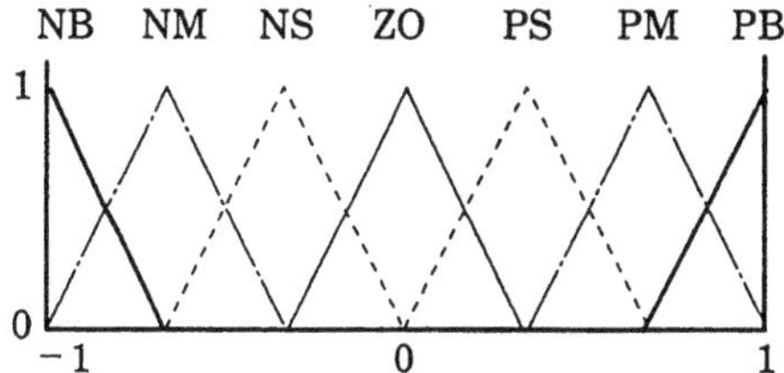

Fig. 11.19 Triangular fuzzy variables. *NB* negative big, *NM* negative medium, *NS* negative small, *ZO* zero, *PS* positive small, *PM* positive medium, *PB* positive big

operator to do so, and performs acceleration and deceleration operations via a computer until a train arrives at the next station.

[1] Basic concept of fuzzy control

Fuzzy control consists of a set of fuzzy control rules and fuzzy inference sections. If plant information (inputs) required for plant control is set to x_1 and x_2 and the output to the target plant is set to y, the "if (antecedent part)—then (consequent part)" based control rule is obtained as follows.

If x_1 is small and x_2 is big, then y is medium
If x_1 is big and x_2 is medium, then y is big

Here, fuzzy control rule is represented generally as follows.

R_i: If x_1 is A_{i1} and x_2 is A_{i2}, then y is B_i, $i = 1,2, \ldots, n$

Here i indicates a control rule number, x_1 and x_2 are variables in the antecedent part, and y is a variable in the consequent part.

A_{i1}, B_i, and other functions are represented by a fuzzy set, so they are referred to as fuzzy variables; however, actually, they are assumed to be fuzzy values of x and y.

Fuzzy variables are classified into two types: continuous and discrete types. The most frequently used continuous-type membership function, which creates a triangular form as shown in Fig. 11.19, is represented by the following equation:

$$A(x) = \frac{1}{a}(-|x - b| + a) \vee 0, \quad a > 0 \tag{11.35}$$

In this equation, parameter b indicates the value of x that sets the grade of a membership function to 1, and a indicates the spread of a membership function. Fuzzy control uses the standardized fuzzy variables. In this example, the underlying set is normalized as division $[-1, 1]$.

[2] Fuzzy control system for feed-water control

In the Fugen reactor, the reactor feed-water control for low output is carried out using an expert operator's knowledge or experience, requiring excessively fine operations. It is, therefore, judged that the application of fuzzy control is appropriate for Fugen, and it was applied to the actual machine in 1992 after the development in a pilot system. The Fugen reactor feed-water control system is used to control the feed-water control valve to keep the water level of the steam drum to the set value. The quantity of feed-water supplied to the steam drum varies extremely (0–460 t/ht/loop) depending on reactor outputs; therefore, it is adjusted using two control values of different quantities: main feed-water control valve (MCV) and low-flow feed-water control valve (LFCV). The LFCV is used in a low-output area to carry out the 1-element control based on only the water level of the steam drum. During rated operation, the MCV is used to carry out the 3-element control based on the water level of the steam drum, main steam flow, and feed-water flow, ensuring the stabilization of the water level of the steam drum.

If it is possible to model Fugen operator's judgments and operations and stabilize the water level of the steam drum with a slight fluctuation at low output, it results in the enhancement of plant operation controllability and reduction of operator's incidence. Therefore, the development of the "fuzzy control system for feed-water control" has been advanced since 1986 in order to <1> accomplish the steam drum water-level stabilization control using the fuzzy online control system, exceeding the control result obtained by the conventional control system, and <2> realize a fully automatic control system to reduce the operator's burden.

The fuzzy control pilot system has been developed to control the LFCV in a low output area in which the reactor output is below approximately 18 %. This system consists of the process data processor, fuzzy inference sections, and workstation as shown in Fig. 11.20.

In this figure, in the fuzzy inference section of the steam drum water level system (1), the operation amount U_1 of the LFCV is inferred based on the difference LE between the actual steam drum water level LV(t) and set value LS(t) as well as the water level percentage change rate CL.

$$\mathrm{LE}(t) = \mathrm{LV}(t) - \mathrm{LS}(t) \tag{11.36}$$

$$\mathrm{CL}(t) = \mathrm{dLV}(t)/\mathrm{d}t \tag{11.37}$$

where t indicates a time.

In the fuzzy inference section for feed-water and steam flow system (2), the operation amount U_2 of the LFCV is inferred based on the difference FE2 between the water inflow to and water outflow from the steam drum. As described above, the low-output area has feed-water flow and steam flow at a low level, so it is impossible to ensure the sufficient measurement accuracy. To solve this problem, these flow values are obtained by the opening of the LFCV and turbine bypass valve (TBV). The flow deviation FE2 is then obtained as shown below using the feed-water flow FD, steam flow TB, and blow-down flow BLOW from the reactor coolant clarification system to the steam condenser.

$$\mathrm{FE2}(t) = \mathrm{FD}(t) - (\mathrm{TB}(t) + \mathrm{BLOW}(t))/2 \tag{11.38}$$

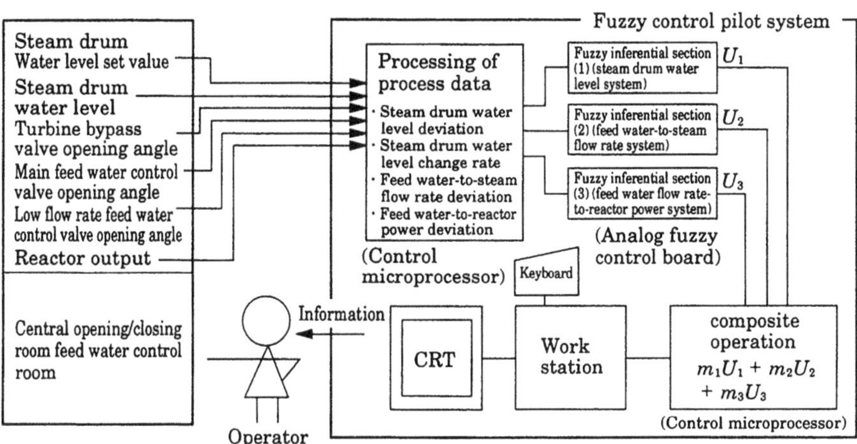

Fig. 11.20 Configuration of fuzzy control system

In the fuzzy inference section for feed-water flow and reactor output system (3), the operation amount U_3 of the LFCV is inferred based on the deviation FE3 between the feed-water flow to the steam drum FD and the appropriate feed-water flow IFD. "IFD" is obtained from the reactor output using the relational expression between the pre-analyzed reactor output and appropriate feed-water flow. The deviation FE3 between the actual feed-water flow and appropriate one is obtained as follows:

$$\mathrm{FE3}(t) = \mathrm{FD}(t) - \mathrm{IFD}(t) \tag{11.39}$$

Fuzzy inference section (1) contains 35 rules, and fuzzy inference sections (2) and (3) contain seven rules. Figure 11.21 shows the rules for each inference section.

The inference method uses the following Min–Max theorem:

$$U_1(x) = \mathrm{Max}\{\{W_{i1}(\mathrm{LE})^* W_{j1}(\mathrm{CL})^* X_{ij}(x)\}, i = 1 \ldots 7,\ \ j = 1 \ldots 5\} \tag{11.40}$$

$$U_2(x) = \mathrm{Max}\{\{W_{i2}(\mathrm{FE}_2)^* X_{i2}(x)\}, i = 1 \ldots 7\} \tag{11.41}$$

$$U_3(x) = \mathrm{Max}\{\{W_{i3}(\mathrm{FE}_3)^* X_{i3}(x)\}, i = 1 \ldots 7\} \tag{11.42}$$

where "$W_{\mathrm{ba}}(c)$" means the fitness of rule b that is evaluated based on input C in fuzzy inference section (a). "$X_{\mathrm{ba}}(x)$" indicates a membership function that represents the conclusion part of rule b in fuzzy inference section (a), and "*" indicates an operator that means the minimum value operation. The fuzzy variables $U_1(x)$, $U_2(x)$, and $U_3(x)$ that indicate the inference results obtained in each fuzzy interference section are converted to non-fuzzy variables U_1, U_2, and U_3 to generate control signals.

Fig. 11.21 Control rule of inferential sections

Water level change rate							
PB	PS	ZO	NS	NB	NB	NB	NB
PS	PB	PM	PS	PS	NM	NB	NB
ZO	PB	PB	PS	ZO	NS	NB	NB
NS	PB	PB	PM	PS	NS	NM	NB
NB	PB	PB	PB	PB	PS	ZO	NS

Water level deviation

Control rule of inferential section (1)

PB	PM	PS	ZO	NS	NM	NB
NB	NM	NS	ZO	PS	PM	PB

Feed water-to-steam flow rate deviation

Control rule of inferential section (2)

PB	PM	PS	ZO	NS	NM	NB
NB	NM	NS	ZO	PS	PM	PB

Feed water flow rate-to-reactor power deviation

Control rule of inferential section (3)

$$U_n = \int U_n(x)x\mathrm{d}x / \int U_n(x)\mathrm{d}x, \; n = 1, \ldots, 3 \tag{11.43}$$

These three inference results are weighted and averaged by the following equation, resulting in the LFCV control signal CA being output.

$$U = m_1 \cdot U_1 + m_2 \cdot U_2 + m_3 \cdot U_3 \tag{11.44}$$

where m_1, m_2, and m_3 indicate the weighted gain for the corresponding fuzzy inference sections (1)–(3).

The functions of the fuzzy control system were verified when the reactor started in October, 1989. Figure 11.22 shows the test results of control characteristics.

At the time of reactor startup, the output is increased slowly until the steam drum reaches the rated pressure (68 kg/cm^2). During this operation, the water level of steam drum A that is controlled using the fuzzy control pilot system is adjusted within ± 5 mm in most operation periods although the difference between the water level and set value has reached approximately ± 10 mm, enabling the best results to be obtained. The water level of steam drum B that is automatically controlled by PI control is adjusted within ± 60 mm. This verified that the fluctuation of the steam drum water level could be reduced approximately 1/2 to 1/6 compared with the PI control results. It then was proven that the fuzzy control system was extremely effective for adjusting the water level of the steam drum.

Fuzzy control advances control processing while inferring the control amount based on process data such as the steam drum water level, reactor output, and feed-water flow in the same way as when an operator performs manual control processing. Fuzzy control realized the better control characteristics that closely match a skilled operator's operations.

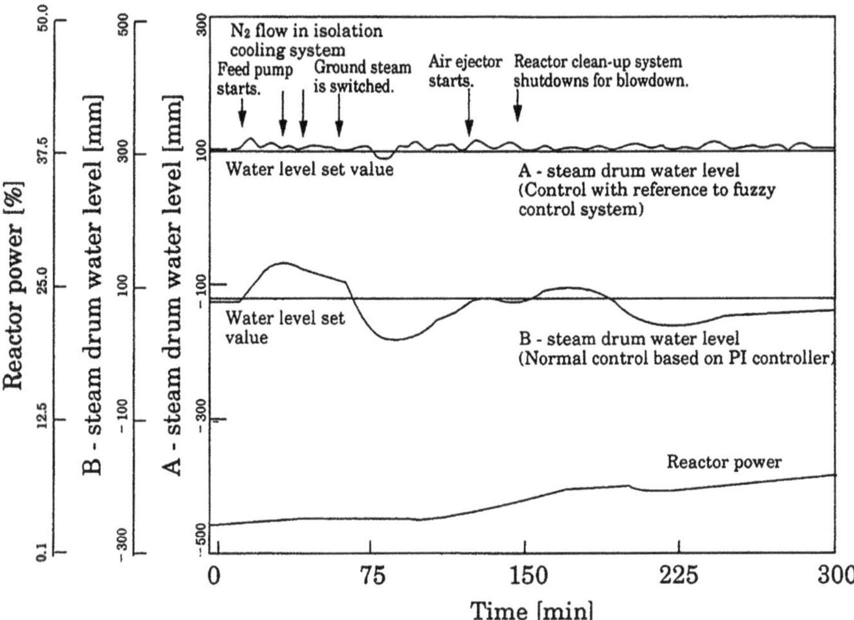

Fig. 11.22 Result of verification test for fuzzy control system

Chapter 11 Exercises

1. Express the Eq. (5.1) shown in Chap. 5 with a state equation; however, the output equation is for current $i(t)$.
2. Prove that the closed loop control system is stabilized by the control input shown in Eq. (11.13).
3. Prove that the Eq. (11.12) of the evaluation function is minimized by feeding back the control input shown in Eq. (11.13).
4. Design an optimal regulator based on the following evaluation function "J" as the minimum, assuming that the control target is

$$\frac{dx(t)}{dt} = \begin{bmatrix} 0 & 1 \\ 0 & -1 \end{bmatrix} x(t) + \begin{bmatrix} 0 \\ 1 \end{bmatrix} u(t)$$

$$J = \int_0^\infty \left\{ x^T(\tau) \begin{bmatrix} 1 & 0 \\ 0 & 1 \end{bmatrix} x(\tau) + u^2(\tau) \right\} d\tau$$

5. Derive Eq. (11.23).

6. Explain that how to correct the weight by learning is expressed in a simple format like Eq. (11.33) if the sigmoid function such as Eq. (11.30) is used as an input–output function for neural network.

Bibliography

1. Ito M, Kimura H, Hosoe S (1978) Design theory of linear control system. The society of instrument and control engineers, Tokyo
2. Mita T (1994) H∞ control. Shokodo, Tokyo
3. March-Leuba J, Cacuci DG, Perez RB (1986) Nucl Sci Eng 93:111–123
4. Suzuki K, Shimazaki J, Shinohara Y (1993) Nucl Sci Eng 115:142–151
5. The Society of Instrument and Control Engineers (1994) Neuro-fuzzy, AI handbook. Ohmsha, Tokyo
6. Nabeshima K (2001) JAERI 1342
7. PNC-TN1410-91-52 (1991)
8. Iguchi M, Isomura K, Ohkawa T, Sakurai N (2003) PNC technical report

ERRATUM

Nuclear Reactor Kinetics and Plant Control

Yoshiaki Oka and Katsuo Suzuki

Editors

Y. Oka and K. Suzuki (eds.), *Nuclear Reactor Kinetics and Plant Control*,
An Advanced Course in Nuclear Engineering, DOI 10.1007/978-4-431-54195-0,

---

DOI 10.1007/978-4-431-54195-0

The publisher regrets that there was an error in a copyright holder name.
The copyright holder of this title is Authors 2013.

The online version of the original book can be found at
http://dx.doi.org/10.1007/978-4-431-54195-0

Answers to Exercises

Part I

Chapter 2

1. If the effective delayed neutron fraction is 0.0076, the decay constant of the delayed neutron precursor is 0.008 s^{-1}, and the prompt neutron lifetime is 0.001 s.

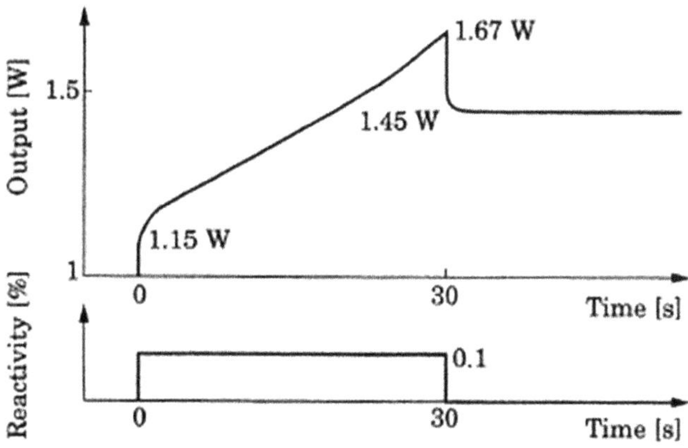

2. If the reactivity has significantly increased in stepwise, the transient effect that appears on the term can be ignored, and the power $P(t)$ of the neutron can be expressed by the following equation.

$$P(t) = P(0)\,\mathrm{e}^{t/T}$$

However, T is expressed by the reactor period. Thus, we obtain the total energy emitted from the following equation.

Y. Oka and K. Suzuki (eds.), *Nuclear Reactor Kinetics and Plant Control*, An Advanced Course in Nuclear Engineering, DOI 10.1007/978-4-431-54195-0, © Springer Japan 2013

$$E = \int_0^t P(t)\,\mathrm{d}t = TP(0)[\mathrm{e}^{t/T}]_0^t = TP(0)(\mathrm{e}^{t/T} - 1) = \{P(t) - P(0)\}T$$

On the other hand, the relationship of the reactivity ρ with the effective multiplication constant k_{eff}, the reactor period T, etc. is expressed by the following equation. If approximation of short time T is used in the inhour equation, it is expressed by the following equation.

$$\rho = \frac{l}{Tk_{\mathrm{eff}}} + \beta$$

If we substitute T from this equation, it is:

$$E = \{P(t) - P(0)\}\frac{l}{k_{\mathrm{eff}}} \cdot \frac{1}{\rho - \beta}$$

3. The equation can be expressed as follows:

$$\frac{\mathrm{d}n}{\mathrm{d}t} = \frac{k_{\mathrm{eff}} - 1}{l} n + S$$

(1) Since S is input in stepwise, the steady-state solution is

$$n = \frac{Sl}{1 - k_{\mathrm{eff}}}$$

The transient solution is obtained by solving $\frac{\mathrm{d}n}{\mathrm{d}t} = \frac{k_{\mathrm{eff}} - 1}{l} n$. If the initial conditions are entered into the equation, it is:

$$n = \frac{Sl}{1 - k_{\mathrm{eff}}}\left(1 - \mathrm{e}^{\frac{k_{\mathrm{eff}} - 1}{l} t}\right)$$

(2) Since $S = 0$, the initial conditions are entered into the equation $\frac{\mathrm{d}n}{\mathrm{d}t} = \frac{k_{\mathrm{eff}} - 1}{l} n$, thus we obtain

$$n = n_0\,\mathrm{e}^{\frac{k_{\mathrm{eff}} - 1}{l} t}$$

(3) Since $k_{\mathrm{eff}} - 1$, $\frac{\mathrm{d}n}{\mathrm{d}t} = S$ is solved to obtain the following.

$$n = n_0 + St$$

It is shown by the following figures.

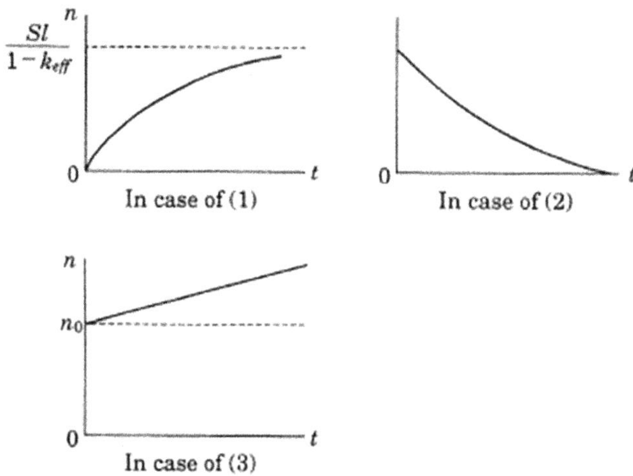

4. Since the positive reactivity is entered in stepwise, if the approximation with delayed neutron constant generation rate is used for the point core dynamic approximation equation, we obtain the following equation.

$$\frac{\mathrm{d}n}{\mathrm{d}t} = \frac{\rho - \beta}{\Lambda} n(t) + \frac{\beta}{\Lambda} n_0$$

If this equation is solved for $n(t)$, we obtain

$$n(t) = n_0 \, \mathrm{e}^{\frac{\rho-\beta}{\Lambda}} + \frac{\beta}{\beta - \rho} n_0 \left(1 - \mathrm{e}^{\frac{\rho-\beta}{\Lambda}t}\right)$$

Since the reactor is operating with the constant output, it $k_{\mathrm{eff}} = 1.0, \Lambda = l^*$. If we rewrite Eq. (1) using $\rho = \frac{k_{\mathrm{eff}}-1}{k_{\mathrm{eff}}} = k_{\mathrm{eff}} - 1 = \delta k$, the equation of the problem is given. By solving with the initial conditions, we can obtain the following solution.

$$n(t) = n_0 \, \mathrm{e}^{\frac{\delta k-\beta}{l^*}t} + \frac{\beta}{\beta - \delta k} n_0 \left(1 - \mathrm{e}^{\frac{\delta k-\beta}{l^*}t}\right)$$

$$\left.\frac{\mathrm{d}n(t)}{\mathrm{d}t}\right|_{t=0} = n_0 \frac{\delta k - \beta}{l^*} - \frac{\beta}{\beta - \delta k} n_0 \frac{\delta k - \beta}{l^*}$$

$$= n_0 \frac{\delta k - \beta}{l^*} + \frac{\beta}{l^*} n_0$$

$$= n_0 \frac{\delta k}{l^*}$$

$$\frac{1}{n} \cdot \left.\frac{\mathrm{d}n}{\mathrm{d}t}\right|_{t=0} = \frac{\delta k}{l^*}$$

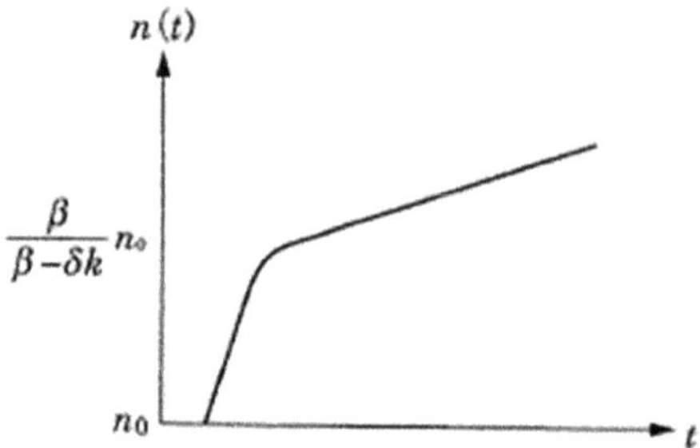

5. The kinetics equation of reactor can be expressed as follows if the delayed neutron precursor C is approximated to be one group.

$$\frac{\mathrm{d}n}{\mathrm{d}t} = \frac{(1-\beta)k_{\mathrm{eff}} - 1}{l} n + \lambda C + S \tag{1}$$

$$\frac{\mathrm{d}C}{\mathrm{d}t} = -\lambda C + \frac{\beta k_{\mathrm{eff}}}{l} n \tag{2}$$

where n is neutron density, C is delayed neutron precursor density, λ is its decay constant, $k_{\mathrm{eff}}\,(1 > k_{\mathrm{eff}} > 0)$ is effective multiplication constant in sub-critical state, β is delayed neutron fraction, S is neutron source, and t is time.

The steady-state value of the neutron density in sub-critical state n_0 is obtained by assigning 0 for the left side of Eqs. (1) and (2).

$$n_0 = \frac{l\lambda C_0}{\beta k_0} = \frac{lS}{1-k_0} \tag{3}$$

where C_0 is the precursor density at n_0, and k_0 is the effective multiplication factor to give n_0.

Now, suppose the reactivity that is equivalent to (δk is given and the neutron flux is abruptly increased by (δn from n_0 temporarily. This period is too short to cause change in generation of the delayed neutron due to decay of the delayed neutron precursor.

More specifically, it is:

$$\frac{\mathrm{d}C}{\mathrm{d}t} = \frac{\mathrm{d}(C_0 + \delta C)}{\mathrm{d}t} = 0$$

Then Eq. (1) is expressed by as follows.

$$\frac{d(n_0 + \delta n)}{dt} = \frac{d\delta n}{dt} = \frac{(1-\beta)(k_0 + \delta k) - 1}{l}(n_0 + \delta n) + \lambda C_0 + S$$

If we expand it and ignore the second-order very small term of $\delta k \times \delta n$, it is:

$$\frac{d\delta n}{dt} = \frac{(1-\beta)k_0 - 1}{l}\delta n + \frac{(1-\beta)\delta k}{l}n_0 + \frac{(1-\beta)k_0 - 1}{l}n_0 + \lambda C_0 + S \quad (4)$$

where, if Eq. (1) is in steady state, it is:

$$\frac{(1-\beta)k_0 - 1}{l}n_0 + \lambda C_0 + S = 0 \quad (5)$$

Then substitute Eq. (5) into Eq. (4).

$$\frac{d\delta n}{dt} = -\frac{\lambda C_0 + S}{n_0}\delta n + \frac{(1-\beta)\delta k}{l}n_0$$

Then we solve this differential equation and assign $\delta n(t = 0) = 0$, the following is obtained.

$$\delta n = \frac{\delta k(1-\beta){n_0}^2}{(\lambda C_0 + S)l}\left(1 - e^{-\frac{\lambda C_0 + S}{n_0}t}\right)$$

From the above equation, the maximum value of δn that has abruptly been raised is

$$\frac{\delta k(1-\beta){n_0}^2}{(\lambda C_0 + S)l}. \quad (6)$$

From Eq. (6), it is seen that even if δk is the same, the smaller n_0 is and the larger S is, the value of δn becomes small.

Chapter 3

1. (1) δk_0: Applied reactivity, ΔT: Rise in temperature, the following equation is satisfied.

$$\frac{1}{\phi} \cdot \frac{d\phi}{dt} = \frac{\delta k}{l} \quad (1)$$

$$\delta k = \delta k_0 - |\alpha|\Delta T \tag{2}$$

$$\frac{\mathrm{d}\Delta T}{\mathrm{d}t} = \frac{\phi A}{C} \tag{3}$$

From Eqs. (1) and (2),

$$\frac{\mathrm{d}\phi}{\mathrm{d}t} = \frac{\delta k_0 - |\alpha|\Delta T}{l}\phi$$

If φ is the maximum, it is $\frac{\mathrm{d}\phi}{\mathrm{d}t} = 0.$

$$\therefore \Delta T' = \frac{\delta k_0}{|\alpha|}$$

(2) From Eqs. (1) and (3),

$$\frac{\mathrm{d}\phi}{\mathrm{d}t} = \frac{C}{lA}(\delta k_0 - |\alpha|\Delta T)\frac{\mathrm{d}\Delta T}{\mathrm{d}t}$$

$$\phi = \phi_0 + \frac{C}{lA}\left(\delta k_0 \Delta T - \frac{|\alpha|}{2}\Delta T^2\right)$$

If we substitute $\Delta T = \frac{\delta k_0}{|\alpha|}$ from this equation, it is :

$$\phi_{\max} = \phi_0 + \frac{C}{2lA}\cdot\frac{\delta k_0{}^2}{|\alpha|}$$

(3) Suppose the rise in temperature is $\Delta T''$ when φ becomes φ_0 again, it is:

$$\Delta T'' = \frac{2\delta k_0}{|\alpha|}$$

In addition, heat dissipation to the outside does not occur; therefore, the temperature rise should finally become constant.

$$\frac{\mathrm{d}\Delta T}{dt} = 0$$

$$\therefore \Delta T_{\max} = \frac{\delta k_0 + \sqrt{\delta k_0 + \frac{2|\alpha|lA}{C}\phi_0}}{|\alpha|}$$

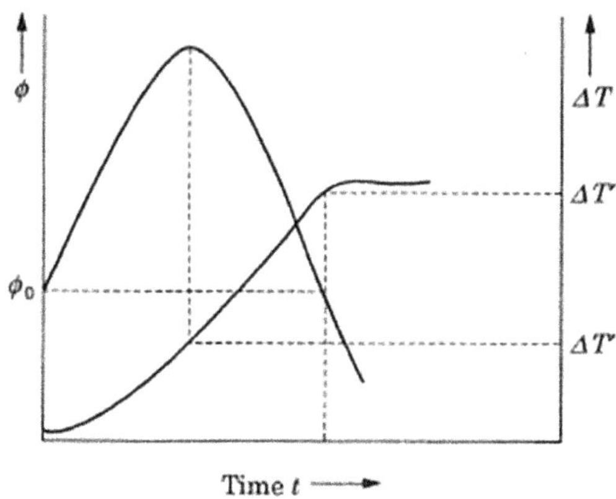

Chapter 4

1.

$$\begin{aligned}\delta k &= \frac{10^{-3}}{30} + \frac{0.0064}{1 + 0.077 \times 30} \\ &= 3.33 \times 10^{-5} + 1.93 \times 10^{-3} \\ &= 1.96 \times 10^{-3}\end{aligned}$$

Thus, the reactivity per 1 cm is

$$\begin{aligned}\frac{1.96 \times 10^{-3}}{15} &= 1.3 \times 10^{-4}\,\text{cm}^{-1} \\ &= 0.013\%/\text{cm}\end{aligned}$$

2. In sub-critical state, the neutron flux φ is

$$\phi = \frac{S}{1 - k_{\text{eff}}}$$

where, S is the intensity of neutron source and k_{eff} is the effective multiplication factor.

Suppose the neutron flux φ is proportional to the measurement C using the neutron counter, it is: $C = \frac{S'}{1-k_{\text{eff}}}$. S' is constant value.

From the definition, the reactivity ρ is expressed as $\rho = \frac{k_{\text{eff}}-1}{k_{\text{eff}}}$.

$$C = -\frac{S'(1-\rho)}{\rho}$$

$$\therefore S' = \frac{-C\rho}{1-\rho} = \frac{3.8728 \times 10^4 \times 0.38 \times 10^{-2}}{1.0038} = 147$$

Also, it is $\rho = \frac{S'}{S'-C} = \frac{147}{147-C}$. Then we obtain the reactivity based on the calibration by substituting measurement value into C.

(a) Insertion position of control rod 60 % $\rho60 = -1.9 \times 10^{-2} = -1.9\ \%$
(b) Insertion position of control rod 40 % $\rho40 = -4.3 \times 10^{-2} = -4.3\ \%$
(c) Insertion position of control rod 20 % $\rho20 = -7.0 \times 10^{-2} = -7.0\ \%$
(d) Insertion position of control rod 0% $\rho0 = -8.6 \times 10^{-2} = -8.6\ \%$

Part II

Chapter 5

1. <1>

$$\frac{\omega}{s^2+\omega^2}$$

<2>

$$\frac{s}{s^2+\omega^2}$$

<3>

$$\frac{1}{a}F\left(\frac{s}{a}\right)$$

2. Use that $y(t)$ is expressed by the sum of 3 functions $f_1(t) = \frac{a}{T}t$, $f_2(t) = -f_1(t-T)$, and $f_3(t) = -a \cdot u(t-T)$.

$$G(s) = \frac{a}{Ts^2}(1-\mathrm{e}^{-Ts}) - \frac{a}{s}\mathrm{e}^{-Ts}$$

3. Suppose $f_1(t) = a\sin\left(\frac{2\pi}{T}t\right) \cdot u(t)$, $f_2(t) = a\sin\left(\frac{2\pi}{T}\left(t-\frac{T}{2}\right)\right) \cdot u\left(t-\frac{T}{2}\right) = f_1\left(t-\frac{T}{2}\right)$, it is expressed by the following

$$f(t) = f_1(t) + f_2(t).$$

Then, we obtain the solution of

$$F(s) = \frac{a\frac{2\pi}{T}\left(1 + e^{\frac{Ts}{2}}\right)}{s^2 + \left(\frac{2\pi}{T}\right)^2}$$

4.

<1>

$$\frac{1}{s}\tanh\left(\frac{Ts}{2}\right)$$

<2>

$$\frac{1}{Ts^2} - \frac{e^{-Ts}}{s(1 - e^{-Ts})}$$

5.

$$f(t) = \frac{k}{a}\left(1 - e^{-\frac{t}{T}}\right)$$

6. It shows the derivation of $G_{xv}(s)$. Focus on the output $U(s)$ at the adder of v and feedback amount. It is only necessary to remove $U(s)$ from $X(s) = C(s)P(s)U(s)$, $U(s) = V(s) - H(s)X(s)$. The same applies to other two equations.
7. (1) Apply Laplace transform under the initial conditions $y(0) = 0$ and $\left(\frac{dy}{dt}\right)_0 = 0$ of the motion equation.

$$G(s) \equiv \frac{Y(s)}{X(s)} = \frac{k}{s(ms + c)}$$

(2) In the block diagram of the feedback control system, the following relational expression is satisfied between the variables.
$Y(s) = G(s)X(s)$
$X(s) = g(U(s) - Y(s) - fsY(s)$
Remove $X(s)$ from the above two equations to obtain the closed loop transfer function as follows.

$$G_c(s) \equiv \frac{Y(s)}{U(s)} = \frac{gk}{s(ms + fk + c) + gk}$$

8. From the block diagram, if the input/output transfer function is obtained as $C(s) = \frac{A}{1+A\left(\frac{1}{K_0}\right)\left(\frac{T_1 s}{T_1 s+1}\right)\left(\frac{1}{T_2 s+1}\right)}$, and the amplifier gain is regarded as $A = \infty$, the approximate expression $C_{\mathrm{A}}(s)$ of $C(s)$ is

$$C_{\mathrm{A}}(s) = K_0\alpha\left(1 + \left(\frac{T_2}{\alpha}\right)s + \frac{1}{\alpha T_1 s}\right)$$

where $\alpha = 1 + \frac{T_2}{T_1}$.

9.

$$C_1(s) = \frac{0.882}{2.65s+1}D_1(s) + \frac{20.8}{2.65s+1}Z(s)$$
$$C_2(s) = \frac{0.5975}{9.39s+1}C_1(s) + \frac{0.4}{9.39s+1}D_2(s) + \frac{0.361}{9.39s+1}Z(s)$$

10. In the Bode diagram, the gain curve of the expression (5.47) and dead time system e^{-Ls} coincide perfectly, and the phase curve varies only in the high-frequency regions. On the other hand, since the high-order system has a low-pass filter property, gain in the high-frequency region becomes small. Consequently, to the dead time system + high-order system, the adverse effect of the approximation in the high-frequency region of the dead time system is reduced. Let's check this using the step response and others in the joint transfer function of the dead time system and second-order system.

11. <1>

$$\left(\frac{5}{s^2+2.5s+5}\right)\frac{1-\frac{1}{2}Ls}{1+\frac{1}{2}Ls}$$

<2>

$$\left(\frac{5}{s^2+2.5s+5}\right)\frac{1-\frac{1}{3}Ls}{1+\frac{2}{3}Ls+\frac{1}{6}(Ls)^2}$$

Chapter 6

1. Substitute Eqs. (6.3) and (6.4) into Eqs. (6.1) and (6.2).
2. If $\rho(t) = 0$, either of $\frac{dN}{dt} = 0$, $\frac{dC_i}{dt} = 0$ is not exactly satisfied for Eqs. (6.1) and (6.2).
3. Omitted
4. Omitted

5. The block diagram using the transfer function of the critical reactor of Eq. (6.12) is as shown below.

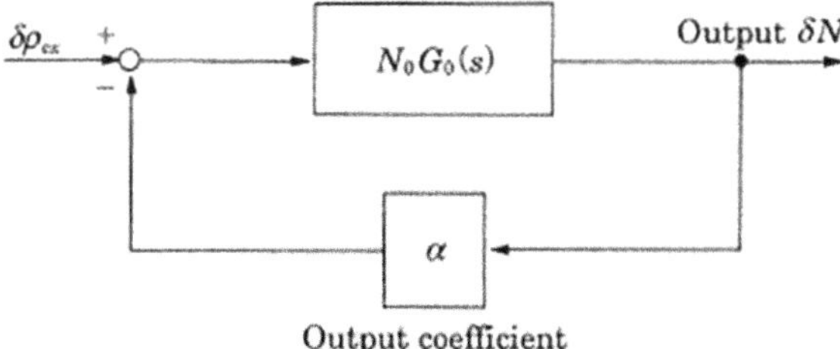

Using the first equation of Eq. (5.6), obtain $G_c(s) = \frac{N_0G_0(s)}{1+\alpha N_0G_0(s)}$.

6. <1> Thermo-hydraulic characteristics, <2> pressure drop, <3> nuclear characteristics, <4> void, <5> nuclear characteristics (or thermo-hydraulic characteristics), <6> thermo-hydraulic characteristics (or nuclear characteristics), <7> high-order mode, <8> high-output and low core flow, <9> selective control rods
7. From the Laplace transformation of Eq. (6.32), obtain $\delta I(s) = \dfrac{\gamma_1\Sigma_f}{s+\lambda_I}$.
   Substitute this to $\delta I(s)$ of Laplace transformation equation of Eq. (6.31) and organize.

Chapter 7

1. Refer to "Sect. 7.3 (Power control)."
   Write any three of the following:

   <1> Log of thermal limit obtained from the core performance computation by the process calculator
   <2> Rod block monitor into which only neutron flux signals around operated control rods are input
   <3> Monitoring on an operating characteristics diagram for the core flow rate and reactor heat output
   <4> Rod worth minimizer for guiding the operation procedure of the control rod

2. Refer to "Sect. 7.2 (Operation Control Technique of BWR)" and "Sect. 7.4 (Operation Control During Steady Operation)."

   <1> (Inserting and pulling out of) Control rod
   <2> (Increase and decrease of) Recirculation flow rate

3. Refer to "Sect. 7.5 (Control When Abnormal Condition Occurs)."

   <1> Reactor scram
   <2> Standby liquid control system (SLC)
   <3> Recirculation pump trip (RPT)

4. Refer to "Sect. 7.5 (Control When an Abnormal Condition Occurs)."

   <1> Turbine bypass valve (TBV)
   <2> Safety relief valve (SR valve)

5. Refer to "Sect. 7.6 (Central Control Panel)".
   Write any three of the following:

   <1> To improve monitoring performance.
   <2> To reduce burden to the operator.
   <3> To prevent human errors

6. Refer to "Sect. 7.6 (Central Control Panel)."
   Write any two of the following:

   <1> First hit display
   <2> Major four events display
   <3> Critical alarm display
   <4> Fixed mimic display
   <5> Large-size screen (or variable display screen)
   <6> Batch alarm display by system

Chapter 8

1. The PWR applies pressure on the core coolant to maintain its single liquid phase and the primary and secondary systems are separated by the steam generator, the effects on the reactivity of the core caused when the reactor pressure changes are minimized. Therefore, time for controlling the control rod control system and the turbine bypass control system is secured.
2. <1> The control rod cluster is used for controlling the power in relatively early stage of the turbine load change. <2> The boron concentration is adjusted to compensate the relatively slow changes in the reactivity of the core for a long time, such as combustion of fuels and changes in xenon concentration. The boron concentration is also adjusted to adjust the position of control rods for the purpose of maintaining the output distribution in the axial direction in the core to be within a targeted range.
3. The steam pressure constant method is ideal for the conditions for designing the turbine system in that a constant steam pressure is obtained regardless of output. It requires, however, excessive primary system equipment such as the capacity or pressurizer and the number of control rods. On the other hand, the T_{avg}

constant method is favorable for the conditions for designing the primary system. It has a defect, however, that requires tight conditions for designing the pressure resistance of the secondary system. Currently, a compromized method is adopted to optimize the primary and secondary system equipment.

4. It is achieved by pulling out the control rod after raising the temperature and pressure by the heat input from the primary coolant pumps and pressurizer heater until the reactor reaches the hot shutdown condition (about 292 °C and 15.41 MPa) and diluting the critical boron concentration near the critical value.
5. The difference in the core life appears as the difference in moderator temperature coefficient. Comparing with EOC, the change in the T_{avg} or the pressurize pressure tends to be larger in BOC when the absolute value of the moderator temperature coefficient is small, because the effect of the negative reactivity feedback caused by the rise in the primary coolant temperature is small.

Chapter 9

1. The reactor power control system is provided to improve quick response and stability at the time of a reactor power change.
   There is a response delay of the reactor vessel outlet Na temperature, which is controlled by the reactor power control system, because the temperature detector fitted on the primary main cooling system piping is far from the reactor outlet and the large heat capacity of the coolant. Therefore, such thermal delays occur in the order of extraction/insertion of the fine control rod, output change and change of reactor vessel outlet sodium temperature when the output command changes, and T_{avg} may deviate from the T_{REF} signal excessively (overshoot). To restrict the deviation, stability is obtained by feeding back a neutron flux signal equivalent to the reactor power in order to compensate thermal delay (refer to Fig. 9.7).
2. The control gauges other than those in the main steam pressure control system are designed to measure one processing volume by 2 units. In this case, in the viewpoint of fail-safe, non-safe side signal is used for the control. Therefore, input signals of "high sorting value" are used.
   On the other hand, the turbine speed and main steam pressure in the main steam pressure control system are designed to measure one processing volume by 3 gauge units. In addition to fail-safe, to improve reliability, it is designed to use "intermediate value" for control so that continuance plant operation is possible even if any one of the gauges malfunctions to high-value side.
3. Since the control system of MONJU is composed of the primary main cooling system (sodium), the secondary main cooling system (sodium) and the water/steam system, dead time due to delay in heat transport is large. To solve this problem, the composition of the plant control equipment of MONJU forms a hierarchical system with the output command device on the top, which controls the reactor output, a primary main cooling system flow rate, a secondary main cooling system flow rate, and a feed water flow rate.

4. There are two methods available for control of fast reactor plant; variable coolant temperature at reactor outlet and constant coolant flow rate, and constant coolant temperature at reactor outlet and variable coolant flow rate. In the former method, control with pump, control valve, etc. is not necessary and the control system is simple. In the latter method, steam conditions can be kept constant even for change of load.
   Since MONJU is a power reactor, such control method is adopted as to proportionate the coolant flow rate to the plant output, to make temperature difference between the reactor vessel outlet and inlet large irrespective of plant output and to make main steam temperature and pressure constant irrespective of plant output as requested for turbine design.
5. Saturated steam (pressure approximately 6 MPa, temperature approximately 280 °C) is used for the steam condition for the high-pressure turbine for the BWR and PWR. On the other hand, for MONJU, since it is possible to make the reactor vessel outlet temperature high, superheated steam (pressure approximately 12.5 MPa, temperature approximately 483 °C) is used for the steam condition.
6. Helical coil once-through separate type. Helical coil-type steam generator tube with a spiral coiled shape is used for MONJU. This design is characterized that it makes it possible to downsize the equipment. Also, the steam condition is high at 483 °C. For this reason, the separate-type superheater made of stainless metal, which has excellent heat resistance, is adopted and used in combination with the evaporator.

Chapter 10

1. To make the output distribution in the radial direction of the core as flat as possible to suppress the output peaking, to maintain the fuel temperature low. For reference purposes, the HTTR adopts various devices to keep the fuel temperature low such as by using 12 fuel bodies with varing concentration and arranging the fuel bodies with high concentration in the upper area and periperial area in the radial direction to make the axial and radial output diesribution flat.
2. To prevent water entry into the primary cooling system, even in case that the heat-transfer pipe of the primary pressurized water cooler is broken. If any water enters, water will reach the core resulting in positive reactivity according to the neutron moderating effect, and causing reactor power to increase. Also, moisture inside the core reacts with the graphite, which is core internal, and may cause adverse effect on structural integrity. For the HTTR, the water concentration in the primary coolant is limited to 0.2 vol ppm or less at reactor outlet temperature 900 °C or more.

Chapter 11

1. Suppose $x(t) = \int i(\tau)\,\mathrm{d}\tau$.

$$\frac{\mathrm{d}}{\mathrm{d}t}\begin{pmatrix} i(t) \\ x(t) \end{pmatrix} = \begin{pmatrix} -\frac{R}{L} & -\frac{1}{LC} \\ 1 & 0 \end{pmatrix}\begin{pmatrix} i(t) \\ x(t) \end{pmatrix} + \begin{pmatrix} \frac{1}{L} \\ 0 \end{pmatrix} e(t), \qquad y(t) = (1 \quad 0)\begin{pmatrix} i(t) \\ x(t) \end{pmatrix}$$

2. The optimal regulator is $\dfrac{\mathrm{d}x(t)}{\mathrm{d}t} = (A - BR^{-1}B^{\mathrm{T}}P)x(t)$.
   Now, suppose that the eigenvalue λ of the matrix $(A - BR^{-1}B^{\mathrm{T}}P)x(t)$ has a nonnegative real part, its eigenvector be $\boldsymbol{\xi}$ and its conjugate transpose be $\boldsymbol{\xi}^{*}$. Multiply $\boldsymbol{\xi}$ on the right side of Eq. (11.14) and $\boldsymbol{\xi}^{*}$ on the left side to obtain

$$2\,Re(\lambda)\xi^{*}P\xi + \xi^{*}PBR - 1BTP\xi + \xi^{*}Q\xi = 0$$

   Each term of this equation is nonnegative. Therefore we obtain $A\xi = \lambda\xi$. By resolving the semi-positive definite symmetric matrix Q to $Q = C^{\mathrm{T}}C$, we obtain $C\xi = 0$ from the third term. From the above two results, $\begin{bmatrix} \lambda I - A \\ C \end{bmatrix}\xi = 0$ is given, which disproves the hypothesis of detectability of (C, A).
3. Obtain Q from Eq. (11.14), substitute it to Eq. (11.12), and square the integrand to obtain

$$J(x_0, u(t)) = \int_0^{\infty} \left[x^{\mathrm{T}}(\tau)(-A^{\mathrm{T}}P - PA + PBR^{-1}B^{\mathrm{T}}P)x(\tau) + u^{\mathrm{T}}(\tau)Ru(\tau)\right]\mathrm{d}\tau$$

$$= x_0^{T}Px_0 + \int_0^{\infty} \left\{u(\tau) + R^{-1}B^{\mathrm{T}}Px(\tau)\right\}^{\mathrm{T}} R\left\{u(\tau) + R^{-1}B^{\mathrm{T}}Px(\tau)\right\}\mathrm{d}\tau$$

4. From Eq. (11.14), obtain the positive definite solution $P_1 = \begin{bmatrix} 2 & 1 \\ 1 & 1 \end{bmatrix}$ of the Riccati equation $P\begin{bmatrix} 0 & 1 \\ 0 & -1 \end{bmatrix} + \begin{bmatrix} 0 & 0 \\ 1 & -1 \end{bmatrix}P - P\begin{bmatrix} 0 & 0 \\ 0 & 1 \end{bmatrix}P + \begin{bmatrix} 1 & 0 \\ 0 & 1 \end{bmatrix} = 0.$
   Using this P_1, calculate the feedback gain K from Eq. (11.13) to obtain $K = [-1\ -1]$.
   Then, the optimal regulator is $\dfrac{\mathrm{d}x(t)}{\mathrm{d}t} = \begin{bmatrix} 0 & 1 \\ -1 & 2 \end{bmatrix}x(t)$, which is asymptotically stable.
5. Suppose $z(s) = \begin{pmatrix} z_1(s) \\ z_2(s) \end{pmatrix}$, from Eq. (11.21) we obtain

$$\begin{aligned} z(s) &= G11(s)w(s) + G12(s)u(s) \\ y(s) &= G21(s)w(s) + G22(s)u(s) \end{aligned}$$

   It is only necessary to substitute $u(s) = K(s)\,y(s)$ and delete $y(s)$.

6. The differentiation of $f(x) = \frac{1}{1+\mathrm{e}^{-(x-a)}}$ is expressed by the following equation:

$$f'(x) = \frac{\mathrm{e}^{-(x-a)}}{\{1+\mathrm{e}^{-(x-a)}\}^2} = \left(1 - \frac{1}{1+\mathrm{e}^{-(x-a)}}\right) \cdot \frac{1}{1+\mathrm{e}^{-(x-a)}} = (1 - f(x)) \cdot f(x)$$

which is expressed in simple form as shown below.

$$f'(I_j^k) = O_j^k \cdot (1 - O_j^k)$$

Index

Y. Oka and K. Suzuki (eds.), *Nuclear Reactor Kinetics and Plant Control*,
An Advanced Course in Nuclear Engineering,
DOI 10.1007/978-4-431-54195-0,

www.ingramcontent.com/pod-product-compliance
Ingram Content Group UK Ltd.
Pitfield, Milton Keynes, MK11 3LW, UK
UKHW021833190726
13853UKWH00003B/1297

* 9 7 8 4 4 3 1 5 4 6 9 3 1 *